# BIOCHEMICAL INSECT CONTROL

# BIOCHEMICAL INSECT CONTROL

## ITS IMPACT ON ECONOMY, ENVIRONMENT, AND NATURAL SELECTION

### M. SAYEED QURAISHI

A WILEY/INTERSCIENCE PUBLICATION

JOHN WILEY & SONS, New York · London · Sydney · Toronto

To the memory of

**IMTIAZ MOHAMMED KHAN (1901–1973)**

my father-in-law

**Library of Congress Cataloging in Publication Data:**

Quraishi. Mohammed Sayeed.
  Biochemical insect control, its impact on economy, environment, and
natural selection.

  "A Wiley-Interscience publication."
  Includes bibliographical references and index.
  1. Insecticides.   2. Insect control.   I. Title.
[DNLM: 1. Insecticides.   2. Insect control.
3. Environmental health.   QX600 Q9b]
SB951.5.Q7       628.9       76-29701
ISBN 0-471-70275-7

Printed in the United States of America

10 9 8 7 6 5 4 3 2 1

The aim of this book is to provide the student of insect toxicology with a balanced account of insecticides and insect control agents, and to provide readers in several disciplines with a broad picture of insecticides in relation to insects, man, and other living organisms. Special emphasis is placed on interactions of insecticidal chemicals and insects, the responses of living organisms, both target and nontarget, to the presence of insecticides in the environment, and the cost benefit equation.

The importance of pesticides in food production and public health cannot be underestimated, nor can some of the hazards involved in the use of pesticides, as practiced today, be ignored. Unfortunately, there has been a polarization of opinions on both sides, and because of this controversy we have relegated second place to two important facts: the ever-present need for better health and greater food production; and the ecological complexities of insect control. Developments in these disciplines have been so rapid and encompass such a wide front that to understand and evaluate the intricacies of the problems involved requires the analyses of data from many areas of scientific endeavor. This is what has been attempted in this book; however, since a number of aspects have been covered, the space allotted to some subjects is understandably brief.

Only a few books on biochemical insect control are available in English, and some of them are rather outdated. During the past few years important advances have been made in our understanding of the various phenomena related to the toxicology of insecticides and their influence on the environment, and scientists have been exploring alternative means of insect control.

v

Dr. C. M. Williams at Harvard has brought "third generation pesticides" into focus. Dr. R. L. Metcalf and his co-workers at the University of Illinois have been doing remarkable work on, among other insecticides, the biodegradable analogs of DDT and on model ecosystems. Dr. J. E. Casida and his group in Berkeley, California, and Dr. M. Elliott and his colleagues in England have made great progress in the metabolism of insecticides, especially pyrethroids. The findings of these groups and others in this country and abroad are exciting. Perhaps the next 10 years might well be the most productive in the history of pest control. Integrated insect control is also receiving more attention now because of the efforts of Dr. R. F. Smith and his group at Berkeley California; a small chapter has been included on the subject. In some states private integrated pest management services are available; I wish more data could be obtained on these.

Before I close, I want to express my gratitude to my wife, Akhtar, whose encouragement has made this effort possible, and to my children, Rana, Naveed, and Sabah for their help. I also want to thank Dr. Walter Ebeling for reviewing chapter 15.

M. SAYEED QURAISHI

*Newtonville, New York*
*July 1976*

# CONTENTS

## CHEMICALS FOR INSECT CONTROL

**Insecticides of Plant Origin**      2

Chapter 1    Pyrethroids      3
Chapter 2    Nicotinoids, Rotenoids, and Other Insecticides of Minor
            Importance      22

**Synthetic Organic Insecticides**      26

Chapter 3    Organophosphorus Compounds      27
Chapter 4    Carbamates      67
Chapter 5    DDT and Related Chemicals      98
Chapter 6    Cyclodiene (Diene-Organochlorine) Insecticides      115
Chapter 7    Lindane      125
Chapter 8    Pharmacodynamics of Chlorinated Hydrocarbons and
            Their Present Status      131

**Other Chemicals**      136

Chapter 9    Organic and Inorganic Insecticides of Minor Importance      137

**Synthetic Insect Control Agents**      141

Chapter 10    Insect Chemosterilants      142
Chapter 11    Attractants, Repellants, and Antifeedants      149

**Insect Control Agents of Insect and Microbial Origin**

Chapter 12    Pheromones and Related Chemicals                         156
Chapter 13    Hormones                                                 160
Chapter 14    Insect Control Agents of Microbial Origin                173

DYNAMICS OF INSECT TOXICOLOGY                                          178

**Insect-Insecticide Interactions**                                    179

Chapter 15    Sorption, Penetration, Distribution Inside the Body and
              Site of Action                                           180
Chapter 16    Mixed Function Oxidases, Synergism                       195
Chapter 17    Activation, Degradation, and Excretion                   204

**Resistance and Genetics of Resistance**                              210

Chapter 18    Natural versus Artificial Selection and the Impact of
              Insecticide Pressure                                     211
Chapter 19    Different Types of Resistance Mechanisms                 218

**Insecticide-Environment Interactions**                               224

Chapter 20    Effects of Insecticides on Non-Target Organisms          225
Chapter 21    Effects of Insecticides on Man                           231
Chapter 22    Carcinogenicity, Mutagenicity, and Teratogenicity        237

INSECTICIDES, FUNGICIDES, AND WORLD ECONOMY                            242

Chapter 23    Role of Insecticides and Fungicides in Food and
              Agriculture, and Public Health                           243

INTEGRATED CONTROL AND PEST MANAGEMENT                                 253

Chapter 24    Economics, Efficacy, and Ethics of Insect Control        254

Author Index                                                           259

Subject Index                                                          271

# CHEMICALS FOR INSECT CONTROL

# INSECTICIDES OF PLANT ORIGIN

# CHAPTER 1 | Pyrethroids

Pyrethrum is one of the oldest insecticides known; it was used for louse control in Iran (Persia) in 400 BC (1). The Chinese and the Japanese may have used it in joss sticks several hundred years ago. Though records of its use in ancient times are scanty, a factory in Amoy (China) has been making pyrethrum joss sticks and coils for nearly 200 years (2). In 1829 pyrethrum was introduced into Europe by an Armenian who thus started a lucrative business. In the beginning several closely related species belonging to the genus *Chrysanthemum* were grown as commercial crops and their powdered flowers were exported; however in 1849 it was discovered that the species *Crysanthemum cinerariaefolium* Vis. had the highest concentration of the insecticidal principles, and since then attention has been centred on this particular plant. In the late nineteenth century, this species was introduced into several countries, including Japan and the United States. In the early twentieth century, Dalmatia (Yugoslavia) acquired a leading place as the producer of pyrethrum. At about the same time Japan also started the commercial cultivation of *C. cinerariaefolium* and soon replaced Dalmatia as a major producer of the insecticide; she enjoyed this position until the beginning of the Second World War. At present (1976) Kenya occupies first place among the producers of pyrethrum; other countries of importance are Tanzania, Uganda, and the Congo.

*Pyrethrins*, the name given to the insecticidal principles in the flowers, are found in highest concentration in the achenes, especially at the time of full bloom. This necessitates a careful handpicking of the flowers, a procedure that makes an industry based on natural flowers unprofitable for countries where

3

the cost of labor is high. Until recently the content of active principles (pyrethrins) in the flowers varied between 0.9 and 1.3%; improved plant breeding techniques and propagation by clones, however, have made yields in the range of 2% feasible. Efforts are presently under way to further improve the pyrethrin content of the flowers, the goal being to achieve a concentration of about 3%.

Although pyrethrum was originally marketed in the form of finely ground dried flowers, dilute (0.06–0.15%) mineral oil extracts of pyrethrins later became more popular. Because of the growth in demand for "aerosols" that dispense solutions of from 1 to 2% pyrethrins, processes for preparing concentrates in the country of origin became more refined. At present flowers are extracted with an organic solvent and the extract is dewaxed and concentrated to contain about 20–25% pyrethrins. By partitioning with nitromethane, the petroleum solution can be further concentrated to a viscous preparation containing about 90% pyrethrins. With increasing concentration, however, the pyrethrins tend to become unstable. A summary of various methods for preparing refined pyrethrum extract is given by Hopkins (3). Since 1950 several analogs of pyrethrins have been synthesized and, of these, allethrins were the first to acquire commercial importance. The natural insecticides and their synthetic analogs are called *pyrethroids*.

The insecticidal activity of pyrethrum is due to six principles shown in Fig. 1.1, namely pyrethrin I (**1**), pyrethrin II (**2**), cinerin I (**3**), cinerin II (**4**), jasmolin

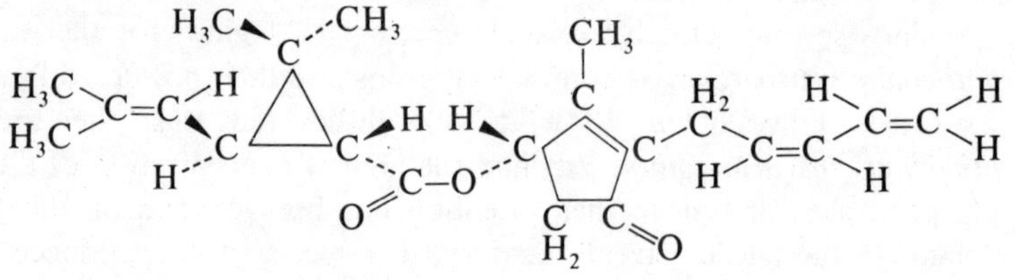

Pyrethrin I

**1**

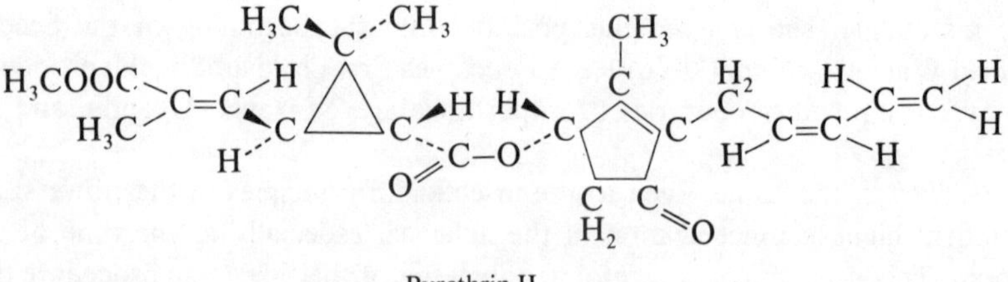

Pyrethrin II

**2**

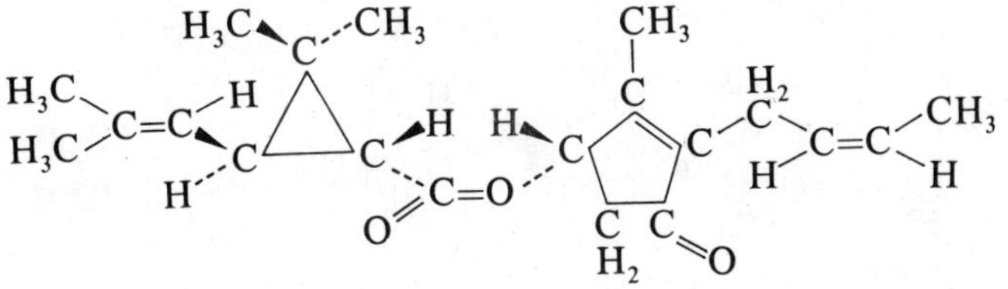

Cinerin I

**3**

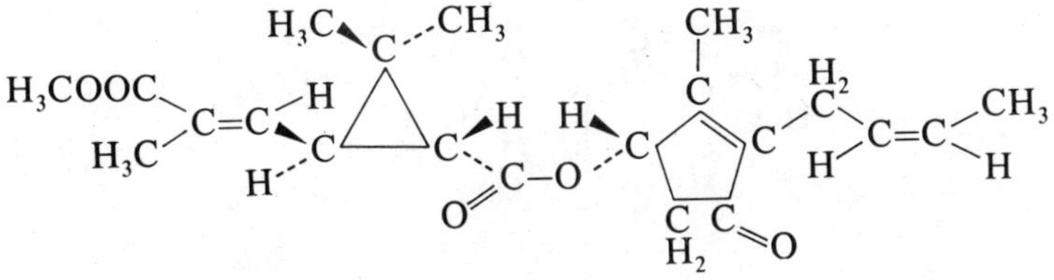

Cinerin II

**4**

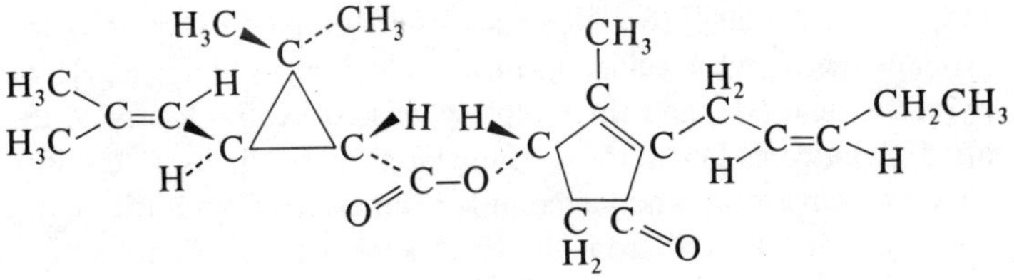

Jasmolin I

**5**

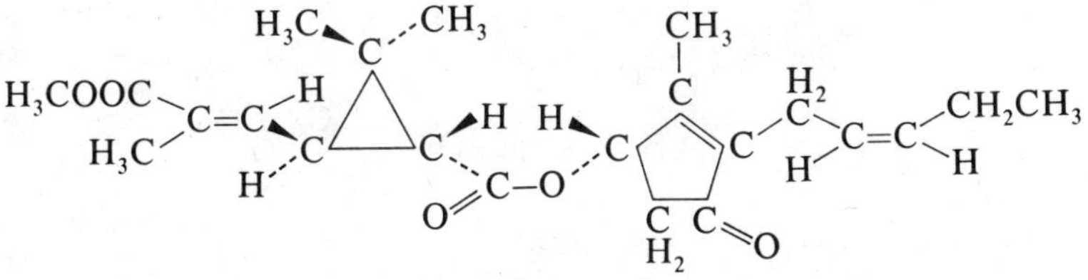

Jasmolin II

**6**

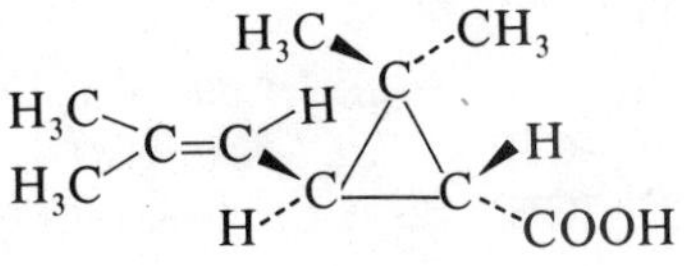

(+)-*trans*-Chrysanthemic acid (I)

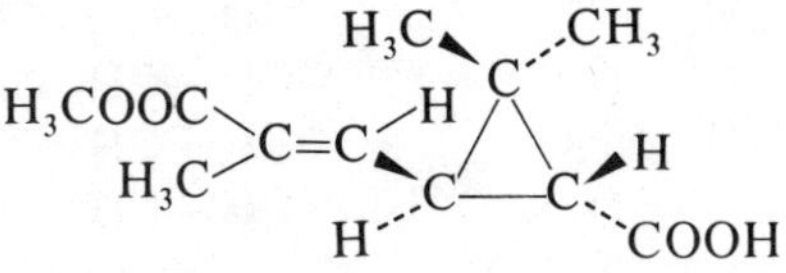

(+)-*trans*-Pyrethric acid (II)

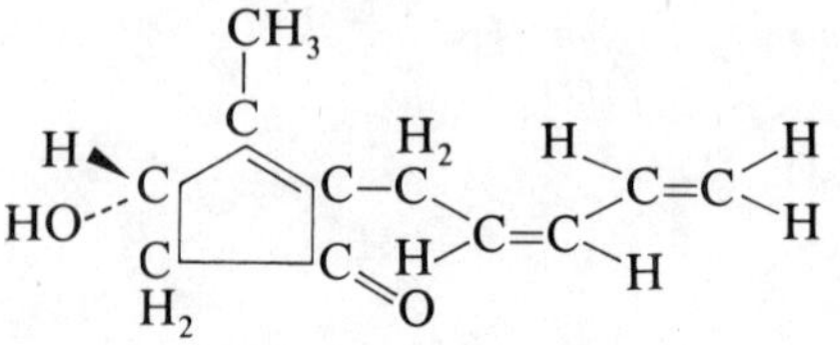

(+)-Pyrethrolone
**7**

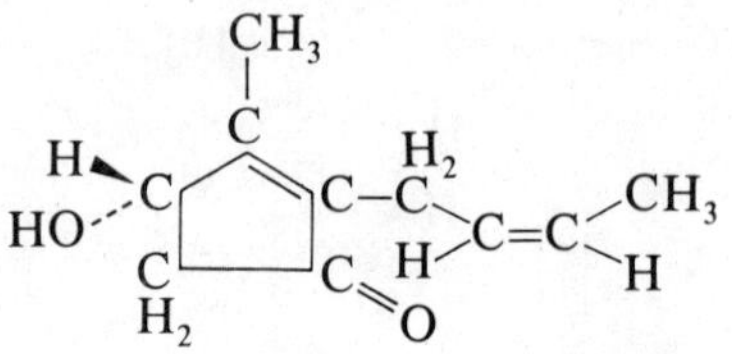

(+)-Cinerolone
**8**

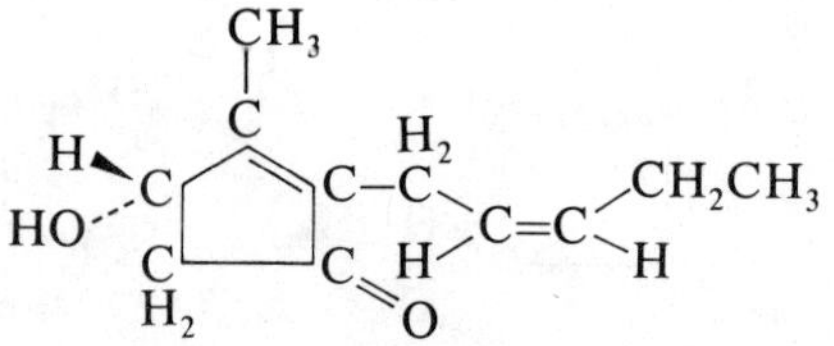

(+)-Jasmolone
**9**

**Fig. 1.1**   The pyrethrins and the acids and alcohols from natural pyrethroids.

I (**5**), and jasmolin II (**6**). All six are esters derived structurally from two cyclo-propane carboxylic acids, (+)-*trans*-chrysanthemic acid (I) and (+)-*trans*-pyrethric acid (II), and three closely related ketols. Esters of the former acid are designated as Is and those of the latter as IIs. Thus pyrethrin I is an ester of chrysanthemic acid, and jasmolin II is an ester of pyrethric acid. The names of the esters are derived from the three keto alcohols (ketols), pyrethrolone (**7**), cinerolone (**8**) and jasmolone (**9**) that form the six esters with the above acids.

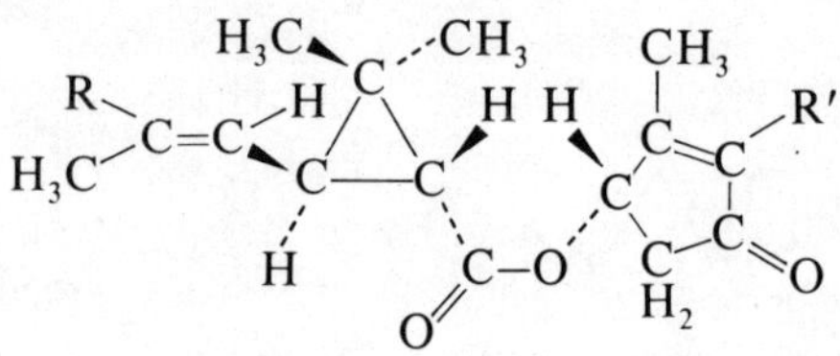

Basic structure

| R | R' | |
|---|---|---|
| $CH_3$ | $-CH_2-CH=CH-CH=CH_2$ | Pyrethrin I |
| $COOCH_3$ | $-CH_2-CH=CH-CH=CH_2$ | Pyrethrin II |
| $CH_3$ | $-CH_2-CH=CH-CH_3$ | Cinerin I |
| $COOCH_3$ | $-CH_2-CH=CH-CH_3$ | Cinerin II |
| $CH_3$ | $-CH_2-CH=CH-CH_2-CH_3$ | Jasmolin I |
| $COOCH_3$ | $-CH_2-CH=CH-CH_2-CH_3$ | Jasmolin II |

**Fig. 1.2**   Pyrethrins.

Harper (4) has suggested an interesting nomenclature for pyrethroids. He designated the esters as *"rethrins"* and their corresponding alcohols as *"rethrolones."* Because insecticidal properties of pyrethrins are dependent on their stereochemistry, structures **1–9** show spatial arrangement of the molecules. These can be simplified and the structures of various pyrethroids explained in terms of a basic structure as shown in Fig. 1.2. Chemists further simplify the drawing of structures of rethrins by often omitting the carbons and hydrogens in the middle. Thus the basic structure in simplified form can be written as

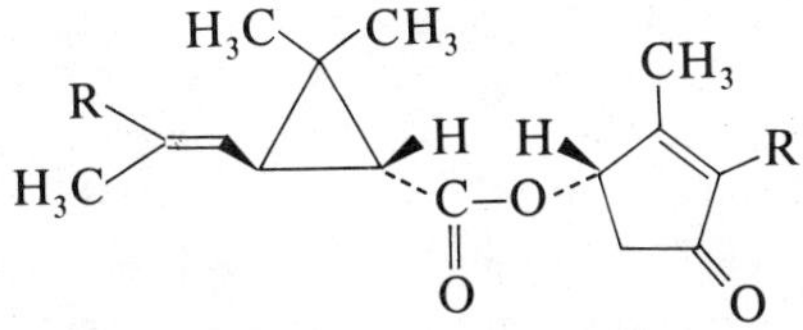

or, if details or stereochemistry are not essential, as

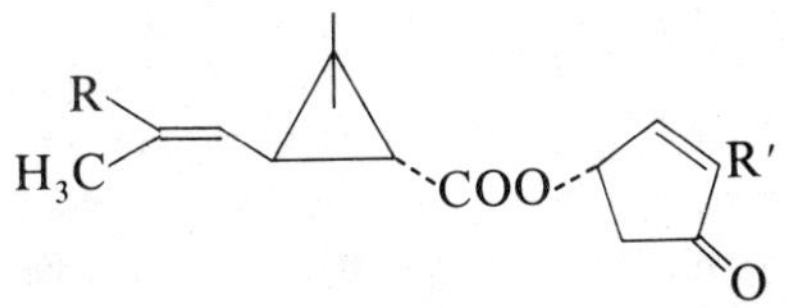

In this short account it is not possible to cover the interesting and, at times, intriguing history of the chemistry of pyrethrum, but the reader is referred to some of the earlier books and reviews (5–9) that have adequately covered the subject. Among the earlier investigators of the chemistry of pyrethrum, mention should be made of Fujitani (10), Staudinger and Ruzicka (11), Yamamoto (12), and LaForge and Barthel (13, 14). Research on the structure of the pyrethrum acids and ketols comprising pyrethrins and cinerins culminated when Crombie and Harper (15) and Katsuda et al. (16, 17) reported on the absolute configurations of these chemicals. Godin and co-workers (18, 19) isolated and characterized two additional pyrethrins, jasmolin I and jasmolin II. Three recent review articles by Crombie and Elliot (20), Barthel (21) and Matsui and Yamamoto (22) and a recent book (9) have summarized the history of recent research on pyrethrins.

Pyrethrins are heat and light sensitive and are easily oxidized. With increasing concentration and purity they tend to become increasingly unstable. In fact, some of the constituents of crude extract act as cosolvents and possibly as preservatives, and their removal during purification and concentration may cause precipitation problems later. The structures of pyrethrins are shown in Figs. 1.1 and 1.2.

## RELATIVE TOXICITIES OF PYRETHRINS

There has been some controversy regarding the relative toxicities of pyrethrin I and pyrethrin II to the house fly (*Musca domestica* L.). Investigators in the United States (23–25) found pyrethrin I to be 1.7 to 4.3 times as toxic to the house fly as pyrethrin II. On the other hand, British workers (26) found that in the case of the house fly, pyrethrin II was more toxic than pyrethrin I. All authors agree that for other insects tested, pyrethrin I is from 1.7 to as much as about 10 times as toxic as pyrethrin II. Pyrethrin I and II are more toxic than the corresponding cinerins (27).

The insecticidal activity of pyrethroids can be improved by mixing them with certain noninsecticidal compounds, a phenomenon known as *synergism* (see Chapter 16). One such chemical, piperonyl butoxide, is used commercially for this purpose. Sawicki (28) studied the synergistic action of piperonyl butoxide on pyrethroids and reported that their resultant toxicities varied with the amounts of synergist used. When the pyrethrins and piperonyl butoxide were used in the ratio of 1:8, the relative toxicities of the four constituents were enhanced as follows: pyrethrum extract, 1.0 (1.0); pyrethrin I, 1.31 (0.85); cinerin I, 1.28 (0.53); pyrethrin II, 0.90 (1.37); and cinerin II, 0.58 (0.49). The relative toxicities of the constituents alone are given in parentheses. At this ratio, according to Sawicki, the synergistic factors of the four constituents were: pyrethrin I, 18; cinerin I, 27; pyretherin II, 6; cinerin II, 14; pyrethrum extract, 10.

## SYNTHETIC PYRETHROIDS

### Modification of Rethronal Moiety:
### Allethrin and Related Compounds

After pyrethrins were characterized, efforts were soon directed toward the synthesis of natural pyrethrins and their analogs. Schechter et al. (29) prepared an ester that they named allethrin from chrysanthemic acid and an allyl analog, allethrolone, of the natural keto alcohol. The ester was found to be as effective against house flies as natural pyrethrins. Allethrin is a mixture of stereoisomers that show a marked difference in their toxicities to house flies. The one that has the same absolute configuration as the natural pyrethrins, that is, (+)-alcohol ester of (+)-*trans*-acid, is the most insecticidal. Allethrins are being produced commercially and are in extensive use in household sprays. Other related pyrethroids include furethrin and cyclethrin. Both of these are insecticidally active compounds. It appears that esters of chrysanthemic acid and rethrolones with a methylene group between a side-chain double bond and the cyclopentenolone ring give esters that are insecticidal in properties.

## Replacement of Alcohol Moiety

Synthetic chrysanthemates in which the rethrolone moiety has been replaced by benzyl or imidomethyl groups have been synthesized. Of these, barthrin, dimethrin, and tetramethrin (phthalthrin) are historically important. The last compound has been developed as an insecticide by Sumitomo Company of Japan under the name Neo-Pynamin®. Recently Nakanishi et al. (30) have reported on another synthetic pyrethroid, kikuthrin, 5-(2-propynyl)-2-methyl-3-furfuryl-methyl chrysanthemate, which they claim to be superior to allethrins in insecticidal properties. Resmethrin is another synthetic pyrethroid that has been developed commercially. It consists of four isomers of which the (+)-*trans*-isomer (bioresmethrin) is insecticidally very active. In resmethrin, the rethronal component of pyrethrins has been replaced by the 5-benzyl-3-furylmethyl alcohol. The structures of synthetic pyrethroids are shown in Fig. 1.3.

Modification of Alcohol Moiety

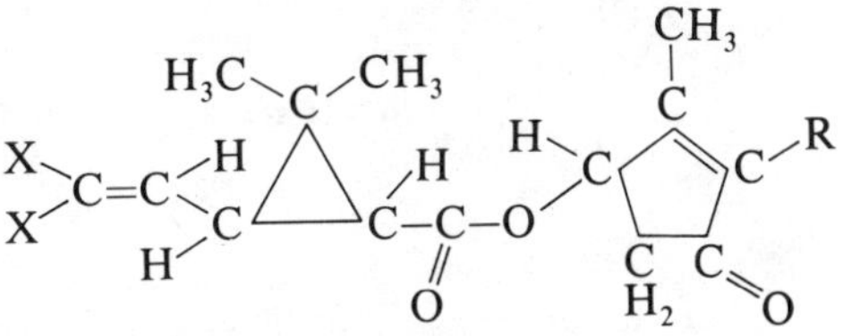

Basic structure

| X | R | Compound |
|---|---|---|
| $CH_3$ | $-CH_2-CH=CH_2$ | Allethrin |
| $CH_3$ | 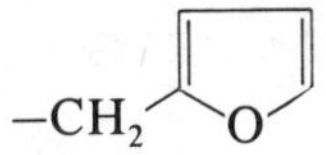 $-CH_2$ (furyl group) | Furethrin |
| $CH_3$ | 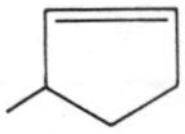 (cyclopentenyl group) | Cyclethrin |

Replacement of alcohol moiety

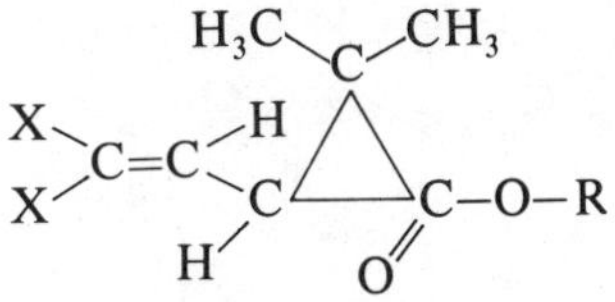

Basic structure

| X | R | Compound |
|---|---|---|
| $CH_3$ | 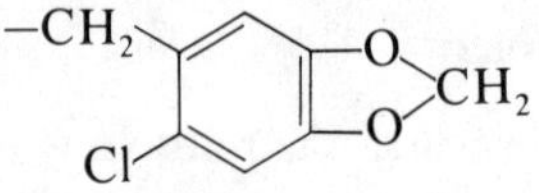 | Barthrin |
| $CH_3$ | 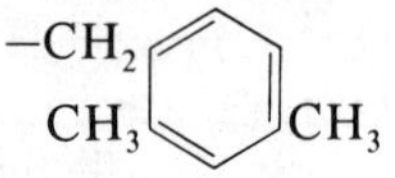 | Dimethrin |
| $CH_3$ | 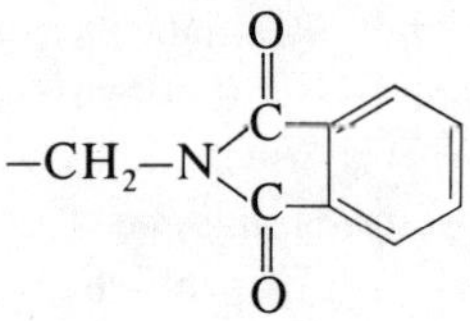 | Phthalthrin (Neopynamin®) |
| $CH_3$ | 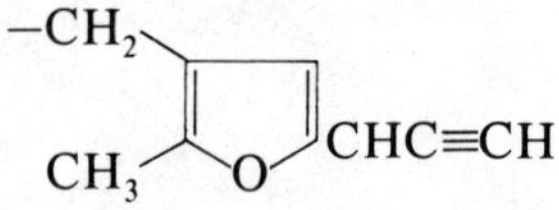 | Kikuthrin |
| $CH_3$ | 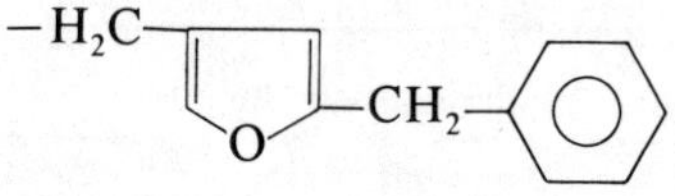 | Resmethrin Mixture of 4 isomers Bioresmethrin (+)-trans isomer |
| $CH_3$ | 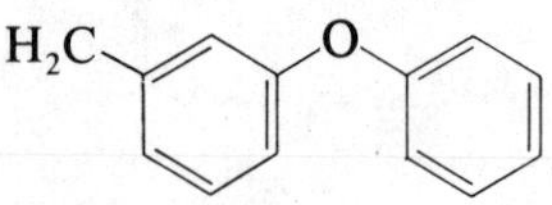 | Phenothrin |

Modification of Acid Moiety and
Replacement of Alcohol Moiety

| Cl | 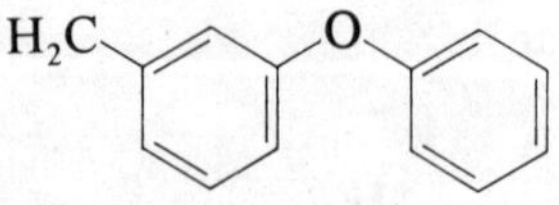 | Permethrin |
|---|---|---|

Fig. 1.3.  Synthetic pyrethroids.

## Recently Synthesized Pyrethroids

Elliott and co-workers, taking advantage of the information that the *trans-(E)-* methyl group of the chrysanthemate side chain in pyrethrin I and related insecticides was the main centre for metabolic attack, synthesized and tested compounds lacking this group. When the *trans-(E)*-methyl group was replaced by a but-1-enyl side chain, compounds of greatest insecticidal activity were obtained (31).

Elliott's group has also synthesized photostable pyrethroids that they claim are from 10 to 100 times more stable in light than previous pyrethroids, possess low mammalian toxicity, and are as insecticidally active as bioresmethrin (32) Fig. 1.3. Phenothrin and permethrin are two of the very recently developed pyrethroids that are more persistent than natural pyrethroids and bioresmethrin (32b). A structurally close dibromovinyl cyano-ester was found to be 23 times as insecticidal for the house fly (32b) as bioresmethrin. These

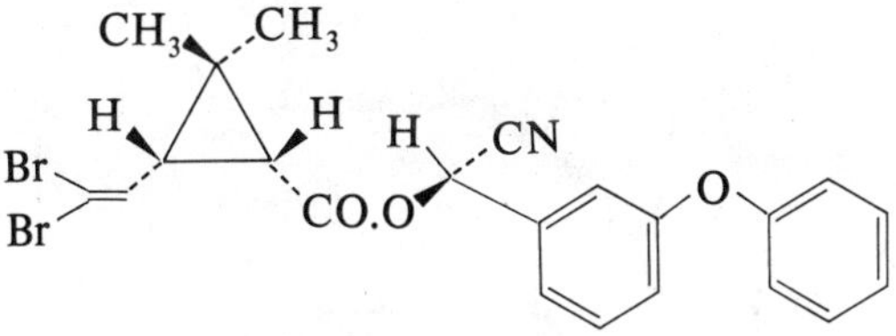

discoveries coupled with the work done by Casida and his group (9) may well revolutionize pyrethroid insecticides.

## STRUCTURE–ACTIVITY RELATIONSHIP

From the structures of insecticidally active pyrethroids, the following requirements for insecticidal activity of the compounds can be inferred. The acid moiety must be a cyclopropanecarboxylic acid with a *gem*-dimethyl attached to it. The alcohol moiety should have a planar or pseudoplanar ring provided with some $\pi$ electrons. A hydroxyl group not coplanar with the ring and located very close to it is essential. Stereoisomerism is also important. The chrysanthemic acid moiety has four stereoisomers, but the natural acid has the (+)-*trans* configuration and, when esterified with alcohols, this isomer produces the most insecticidal compounds. Similarly, allethrin has eight possible isomers, but the most insecticidal, allethrin, (+)-allethronyl (+)-*trans* chrysanthemate, has the same absolute configuration as natural pyrethrins. For more details see Elliott and Janes (32a), and Elliott et al. (32b).

## MODE OF ACTION OF PYRETHROIDS

Pyrethrins have a rapid paralytic action on insects; even sublethal dosages result in quick knockdown, though the effect is usually temporary and recovery is complete. Insects treated with lethal dosages of pyrethrins exhibit typical symptoms of nerve poisoning. Initially there is excitation, followed by convulsions, paralysis, and finally death. In the case of sublethal dosages, however, recovery after initial stages of paralysis is complete. This total recovery would indicate a reversible mechanism of action and also the ability of the insect to effectively metabolize small amounts of pyrethrins. The exact site of action in the central nervous system has defied identification. Insects killed by pyrethrum have shown indications of histopathological changes, but whether

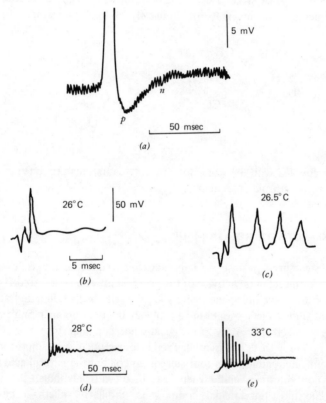

**Fig. 1.4.** Action potentials recorded intracellularly from the giant axon of the American cockroach. (a) Produced by a brief pulse of current; $p$, positive phase; $n$, prolonged negative afterpotential. (b), (c), (d), and (e) At different temperatures, resulting from a single shock by allethrin ($3 \times 10^{-7}$ gram/ml). Redrawn after Narahashi (33–35).

these are cause of lethal paralysis or its effects cannot be firmly established. The latter explanation may prove to be more accurate.

Blum and Kearns (36) reported that in the American cockroach, topically applied pyrethrum showed a negative coefficient of action between 15 and 35°C. The $LD_{50}$ at 35°C (6 $\mu$g) was six times that at 15°C (1 $\mu$g). Also, treated cockroaches that were prostrate at 15°C returned to normal when the temperature was raised to 35°C. However the rate of penetration of pyrethrum is greater at 35 than at 15°C. Blum and Kearns (36) have further observed that a neuroactive toxin is released into the hymolymph of pyrethrum-treated cockroaches. This neurotoxin is neither a pyrethroid metabolite nor a sodium or potassium ion. It has been suggested that it may be secondarily responsible for further stimulation and eventual paralysis of the insect nerve. Sullivan et al. (37) reported that the house fly and the Madeira cockroach (*Leucophaea maderae* Fabricius) demonstrate a circadian rhythm in their susceptibility to pyrethrum aerosols.

Interactions of pyrethroids with insect nerves have been studied recently and a few publications are available on the subject (33–36, 38). It was found that pyrethroids stimulate the nerve to discharge repetitively and then paralyze it. Narahashi (34) studied the interaction of allethrin with nerve axons and, with the aid of intracellular microelectrodes, made external recordings of action potentials. It was found that in low concentrations $3 \times 10^{-7}$ gram/ml (Fig. 1.4) allethrin increased and prolonged the negative afterpotential. At this concentration, repetitive afterdischarges may also be elicited by a single shock. This increased negative afterpotential in allethrin-poisoned axons, according to Narahashi, is possibly caused by accumulation of some depolarizing substances, other than sodium or potassium, either inside or outside the nerve membrane. The increased negative afterpotential probably initiates repetitive discharges, resulting in the hyperactivity of the poisoned insect. At concentration of $10^{-6}$ gram/ml or more, allethrin caused a slight fall of resting potential and an increase in negative afterpotential followed by a decrease in spike height and an eventual conduction block causing paralysis (34). The same author found that there is a profound effect of temperature on the initial height of the negative afterpotential that increased with a rise in temperature. However somewhere between 26 and 26.5°C there is a critical temperature above which repetitive excitation is produced by a single shock (compare DDT, Chapter 5).

The major site of excitation is the nerve membrane that reacts to a stimulus by increasing its conductance to sodium and potassium ions; this accounts for excitation or production of action potentials. Under normal circumstances, the nerve returns to resting potential by restoring the sodium and potassium concentration gradient across the membrane by physicochemical processes. However Narahashi (36) has found evidence to show that nerve paralysis in

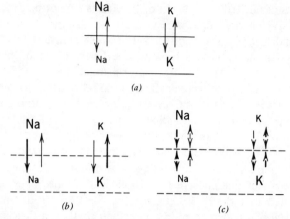

**Fig. 1.5.** Diagrammatic representation of the flow of Na and K ions across nerve membrane. The thickness of the arrows indicates the rate of flow and the size of the letters the concentration gradients across the membrane. (*a*) Resting stage; (*b*) active stage; (*c*) conductances blocked by the chemical. (*a*)and (*b*) redrawn after Narahashi (35).

allethrin-treated insects may be due to changes in nerve membrane that block its sodium and potassium conductance (Fig. 1.5). Narahashi (36), referring to a previous work with Anderson (38) on lobster axon, has further suggested that the allethrin molecule directly plugs the nerve membrane through intramolecular forces.

Though of relatively low oral and dermal toxicity to mammals, pyrethrins are highly toxic by intravenous injection; however an interesting feature is their relative ineffectiveness on isolated vertebrate nerve. This lack of activity cannot be explained solely on the basis of penetration through the nerve membrane of vertebrates. First, the nerve membrane should not provide a barrier to the apolar pyrethrins, and, second, even when the vertebrate nerve is desheathed the activity of the pyrethrins is not considerably increased (39).

In the case of insects, recovery from sublethal doses is complete, indicating metabolism of pyrethroids to nontoxic products. The metabolism of pyrethroids by animal tissue has been worked out but what happens to the molecules that presumably plug the insect nerve membrane is not known. However recovery in isolated preparations is very poor.

## METABOLISM OF PYRETHROIDS

### Metabolism in Insects

The ester linkage in pyrethroids appears to be a vulnerable site in the molecule, and earlier workers who started the studies on pyrethrum metabolism supported this contention. Chamberlain (40) incubated pyrethrins *in vitro* with acetone preparations of whole American cockroaches and house flies and obtained an unidentified acid. The production of acid lent further support to the assumption that the ester linkage was labile. Later investigations by Zeid et al. (41) appeared to support this view; their results, however, could not be confirmed: it was later discovered that their $^{14}C$ pyrethrins were impure. Bridges (42) and Winteringham et al. (43) also attacked the problem but could not provide definitive answers. Hopkins and Robbins (44) did not find evidence for hydrolysis. Chang and Kearns (45) studied the metabolism of randomly labeled pyrethrin I and cinerin I in the house fly and concluded that hydrolysis of the ester linkage was not a major detoxication mechanism; they surmised that the detoxication process started on the keto alcohol moiety. The

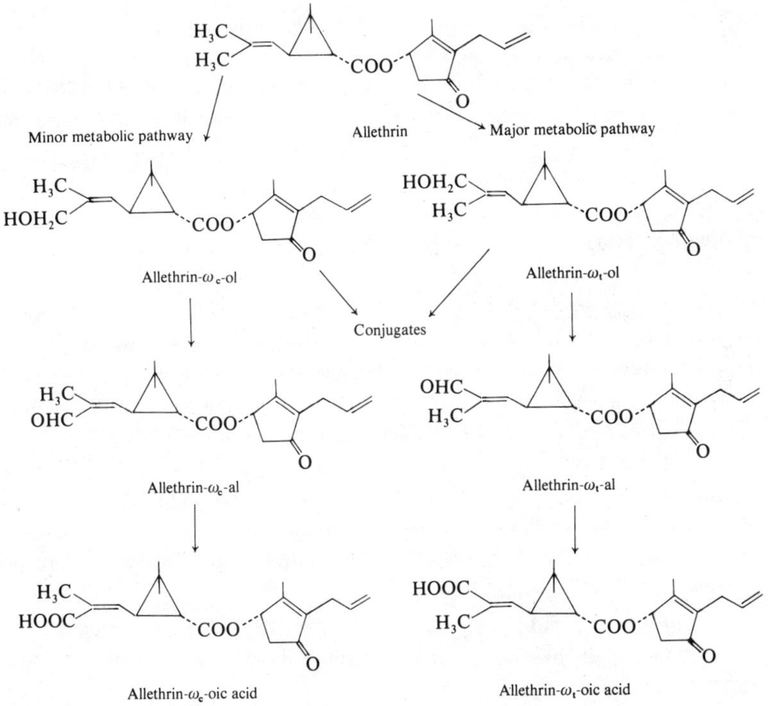

**Fig 1.6** Tentative metabolic pathway for allethrin in house flies. Redrawn after Yamamoto et al. (46). Reprinted with permission from *J. Agri. Food Chem.* **17**: 1227 (1969). Copyright by the American Chemical Society.

metabolism of pyrethroids was finally worked out by Yamamoto et al. (46) who synthesized $^{14}$C-labeled pyrethrin I, allethrin, dimethrin, and phthalthrin. These chemicals were allowed to interact *in vitro* with resistant house fly abdomen homogenates fortified with NADPH$_2$ (reduced nicotine adenine dinucleotide phosphate). In the case of allethrin, more than 10 metabolites were found and in each metabolite, the ester linkage and the allethrolone moiety remained unchanged. The major metabolic pathway was through hydroxylation of the *trans*-methyl group of the isobutenyl side chain in the acid moiety. This resulted in the formation of the alcohol allethrin-$\omega_t$-ol, which, by the action of other enzymes, formed a conjugate or was oxidized to the cor responding aldehyde, allethrin-$\omega_t$-al, and acid, allethrin-$\omega_t$-oic acid. A minor pathway that corresponds to the above involved a similar series of reactions at the *cis*-methyl group. In all probability the actual detoxication steps were the ones succeeding the formation of alcohols. At low enzyme rates the $\omega$-ol metabolites predominated, but at high enzyme level they were mainly oxidized to $\omega$-oic acids. In the case of living house flies, the same metabolites were formed, though the intermediate alcohol was the major metabolite and was apparently conjugated and excreted as glucosides. Other unidentified metabolites of a polar nature were also found in small quantities. These were produced, possibly, by further oxidation and conjugation of the alcohol and acid moieties (Fig. 1.6).

## Metabolism in Rats

Metabolism of pyrethrin I, pyrethrin II, and allethrin in rats has recently been studied by Yamamoto et al. (47) with the aid of $^{14}$C and tritium ($^3$H)-labeled compounds. The metabolism involves oxidative modifications in both the acid and alcohol moieties, hydrolysis of the methoxycarbonyl group of pyrethrin II, and hydrolysis of the ester linkage. The last reaction is not as important as the first two. These reactions, according to the authors "contribute to, or account for, the low toxicity of these compounds for mammals."

Metabolites of pyrethrins appeared in the expired air, urine, and feces of the treated rats. There were differences in the metabolic pathways of pyrethrins and allethrins (Figs. 1.7 and 1.8).

In rats, pyrethrin I and pyrethrin II both gave the four metabolites shown in Fig. 1.7. (47). Allethrin also gave four metabolites which are represented in Fig. 1.8.

Recent investigations have shown that there are differences in the metabolism of allethrins and pyrethrins in mammals (48); rats cleave the cyclopropane carboxylic ester in allethrins that are chrysanthemates of primary alcohols to a small extent, but this cleavage does not occur in the case

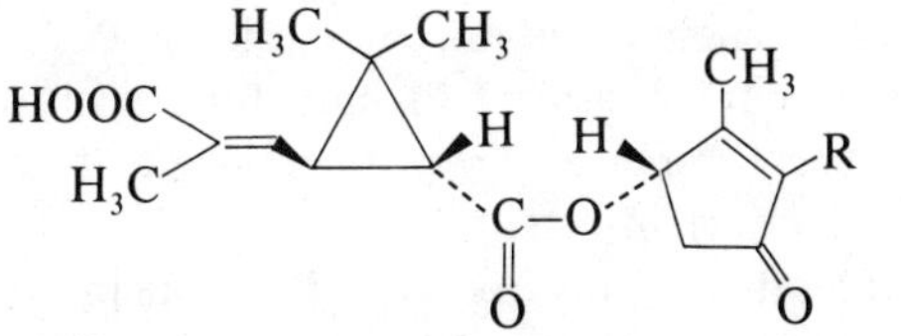

Fig. 1.7  Basic structure of the metabolites of pyrethrins.

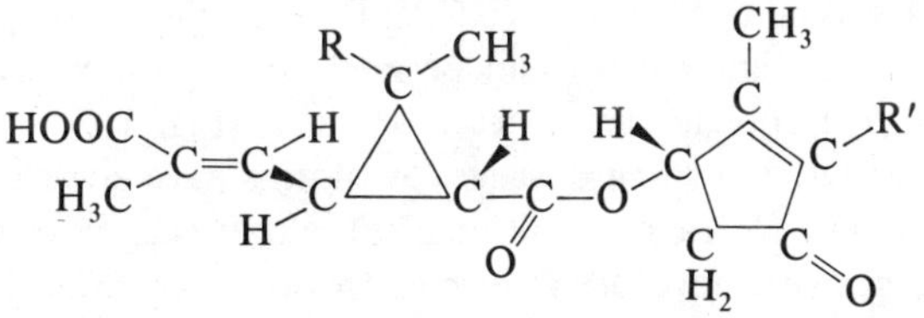

| Metabolite | R |
|---|---|
| A | $-CH_2-CH\stackrel{c}{=}CH-CHOH-CH_2OH$ |
| B | $-CH_2-CHOH-CH\stackrel{t}{=}CH-CH_2OH$ |
| C | $-CH_2-CH\stackrel{c}{=}CH-CH-O-$conj.<br>　　　　　　　　　$\mid$<br>　　　　　　　　　$CH_2OH$ |
| D | $-CH_2-CH\stackrel{c}{=}Ch-CH=CH_2$ |

Fig. 1.8  Basic structure of the metabolites of allethrins.

| Metabolite | R | R' |
|---|---|---|
| A' | $CH_3$ | $-CH_2-CHOH-CH_2OH$ |
| B' | $CH_3$ | $-CHOH-CH=CH_2$ |
| C' | $CH_2OH$ | $-CH_2-CH=CH_2$ |
| D' | $CH_3$ | $-CH_2-CH=CH_2$ |

of pyrethrins that are cyclopentenonyl esters. The synthetic pyrethroids that are esters of *trans*-chrysanthemates are extensively hydrolyzed by rats and the *cis* esters are readily cleaved by an esterase in mouse liver microsomes (48–50). According to Elliott et al. (48–50), the low mammalian toxicity of the pyrethrates is due to the *in vivo* hydrolysis of the methoxycarbonyl group, and the rapid oxidation of the isobutenyl moiety probably reduces the toxicity of the pyrethrins. The ease of oxidation of the pentadienyl side chain also plays an important part in reducing the toxicity of the pyrethrins to mammals.

Though pyrethrins are easily decomposed by oxidation and ultraviolet light (UV), their life can be extended by adding antioxidants and UV absorbers to the synergists. Prolonged activity of the pyrethrins is desirable for outdoor spraying and insect-resistant treatment of bags for ocean shipment of food stuffs. This is particularly important because resistance to malathion is now widespread in stored-product insects.

As stated earlier Elliott and co-workers at the Rothamstead Experimental Station in England and Japanese scientists at Sumitomo Chemical Company have synthesized and tested numerous synthetic pyrethroids, some of which are not easily decomposed by light and are more effective than the natural pyrethrins. Resmethrin, 5 benzyl-3-furylmethyl-*dl-cis,trans*-chrysanthemate, is one of the synthetic pyrethroids that consists of two highly insecticidal components, the (+)-*trans* isomer; and the (+)-*cis* isomers of bioresmethrin, and the two corresponding (−)isomers which are noninsecticidal. Bioresmethrin is 50 times more insecticidal to the house fly by topical application than pyrethrin I and yet it is less toxic to mammals (32). Metabolism of insecticidal components of resmethrin recently has been studied (51).

Pyrethroids even more persistent to light than resmethrin have been synthesized by Elliot and co-workers (52). Two of these are phenothrin (3-phenoxybenzyl chrysanthemate) and permethrin (3-phenoxybenzyl dichloro-vinyldimethylcyclopropane carboxylic acid ester) (Fig. 1.3).

When administered in small dosages, pyrethrins have no deleterious effects on mammals ($LD_{50}$ for rats is 1500 mg/kg). No effects on cholinesterase or other enzyme systems have been detected in mammals treated with dosages amounting to several milligrams per kilogram of body weight. On the whole, they are safe chemicals to use as household insecticides and can be used on livestock and for treating mills, warehouses, grain sacks, and even grain in storage.

Though resistance to pyrethrins has been reported in several insect species, it is not widespread. Some insects resistant to pyrethrins also develop some cross resistance to DDT (53). This would indicate more than one resistance mechanism in such insects. For further discussion of resistance and synergism see Chapters 15–19.

Recently pyrethrins were aerially applied to a forest ecosystem (54) at the rate of 0.1 and 0.2 lb/acre and short- and long-term effects on nontarget organisms were studied; surprisingly, fish were not affected (pyrethrum is toxic to fish and aquatic insects and crustaceans on which the fish feed). As expected, no adverse effects were found on birds and wildlife. In a recent study Kaya et al. (55) tested two pyrethroids, namely, resmethrin (SBP-1382, S. B. Penick Co., New York, New York), and Bioethanomethrin (RU 11679, MGK Corp., Minneapolis, Minn.), by aerial application against the gypsy moth, *Porthetria dispar* (L.), in Connecticut. Active ingredients at 0.05 lb/gal/acre were sprayed; knockdown rate of the larvae was high, but after 5 hours larvae started recovering and foliage protection was not obtained.

## REFERENCES

1. J. B. Moore. *Pyrethrum Post.* **7**: 15 (1964).
2. F. Sham. *op. cit.* **11**: 50 (1971).
3. L. O. Hopkins. *op. cit.* **7**: 41 (1964).
4. S. H. Harper. *Chem. Ind. (Lond.)* **1949**: 639.
5. C. B. Gnadinger. *Pyrethrum Flowers*, 2nd ed., McLaughlin Gormley King Co., Minneapolis, 1936.
6. C. B. Gnadinger. *Pyrethrum Flowers*, Suppl. to 2nd ed. (1936–1945), McLaughlin Gormley King Co., Minneapolis, 1945.
7. C. C. McDonnell, R. C. Roark, F. B. LaForge, and G. L. Keenan. U.S. Department of Agriculture Bulletin No. 284, *U.S. Govt. Printing Office*, 1962.
8. H. H. Shepard. *The Chemistry and Action of Insecticides*, McGraw-Hill, New York, 1951.
9. J. E. Casida. (ed.). *Pyrethrum, the Natural Insecticide*, Academic Press, New York, 1973.
10. J. Fujitani. *J. Arch. Exp. Pathol. Pharmakol.* **61**: 47 (1909).
11. H. Staudinger and L. Ruzicka. *Helv. Chim. Acta* **7**: 177 (1924).
12. R. Yamamoto. *J. Chem. Soc. Japan* **44**: 311 (1923).
13. F. B. LaForge and F. W. Barthel. *J. Org. Chem.* **9**: 242 (1944).
14. F. B. LaForge and W. F. Barthel. *J. Org. Chem.* **10**: 222 (1945).
15. L. Crombie and S. H. Harper, *J. Chem. Soc.* **1954**: 470.
16. Y. Katsuda, T. Chikamoto, and Y. Inouye. *Bull. Agr. Chem. Soc. Jap.* **22**: 427 (1958).
17. Y. Katsuda, T. Chikamoto, and Y. Inouye. *Bull. Agr. Chem. Soc. Jap.* **23**: 174 (1959).
18. P. J. Godin, J. H. Stevenson, and R. M. Sawicki. *J. Econ. Entomol.* **58**: 548 (1965).
19. P. J. Godin, R. L. Sleeman, M. Snarey, and E. M. Thain. *J. Chem. Soc. (C)* **1966**: 322.
20. L. Crombie and M. Elliott. In *Progress in the Chemistry of Organic Products* (L. Zechmeister, ed.), Springer, Vienna, 1961, p. 120.
21. W. F. Barthel. *World Rev. Pest Control* **6**: 59 (1967).
22. M. Matsui and I. Yamamoto. In *Naturally Occurring Insecticides* (M. Jacobson, and D. G. Crosby, eds.), Dekker, New York. 1971, p. 3.
23. S. C. Chang and C. W. Kearns. *J. Econ. Entomol.* **55**: 919 (1962).
24. W. A. Gersdorff. *J. Econ. Entomol.* **40**: 878 (1947).
25. H. H. Incho and H. W. Greenberg. *J. Econ. Entomol.* **45**: 794 (1952).
26. R. M. Sawicki and M. Elliott. *J. Sci. Food Agr.* **16**: 85 (1965).
27. M. Elliott, P. H. Needham, and C. Potter. *J. Sci. Food Agr.* **20**: 561 (1969).
28. R. M. Sawicki. *J. Sci. Food Agr.* **13**: 260 (1962).
29. M. S. Schechter, N. Green, and F. B. LaForge. *J. Amer. Chem. Soc.* **71**: 3165 (1949).
30. M. Nakanishi, A. Tsuda, K. Abe, S. Inamasu, and T. Mukai. *Botyu-Kagaku* **35**: 91 (1970).
31. M. Elliott, A. W. Farnham, N. F. Janes, P. H. Needham, and D. A. Pulman, *Nature (Lond.)* **244**: 456 (1973).1973).
32. M. Elliott, A. W. Farnham, N. F. Janes, P. H. Needham, D. A. Pulman, and J. H. Stevenson. *Nature (Lond.)* **246**: 169
32a. M. Elliott and N. F. Janes. *In "Pyrethrum"* (J. E. Casida, ed.), Academic Press, New York, 1973, p. 56.
32b. M. Elliott, A. W. Farnham, N. F. Janes, P. H. Needham, and D. A. Pulman. In *Mechanism of Pesticide Action* (G. K. Kohn, ed.), American Chemical Society, Washington, 1974, p. 80.
33. T. Narahashi. *J. Cell. Comp. Physiol.* **59**: 61 (1962).
34. T. Narahashi. *J. Cell. Comp. Physiol.* **59**: 67 (1962).
35. T. Narahashi. In *The Physiology of the Insect Nervous System*, 12th Internation Congress of Entomology, London (J. E. Treherne and J. W. L. Beament, eds.), Academic Press, New York. 1965, p. 1.

36. M. S. Blum and C. W. Kearns. *J. Econ. Entomol.* **49**: 862 (1956).
37. W. N. Sullivan, C. Bryant, D. K. Hayes, and J. Rosenthal. *J. Econ. Entomol.* **63**: 159 (1970).
38. T. Narahashi and N. C. Anderson. *Toxicol. Appl. Pharmacol.* **10**: 529 (1967).
39. G. Camougis. In *Pyrethrum, the Natural Insecticide* (J. E. Casida, ed.), Academic Press, New York, 1973, p. 211.
40. R. W. Chamberlain. *Amer. J. Hyg.* **52**: 153 (1950).
41. M. M. I. Zeid, P. A. Dahm, R. E. Hein, and R. H. McFarland. *J. Econ. Entomol.* **46**: 324 (1953).
42. P. M. Bridges. *Biochem. J.* **66**: 316 (1957).
43. F. P. W. Wintcringham, A. Harrison, and P. M. Bridges. *Biochem. J.* **61**: 359 (1959).
44. T. L. Hopkins. and W. E. Robbins. *J. Econ. Entomol.* **50**: 684 (1957).
45. S. C. Chang and C. W. Kearns. *J. Econ. Entomol.* **57**: 397 (1964).
46. I. Yamamoto, E. C. Kimmel, and J. E. Casida. *J. Agr. Food Chem.* **17**: 1227 (1969).
47. I. Yamamoto, M. Elliott, and J. E. Casida. *Bull. WHO* **44**: 347 (1971).
48. M. Elliott, N. F. Janes, E. C. Kimmel, and J. E. Casida. *J. Agr. Food Chem.* **20**: 300 (1972).
49. C. O. Abernathy and J. E. Casida. *Science* **179**: 1235 (1973).
50. M. Elliott, N. F. Janes, E. C. Kimmel, and J. E. Casida. In *Insecticides* (A. S. Tahori, ed.), Vol. I, Gordon and Breach, New York, p. 141.
51. K. Ueda, L. C. Gaughan, and J. E. Casida. *J. Agr. Food Chem.* **23**: 106 (1975).
52. M. Elliott, A. W. Farnham, N. F. Janes, P. H. Needham, D. A. Pulman, and J. H. Stevenson. *Nature* **246**: 169 (1973).
53. C. J. Llloyd. *J. Stored Prod. Res.* **5**: 337 (1969).
54. R. E. Pillmore. *In Pyrethrum* (J. E. Casida, ed.), Academic Press, New York, 1973, p. 143.
55. H. Kaya, D. Dunbar, C. Doane, R. Weseloh, and J. Anderson. "Gypsy moth: Aerial Tests with *Bacillus thuringiensis* and Pyrethroids," Bull. No. 744, The Connecticut Agricultural Station, New Haven, 1974.

# INSECTICIDES OF PLANT ORIGIN

| | Nicotinoids, Rotenoids, and Other |
|---|---|
| **CHAPTER 2** | Insecticides of Minor Importance |

## NICOTINOIDS

As early as 1690 tobacco extract was suggested as plant spray in parts of Europe; use of water extracts of tobacco for the control of plum curculio, *Conotrachelus nenuphar* (Herbst), was suggested in 1734. The alkaloid nicotine from tobacco was named after Jean Nicot in 1829. Nicotine, *l*-1-methyl-2-(3-pyridyl)-pyrrolidine, is the principal alkaloid in the leaves of *Nicotiana tabacum* L. (2–5%) and *N. rustica* L. (5–14% and up to 20% in

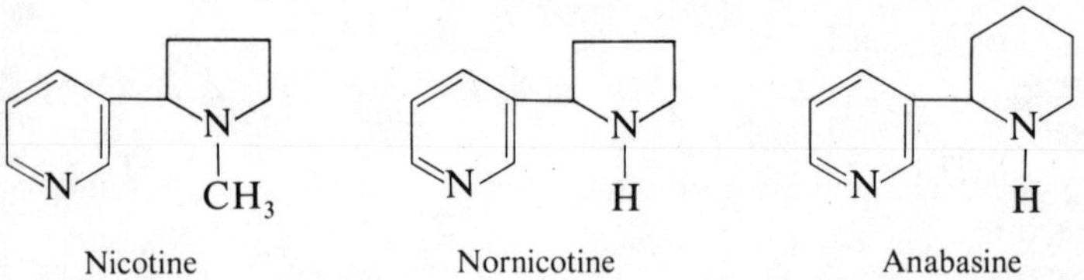

| Nicotine | Nornicotine | Anabasine |
|---|---|---|

some varieties); other alkaloids of insecticidal importance are nornicotine and anabasine. An excellent review on nicotine and other tobacco alkaloids is available (1).

Nicotine is highly toxic, its oral $LD_{50}$ for rat being 30 mg/kg, but it is rapidly detoxified and eliminated from animal body. In the presence of light it decomposes easily. Nicotine is often used as the sulfate, representing 40% of the actual nicotine. It is not volatile and is less toxic to man than the parent compound. Nicotine is also used as a contact insecticide for aphids, as a fumigant in greenhouses, and against poultry mites.

Yamamoto et al. (2) and Yamamoto (3) conducted detailed investigations on the structure–activity relationships of 26 synthetic nicotine analogs and discovered that those resembling acetylcholine in configuration and charge distribution were toxic to several insect species. They consider the following requirements as essential for insecticidal properties: (*a*) a pyridine ring; (*b*) a highly basic nitrogen on the pyrrolidine ring; (*c*) an optimum distance of about 4.2 Å between the two nitrogen atoms; and (*d*) an unsubstituted α-position of the pyridine ring. However the ionic barrier surrounding the insect nerve synapse prevents the effective penetration of highly ionized compounds. In the case of nicotine the N on the aromatic ring is weakly basic ($pK_a \simeq 3$); the N on the pyrrolidine ring is moderately basic ($pK_a$ about 8). Therefore at physiological pH about 90% of the compound is protonated at the pyrrolidine nitrogen. Those nicotinoids that have highest insecticidal action usually exist largely in the ionized form at physiological pH.

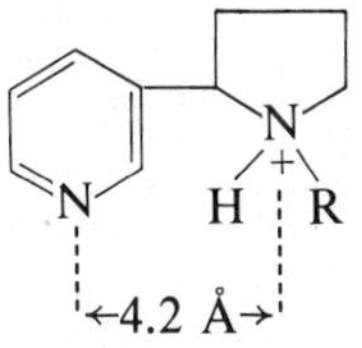

Nicotinoids
(insecticidal)

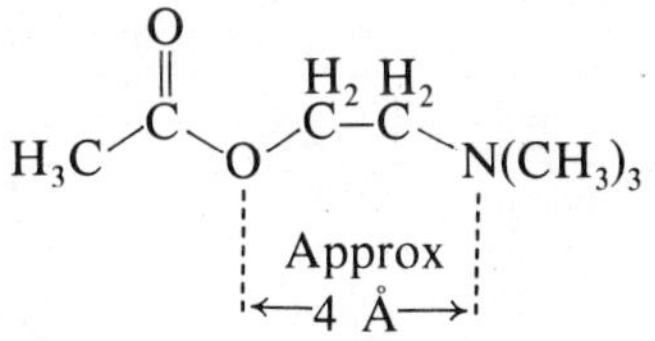

Acetylcholine

## Mode of Action

Yamamoto feels that the free base penetrates easily through the integument, the nicotinium ion, with the positive charge on the pyrrolidine N is slow to penetrate (3). After reaching the body fluid an equilibrium is established between the two at the pH of the insect body hemolymph. Only free bases pass through the synaptic barrier, and the nicotinium ion is metabolized and excreted. After penetration through the nerve barrier a new equilibrium is established between the ion and the free base. The nicotinium ion is the physiologically active entity. It interacts with acetylcholine, which it resembles, and hence the insecticidal action.

Eldefrawi et al. (4) studied the binding of nicotine-$H^3$ and five analogs to extracts of house fly head and compared it with the binding of $H^3$-muscarone. There was reversible binding in both cases with the same protein, probably acetylcholine, receptors. Good correlation was found between the toxicity of nicotine and five analogs to house fly and their ability to bind with the receptor(s); this binding was surmised to be the cause of toxic action of nicotine and its analogs.

## ROTENOIDS

Plants containing rotenone have been used as fish poisons for a long time, but as insecticides rotenoids were first used in 1848 against leaf-eating caterpillars. Use of rotenone in the United States was encouraged because of lead residues. The plants of economic importance are *Derris* spp. of Malaya and the East Indies and *Lonchocarpus* of South America. There are several insecticidal principles in the roots and seeds of these plants, but rotenone, which is found principally in the roots, is the most important. It is toxic, the oral $LD_{50}$ for rat being 132 mg/kg. The chemistry and insecticidal activities of rotenoids have recently been reviewed (5, 6).

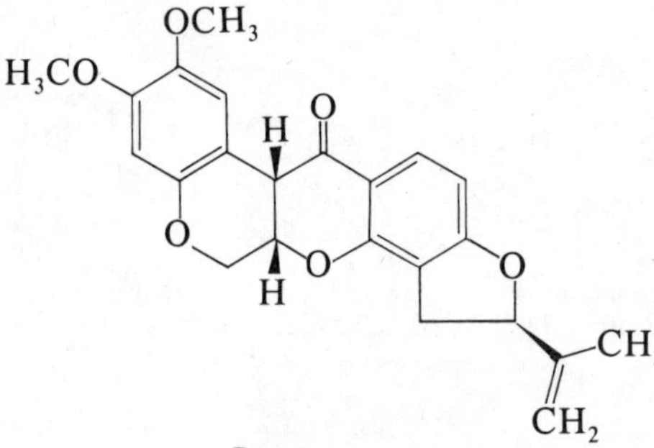

Rotenone

Rotenone is easily decomposed in the presence of sunlight and air and can therefore be used on edible produce just before harvesting. Photodecomposition of rotenone has been studied by Cheng et al. (7) who found that at least 20 compounds were formed when rotenone was exposed to sunlight.

The toxicity of rotenone is possibly due to its inactivation of the respiratory enzyme glutamic acid oxidase.

Other insecticides of plant origin include Sabadilla, which is obtained from the ground seeds of *Schoenocaulon officinale* A. Gray (Liliaceae) of South and Central America, and Ryania which is ground roots and stem of *Ryania speciosa* Vahl (Flacourtiaceae) of South America. Sabadilla is used to control thrips and hemipterous insects, and ryania is effective against lepidopterous larvae.

## REFERENCES

1. I. Schmeltz. In *Naturally Occurring Insecticides*. (M. Jacobson and D. G. Crosby, eds.), Dekker, New York, 1971, p. 99.
2. I. Yamamoto, H. Kamimura, R. Yamamoto, S. Sakai, and M. Goda. *Agr. Biol. Chem.* **26:** 709 (1962).
3. I. Yamamoto. In *Adv. Pest Control Res.* **6:** 231 (1965).
4. M. E. Eldefrawi, A. T. Eldefrawi, and R. D. O'Brien. *J. Agr. Food. Chem.* **18:** 1113 (1970).
5. H. Fukami and M. Nakajima. In *Naturally Occurring Insecticides* (M. Jacobson and D. G. Crosby, eds.), Dekker, New York, 1971, p. 71.
6. L. Crombie. In *Insecticides,* Proceedings of the Second International IUPAC Congress of Pesticide Chemistry. (A. S. Tahori, ed.), Vol. I Gordon and Breach, New York, 1972, p. 101.
7. H.-M. Cheng, I. Yamamoto, and J. E. Casida. *J. Agr. Food Chem.* **20**: 850 (1972).

# SYNTHETIC ORGANIC INSECTICIDES

# CHAPTER 3 | Organdphosphorus Compounds

The organophosphorus compounds first developed by Gerhard Schrader in Germany during the 1930s are one of the most important classes of pesticidal chemicals. In addition to being good insecticides and acaricides, chemicals belonging to this group have provided defoliants, fungicides, herbicides and nematicides. Some of the desirable characteristics of organophosphorus insecticides are discussed below.

## BROAD-SPECTRUM INSECTICIDES

Parathion and other related chemicals are general purpose broad-spectrum insecticides and are used when quick control of insects is desired.

## HIGHLY SELECTIVE SYSTEMIC INSECTICIDES FOR PLANTS

Schradan (OMPA, Pestox III, Systam), octamethyl pyrophophoramide, and other similar pesticides are highly selective in their toxic action. Schradan is almost nontoxic to chewing insects, but it is an effective systemic insecticide for the control of sucking insects; it is, however, very toxic to mammals.

## SYSTEMIC INSECTICIDES FOR PLANTS

Among systemic insecticides, mention should be made of demeton (Systox®), an acaricide and an insecticide, and phorate (Thimet®). The latter is absorbed

27

and translocated in the entire seedling when applied as seed dressing at the time of planting and protects the plant during the critical period of its existence (20–25 days).

## ANIMAL SYSTEMICS

Organophosphorus compounds, such as ronnel (Korlan®), that exhibit low mammalian toxicity but effective insecticidal action can either be fed to the animals for grub control, as in the case of trichlorfon (Neguvon®), or can be applied externally for eventual systemic action.

## MATERIALS WITH SHORT RESIDUAL ACTION

These insecticides have a very short "life" and are readily hydrolyzed to non-toxic products. Their use is indicated when insect control is desired shortly before harvesting. Among chemicals of this type are tetraethyl pyrophosphate (tepp or Tetron®) and mevinphos (Phosdrin®), though the latter insecticide also has high vapour pressure and easily volatilizes.

## MATERIALS WITH PROLONGED ACTIVITY FOR USE IN AGRICULTURE

Where persistent insecticidal action is a desirable quality in agriculture, the use of azinphosmethyl (Guthion®) is recommended. This chemical is stable and can be stored for prolonged periods without deterioration. It is, however, very toxic to mammals.

## MATERIALS WITH PROLONGED ACTIVITY FOR THE CONTROL OF INSECTS OF MEDICAL IMPORTANCE

Where materials with low mammalian toxicity and persistent qualities are required, organophosphorus compounds have not been found lacking. Baytex® or fenthion, unlike azinphosmethyl, is not very toxic to mammals; but like azinphosmethyl, it is stable and can be stored for prolonged periods without decomposition. It is used for the control of insect vectors of diseases that have acquired resistance to chlorinated hydrocarbons. Fenthion has also been used for the purpose of malaria control in the form of an indoor residual spray on the walls of houses. Sumithion® (fenitrothion) as a residual indoor spray (2 grams/m²) has given effective kills of *Anopheles gambiae* for 2 months or more in Africa.

## CHEMICALS WITH EXTREMELY LOW MAMMALIAN TOXICITY

Though the initial interest in organophosphorus compounds started because of their neurotoxicity, this group has also yielded compounds showing extremely low toxicity to mammals while possessing very good pesticidal properties. Among such compounds are malathion (Cythion®) and Abate®. Malathion has remained effective as an indoor residual spray for a period of about 3 months and has been used by the World Health Organization (WHO) for this purpose.

Among some of the more desirable properties of organophosphorus compounds are: degradation to products nontoxic to higher forms of life; low dosage with rapid action on plant pests; relatively rapid metabolism in and low chronic toxicity to mammals. Some of the organophosphorus insecticides, however, are highly toxic to mammals and many fatalities have been reported because of the improper handling or use of these chemicals.

## MODE OF ACTION

It is generally accepted that organophosphorus compounds are toxic because they phosphorylate vital esterases, thus forming complexes that are either irreversible or do not readily release the enzymes. The enzyme mainly affected its accepted to be cholinesterase, an enzyme that plays a vital role in hydrolyzing acetylcholine. In order to understand the function of acetylcholine and

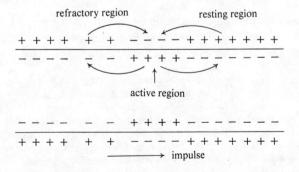

**Fig. 3.1** The conduction of impulse in an axon. Redrawn after Narahashi (6). From *The Physiology of Insect Central Nervous System*, Academic Press, London, 1965.

cholinesterase, it is necessary to discuss briefly the physiological processes involved. At myoneural junctions in mammals and in insects, the terminal portion of the muscle and the nerve fibre are at a slightly negative potential with respect to the outside. If this gradient is altered so that the inside becomes less negative, an action potential moves down the nerve (Fig. 3.1); when this potential reaches the nerve ending at the myoneural junction, acetylcholine is released and diffuses through the gap between the nerve and the muscle (the gap is about 100 Å wide) and is adsorbed onto the muscle ending. This interaction of acetylcholine with the receptor end of the muscle gives rise to an action potential in the muscle that responds by performing work, that is, contracting. After its purpose is accomplished, the acetylcholine is hydrolyzed by acetylcholinesterase (cholinesterase)* and the muscle returns to its original state, ready to respond to future stimuli. Reviews on the biochemistry of these enzymes and their interactions with organophosphorus esters are available (1–5).

The reaction between acetylcholine (ACH) and cholinesterase (HE) takes place in three stages and can be represented by the following sequence:

$$(CH_3)_3\overset{+}{N}C_2H_4O-\overset{\overset{\displaystyle O}{\|}}{C}-CH_3 \; + \; HE \; \rightleftharpoons \; (CH_3)_3NC_2H_4O-\overset{\overset{\displaystyle O}{\|}}{C}-CH_3/HE$$

Acetylcholine                                                Enzyme complex

At this stage there is an equilibrium between the enzyme and its substrate on the one hand and a complex of the two on the other.

$$(CH_3)_3NC_2H_4O\overset{\overset{\displaystyle O}{\|}}{C}-CH_3/HE \; \longrightarrow \; (CH_3)_3\overset{+}{N}C_2H_4OH \; + \; CH_3CO\cdot E$$

Choline                          Acetylated
enzyme

The complex yields choline and acetylated enzyme in the second stage. The final stage is the deacetylation reaction in which the acetylated enzyme is hydrolyzed to give the free enzyme and acetic acid.

---

* Any of the several enzymes that hydrolyze choline esters is called cholinesterase. The enzyme that hydrolyzes the choline ester, acetylcholine, is called acetylcholinesterase. It is the main cholinesterase of vertebrate nervous systems and is also found in vertebrate erythrocytes. Toxic organophosphates and carbamates inhibit this enzyme. Acetylcholinesterase is also called "true cholinesterase" to distinguish it from a similar enzyme, pseudocholinesterase, that is found primarily in vertebrate blood plasma. The importance of this enzyme has not been investigated in detail, it can be totally inhibited without any apparent harmful effects. The model depicting two point landing of the insecticide or acetylcholine on the anionic and esteratic site of the enzyme is a simplified explanation. In fact acetylcholinesterase occurs with multiple forms and the inhibition of this enzyme may be a complex phenomenon involving several steps.

$$CH_3CO.E \xrightarrow{H_2O} CH_3COOH + HE$$

The active center of acetylcholinesterase (AChE) is structurally complementary to its substrate acetylcholine. Acetylcholine contains a trimethylammonium group with a positive charge on N and an ester linkage. The enzyme's active centre contains a negatively charged anionic site, which binds the trimethylammonium group, and a relatively "nonspecific" esteratic site, which catalyzes the hydrolysis of the ester linkage. In the esteratic site there are basic (histidine imidazole), serine hydroxyl, and acidic (tyrosine hydroxyl) groups. This model postulated by Krupka (4) is represented in Fig. 3.2.

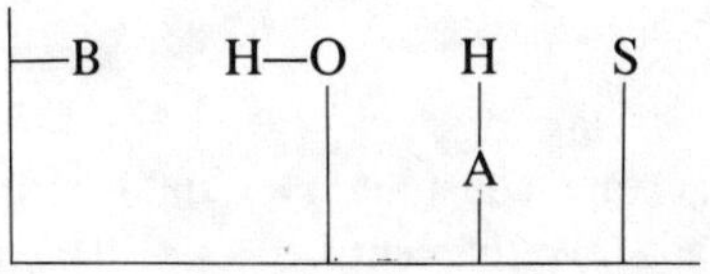

**Fig. 3.2** Proposed groupings in acetylcholinesterase. AH, acidic group; B, basic group; OH, serine hydroxyl; S, anionic site. Redrawn after Krupka (4). Reproduced by permission of the National Research Council of Canada from the *Can. J. Biochem.*, **42**: 677–693 (1964).

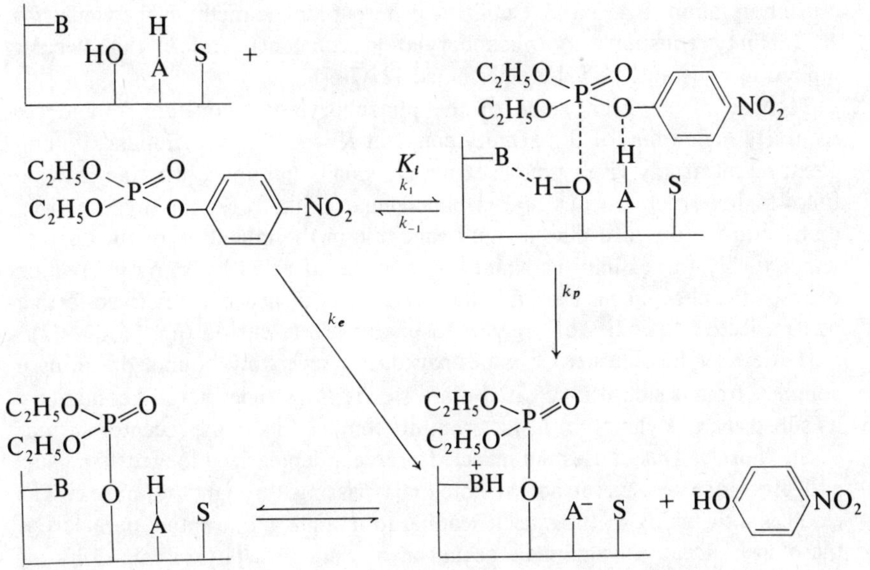

**Fig. 3.3** Reaction between cholinesterase and paraoxon (Redrawn with some changes from Fukuto (2)).

The reaction between toxic phosphorus esters, using paraoxon as an example, and cholinesterase is represented in Fig. 3.3. When the two chemicals interact there is a nucleophilic attack of the serine hydroxyl on the phosphorus atom that is aided by the acidic and basic groups present in the esteratic site of the enzyme. This results in the formation of a "reversible" complex that finally yields phosphorylated enzyme and $p$-nitrophenol. Aldridge (7) investigated the inhibition of cholinesterase by parathion and related chemicals and found that the complex did not show "significant reversibility." In other words, the inhibition of cholinesterase in this case followed first-order kinetics and was bimolecular, that is,

$$K = \frac{1}{tI} \ln \frac{100}{b}$$

where $K$ = bimolecular rate constant, $t$ = time in minutes, $I$ =molar inhibitor concentration, and $b$ = percentage residual activity.

Correlation between the reactivity of a phosphorus ester and its inhibition of cholinesterase, however, has not always been ideal, and Main (8) introduced a kinetic treatment for the reaction that takes into account the reversibility of the complex. This reversibility is dependent on the affinity of the inhibiting compound for the active site of cholinesterase as well as on the rate of phosphorylation (Fig. 3.3). By utilizing different kinetic methods the values for $K_i$ (affinity constant), $k_p$ (phosphorylation constant), and $k_e$ (bimolecular inhibition constant) may be determined (2, 7–9).

Because of the high value of the phosphorylation constant $k_p$ and the relatively high value of the affinity constant $K = k_{-1}/k_1$, the amount of complex present at any given time is extremely small, and this is why the reaction follows first-order kinetics and is bimolecular. Fukuto (2) has hypothesized that steric factors also play a significant role in the inhibition of the enzyme and possibly the affinity constant $K_i$ "may be affected by steric interaction between the phosphorus ester and the enzyme." This aspect is reviewed further when selective toxicities of organophosphates are discussed (pp. 36, 54–62).

If the acetylcholinesterase is destroyed, is irreversibly bound, or forms a complex from which it is released more slowly than under natural conditions, its substrate, acetylcholine, is not promptly removed from the receptor surface of the muscle. This causes the muscle to remain depolarized longer than usual and gives rise to several action potentials passing through the muscle. The result is a twitching of the muscle leading to tetanus and eventual paralysis of the muscle. Death in mammals occurs as a result of asphyxia caused by the paralysis of respiratory muscles.

## NOMENCLATURE OF ORGANOPHOSPHORUS COMPOUNDS

There has never been a universal agreement on the naming of organophosphorus compounds. Unfortunately, different countries more often than not employ their own systems of nomenclature. There are some guidelines that one can follow. In general, organophosphorus compounds can be regarded as derivatives of corresponding acids. Chemical names of compounds formed from phosphorus acid have the ending "ite" and those from phosphoric acid are given the ending "ate." If the chemical has the general formula

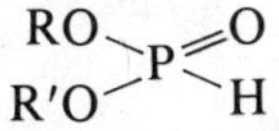

$$\begin{array}{c} RO \diagdown \\ R'O \diagup \end{array} P \begin{array}{c} \diagup O \\ \diagdown H \end{array}$$

it is called a phosphite. If it has the formula

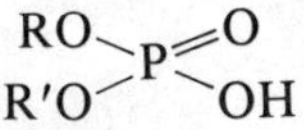

$$\begin{array}{c} RO \diagdown \\ R'O \diagup \end{array} P \begin{array}{c} \diagup O \\ \diagdown OH \end{array}$$

it is called a phosphate, for example,

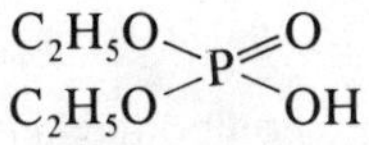

$$\begin{array}{c} C_2H_5O \diagdown \\ C_2H_5O \diagup \end{array} P \begin{array}{c} \diagup O \\ \diagdown OH \end{array}$$

is diethyl phosphate.

When there is a sulfur atom attached to phosphorus in the molecule, the chemical can be designated by the ending "thiophosphate," and the position of the S atom is indicated. Thus

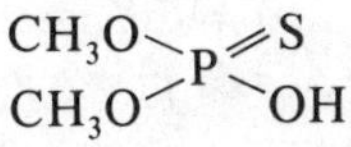

$$\begin{array}{c} CH_3O \diagdown \\ CH_3O \diagup \end{array} P \begin{array}{c} \diagup S \\ \diagdown OH \end{array}$$

is $O,O$-dimethyl thiophosphate, or $O,O$-dimethyl phosphorothionate. The compound

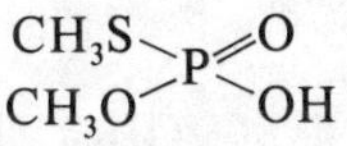

$$\begin{array}{c} CH_3S \diagdown \\ CH_3O \diagup \end{array} P \begin{array}{c} \diagup O \\ \diagdown OH \end{array}$$

is $O,S$-dimethyl thiophosphate or $O$-methyl $S$-methyl phosphorothiolate. Both these compounds can have the ending "phosphorothioate" in place of "thiophosphate"; those who use these two terms do not distinguish between a thiono (=S) sulfur (thionate) and thiolo (—S—) sulfur (thiolate).

Derivatives of alkyl phosphonic acid, in which one carbon atom is attached

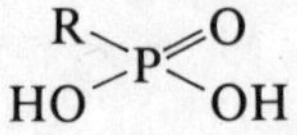

$$\begin{array}{c} R \diagdown \quad \diagup O \\ \quad P \\ HO \diagup \quad \diagdown OH \end{array}$$

to phosphorus are called phosphonates, and the derivatives of dithiophosphoric acid of the general formula

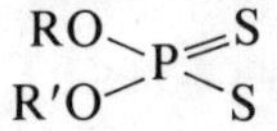

$$\begin{array}{c} RO \diagdown \quad \diagup S \\ \quad P \\ R'O \diagup \quad \diagdown S \end{array}$$

are called dithiophosphates or phosphorodithioates.

Derivatives of pyrophosphoric acid

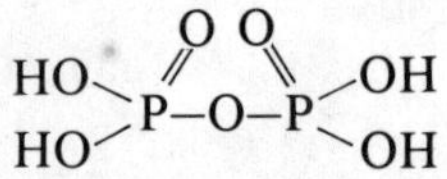

$$\begin{array}{c} \quad\quad O\ \ O \\ HO\diagdown\ \ \| \ \ \| \ \diagup OH \\ \quad\ \ P{-}O{-}P \\ HO\diagup \quad\quad\quad \diagdown OH \end{array}$$

are called pyrophosphates.

Compounds of the type

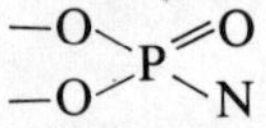

$$\begin{array}{c} {-}O \diagdown \quad \diagup O \\ \quad P \\ {-}O \diagup \quad \diagdown N \end{array}$$

are called phosphoramidates, and compounds in which two nitrogen atoms are attached to phosphorus are called phosphorodiamidates.

This nomenclature does not appear to be difficult, but complications arise when the alcohol side chain is complicated. However the side chain is usually separated from the parent molecule as a result of hydrolysis. This is explained later when reactions of organophosphates are discussed (p. 35). Obviously the word organophosphate as it is used in the preceding sentence is a generic term covering all insecticidal compounds of phosphorus, despite the fact that it has the ending "ate." This generic usage is now generally accepted.

Even if the general principles discussed above are followed, a chemical can still be named in more than one way. For example, malathion

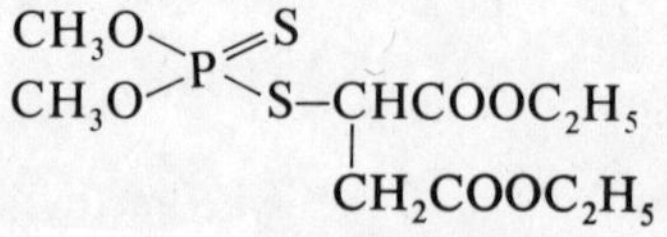

$$\begin{array}{c} CH_3O \diagdown \quad \diagup S \\ \quad P \\ CH_3O \diagup \quad \diagdown S{-}CHCOOC_2H_5 \\ \quad\quad\quad\quad\quad\quad | \\ \quad\quad\quad\quad\quad CH_2COOC_2H_5 \end{array}$$

can be called diethyl mercaptosuccinate, *S*-ester with *O,O*-dimethyl phosphorodithioate or it can be named *O,O*-dimethyl dithiophosphate of diethyl mercaptosuccinate. The former name is in accordance with the principles of *Chemical Abstracts* nomenclature. The latter name is the one used by its manufacturers, American Cyanamid Co., Princeton, N. J. Additionally, most of the chemicals (and this includes all classes of pesticides) that have been registered for use are given common names. Approved common names are required for use in most publications. The *Entomological Society of America* publishes these names from time to time.* Usually registered trade names are capitalized, whereas common names are not.†

## SOME REACTIONS OF ORGANOPHOSPHATES

The organophosphates conform to the general formula

$$\begin{array}{c} RO \\ \diagdown \\ R'O \end{array} P \begin{array}{c} \diagup A \\ \diagdown \\ X \end{array}$$

In most of the commercial insecticides the alkoxy groups, that is, OR and OR', are identical: they are basic and are usually $O,O$-dimethyl or $O,O$-diethyl. A is oxygen or sulfur, and X is the acidic group that is especially subject to chemical attack because of the positive charge developing on the phosphorus atom. The extent of this positive charge on P or, in other words, the electrophilic nature of phosphorus, depends on the nature of group X. If the group is electrophilic, that is, if it has affinity for electrons, it tends to draw electrons away from P thereby creating a positive site in the vicinity of the phosphorus atom. The more positive the nature of this site, the more easily is the compound hydrolyzed. Thus the reaction between paraoxon and hydroxyl ion can be written as

$$\begin{array}{c} C_2H_5O \\ \diagdown \\ C_2H_5O \end{array} P \begin{array}{c} \diagup O \\ \diagdown \\ O \end{array} \!\!-\!\! \bigcirc \!\!-\!\! NO_2 + O\overline{H} = \begin{array}{c} C_2H_5O \\ \diagdown \\ C_2H_5O \end{array} P \begin{array}{c} \diagup O \\ \diagdown \\ O^- \end{array} + HO \!\!-\!\! \bigcirc \!\!-\!\! NO_2$$

                   Diethyl phosphate    $p$-nitrophenol

* Commercial and Experimental Organic Insecticides by E. E. Kenega, and C. S. End is the latest revision published in 1974 by the Entomological Society of America as a special publication 74 (1).

† In this publication, the chemical names used by Chemical Abstracts and the common names approved by the Entomological Society of America have been given. Other names have also been included. Registered trademarks have been followed by a superscript Ⓡ. Names used by the manufacturers have been given. Mammalian toxicology and chemical structures have been included.

  Other useful publications are Farm Chemicals Handbook, an annual publication by Meister Publishing Co., Willoughby, Ohio; and Pesticide Handbook-Entoma, published annually by the Entomological Society of America, College Park, Maryland.

The group O⁻—⟨benzene ring⟩—NO₂ is termed the leaving group.

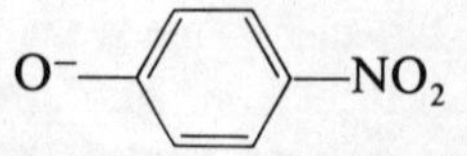

Though the nature of group X is principally responsible for the activity of organophosphorus compounds, each of the other radicals cannot be totally ignored. For example, $O,O$-diisopropyl $O$-nitrophenyl thiophosphate (isopropyl analog of parathion)

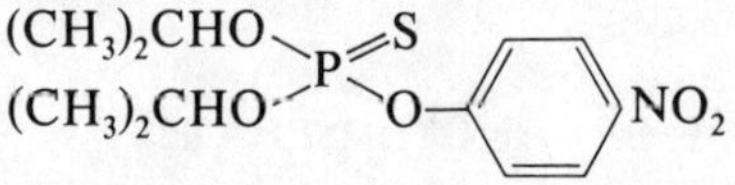

$$(CH_3)_2CHO \diagdown \underset{(CH_3)_2CHO \diagup}{P} \diagup \overset{S}{\diagdown} O—⟨benzene ring⟩NO_2$$

is usually biologically much less active than parathion. Both have the same leaving or acidic group. Also, the biological activity of organophosphates depends not only on their chemical properties but on steric factors, since the biochemical interactions of the molecule are more pronounced if the molecule can effectively block the active centres of the esterases by fitting "irreversibly" into them.

Some of the organophosphorus compounds are in themselves poor inhibitors of cholinesterase *in vitro* and yet they completely suppress cholinesterase *in vivo*. This difference is due to the conversion of the compound, in the living system, to another compound that effectively binds cholinesterase. Parathion,

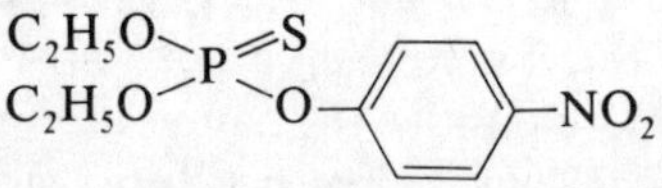

$$C_2H_5O \diagdown \underset{C_2H_5O \diagup}{P} \diagup \overset{S}{\diagdown} O—⟨benzene ring⟩—NO_2$$

which *in vitro* is a poor inhibitor of cholinesterase, is converted *in vivo* into paraoxon, a very efficient inhibitor of the enzyme.

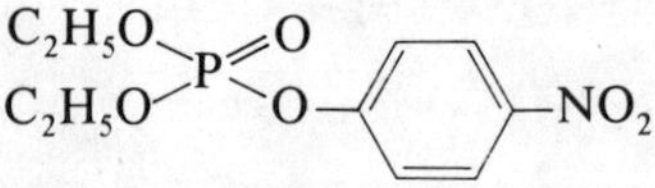

$$C_2H_5O \diagdown \underset{C_2H_5O \diagup}{P} \diagup \overset{O}{\diagdown} O—⟨benzene ring⟩—NO_2$$

This conversion is often termed oxidation, but since both oxygen and sulfur belong to the same group in the periodic table, no change of valence state takes place. The term desulfuration is therefore more appropriate than oxidation (10). Biologically speaking, there is a conversion to a more active compound; the change is therefore termed bioactivation or simply activation. The anticholinesterase activity is enhanced because =O is more electrophilic than =S. Consequently, P in P=O compounds acquires sufficient positive charge to interact rapidly with cholinesterase. Other types of activations are discussed later (p. 55).

## ORGANOPHOSPHORUS INSECTICIDES

### Derivatives of Phosphonic Acid

Several derivatives of phosphonic acid have been prepared and tested. They possess acaricidal, fungicidal, herbicidal, and insecticidal properties, but only a few of these compounds are being used as insecticides in agriculture; several others are under investigation. One of the best known insecticides of this group is *trichlorfon* or Dipterex®, dimethyl, (2,2,2-trichloro-1-hydroxyethyl)-phosphonate:

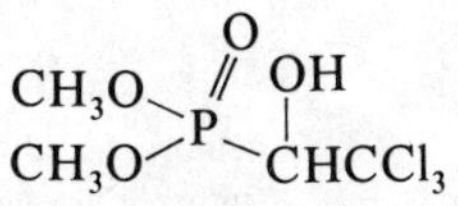

$$
\begin{array}{c}
\text{CH}_3\text{O} \quad \overset{\displaystyle O}{\underset{\displaystyle}{\parallel}} \quad \text{OH} \\
\phantom{\text{CH}_3\text{O}}\diagdown \text{P} \diagup \overset{\displaystyle |}{\phantom{P}} \\
\text{CH}_3\text{O} \diagup \phantom{\text{P}} \diagdown \text{CHCCl}_3
\end{array}
$$

It is a white crystalline solid, mp 83–84°C, soluble in water (12% at 26°C), alcohols, and ketones. Trichlorfon is moderately toxic to mammals and birds; its acute oral $LD_{50}$ for rat is 450–500 mg/kg.

In mammals trichlorfon rapidly undergoes cleavage at two places to yield dimethyl and methyl phosphoric acids, trichloroethanol, and desmethyl trichlorfon (2,2,2-trichloro-1-hydroxyethyl phosphonic acid). The trichloro-ethanol is excreted as glucuronide.

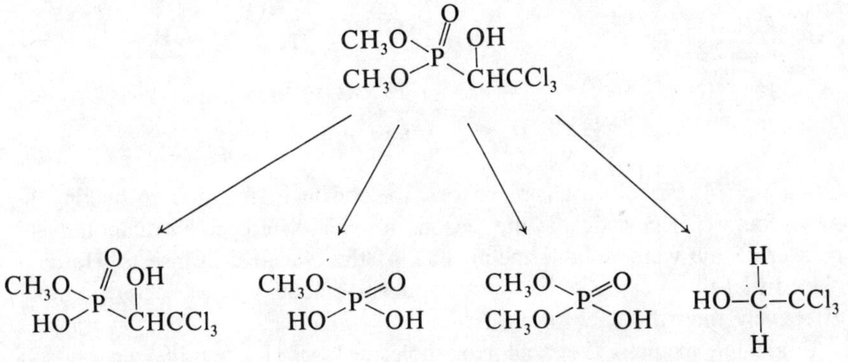

| Desmethyl trichlorfon | Methyl phosphoric acid | Dimethyl phosphoric acid | Trichlorethanol |

A small amount of trichlorfon also dehydrochlorinates and rearranges itself to form dichlorvos. Degradation of this chemical in insects is similar to that in mammals. Its metabolism has been studied by Arthur and Casida (11), Hassan et al (12), Metcalf et al. (13), Robbins et al. (14), Saito (15), and Zayed et al.

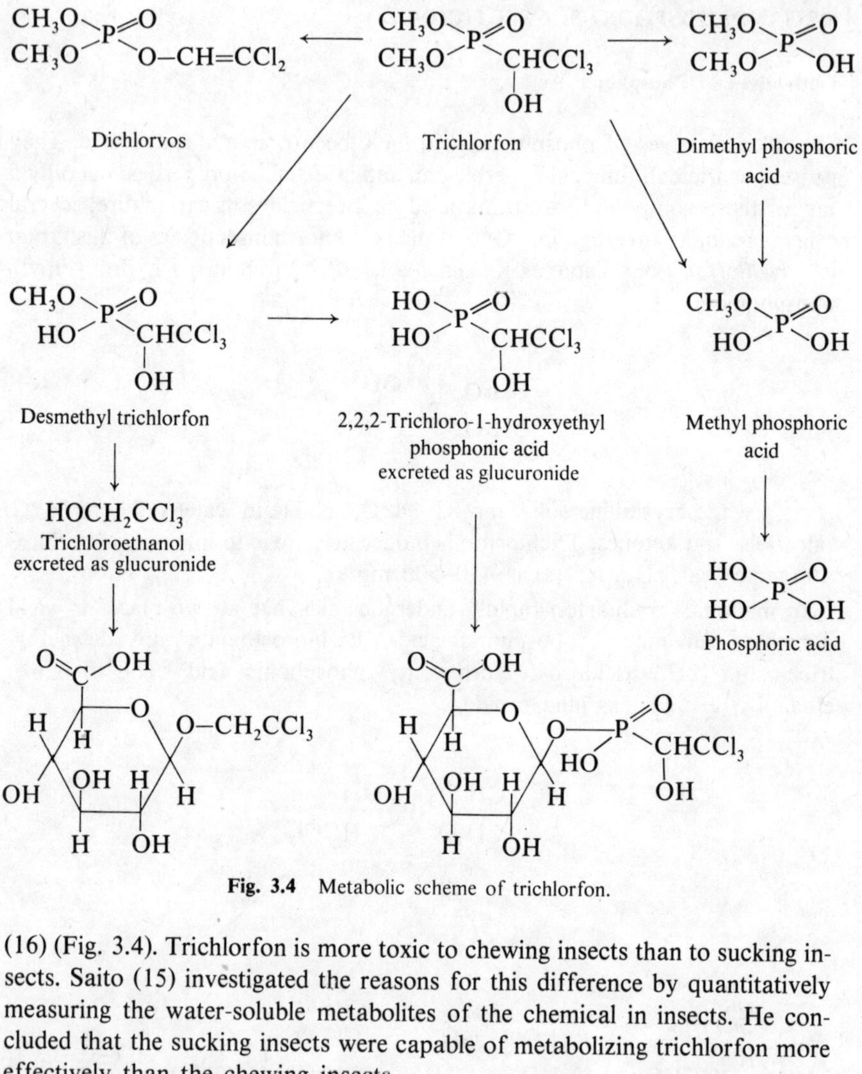

$$CH_3O\diagdown P \diagup O$$
$$CH_3O \diagup \quad O-CH=CCl_2$$

Dichlorvos

$$CH_3O\diagdown P \diagup O$$
$$CH_3O \diagup \quad CHCCl_3$$
$$\qquad\qquad OH$$

Trichlorfon

$$CH_3O\diagdown P \diagup O$$
$$CH_3O \diagup \quad OH$$

Dimethyl phosphoric
acid

$$CH_3O\diagdown P \diagup O$$
$$HO \diagup \quad CHCCl_3$$
$$\qquad\qquad OH$$

Desmethyl trichlorfon

$$HO\diagdown P \diagup O$$
$$HO \diagup \quad CHCCl_3$$
$$\qquad\qquad OH$$

2,2,2-Trichloro-1-hydroxyethyl
phosphonic acid
excreted as glucuronide

$$CH_3O\diagdown P \diagup O$$
$$HO \diagup \quad OH$$

Methyl phosphoric
acid

$HOCH_2CCl_3$
Trichloroethanol
excreted as glucuronide

$$HO\diagdown P \diagup O$$
$$HO \diagup \quad OH$$

Phosphoric acid

Fig. 3.4   Metabolic scheme of trichlorfon.

(16) (Fig. 3.4). Trichlorfon is more toxic to chewing insects than to sucking insects. Saito (15) investigated the reasons for this difference by quantitatively measuring the water-soluble metabolites of the chemical in insects. He concluded that the sucking insects were capable of metabolizing trichlorfon more effectively than the chewing insects.

In alkaline medium, the trichlorfon molecule loses HCl and rearranges itself to yield a phosphate, dichlorvos, which in itself is a potent insecticide. As stated earlier, this conversion also takes place in insects and mammals.

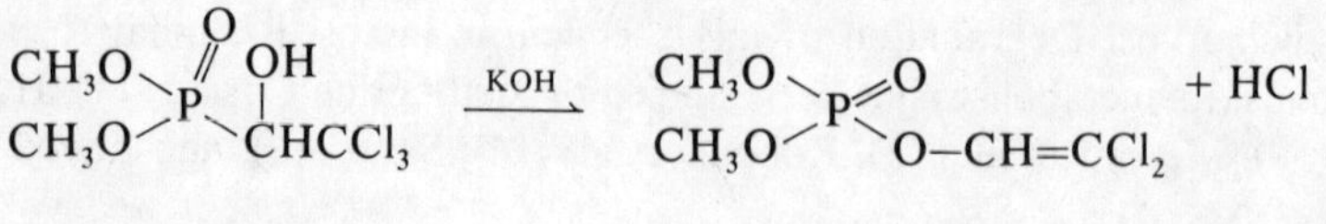

$$CH_3O\diagdown \overset{O}{\underset{P}{\diagdown}} OH \qquad \xrightarrow{KOH} \qquad CH_3O\diagdown P \diagup O \qquad + HCl$$
$$CH_3O \diagup \quad CHCCl_3 \qquad\qquad\qquad CH_3O \diagup \quad O-CH=CCl_2$$

*EPN*, *O*-ethyl *O-p*-nitrophenyl phenylphosphonothioate,

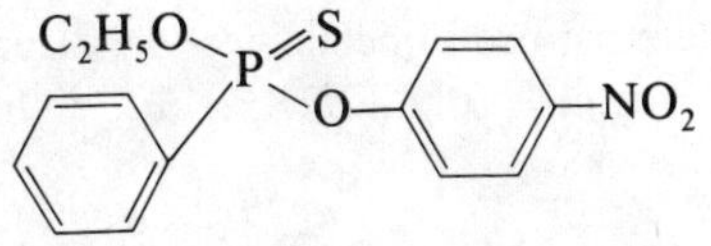

is a white crystalline material, mp 36°C that is soluble in most organic solvents but insoluble in water. The technical insecticide, however, is a liquid with an unpleasant odor. It is a good acaricide and insecticide. EPN is highly toxic to mammals (acute oral $LD_{50}$ for rat is 14–42 mg/kg). In mammals the P=S in EPN is converted to P=O; hydrolysis of the parent compound and its oxygen analog results in the formation of *p*-nitrophenol and other metabolites (17). Dahm (18) studied the microsomal degradation of EPN and concluded that it was metabolized oxidatively (p. 60), the aryl phosphate bond being cleaved to yield *p*-nitrophenol. EPN potentiates the toxicity of malathion. The importance of these findings is discussed in Chapter 16.

*Cyanofenphos*, *Surecide*®, *O-p*-cyanophenyl *O*-ethyl phenylphosphonothioate,

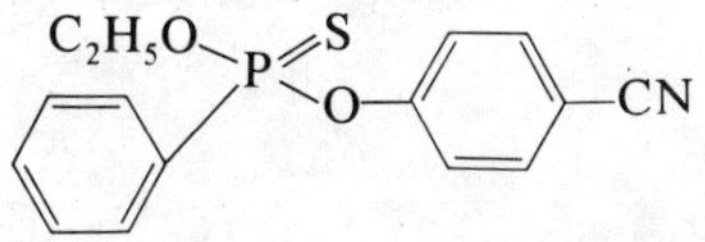

has been developed by the Sumitomo Company of Japan and it is being used experimentally in that country. Surecide is toxic to mammals, its acute oral $LD_{50}$ for rat is 1000 mg/Kg.

*Fonofos*, *Dyfonate*®, *O*-ethyl *S*-phenyl ethylphosphonodithioate,

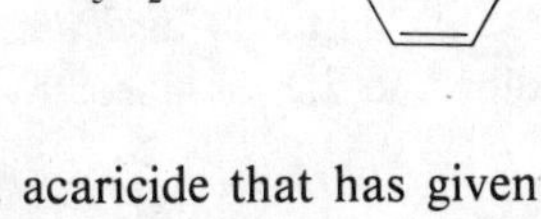

is a soil insecticide and acaricide that has given satisfactory control of corn rootworms and wireworms. It is highly toxic to mammals. Metabolism of Dyfonate has been studied by Hoffman et al. (19) and McBain et al. (20, 21). It undergoes a variety of biotransformations in plants and in mammals.

## Derivatives of Phosphoric Acid

Among insecticides belonging to this group of organophosphorus compounds are found some of the most useful pesticides; we discuss some of these below.

Dichlorvos or Vapona® or DDVP, 2,2-dichlorovinyl O,O-dimethyl phosphate,

$$
\begin{array}{c}
CH_3O \\
CH_3O
\end{array}
P
\begin{array}{c}
O \\
O-CH=CCl_2
\end{array}
$$

which has been mentioned earlier, is a colorless liquid, only slightly (1% at 25°C) soluble in water but highly soluble in organic solvents. It has considerable vapor pressure, 0.032 mm Hg at 31°C, and the vapor has rapid insecticidal action. Although acute oral toxicity of dichlorvos to mammals is high (LD$_{50}$ for rat is 56–80 mg/kg), mammalian systems rapidly detoxify sublethal dosages. Thus there is a significant difference between the amount that causes toxic symptoms in a single acute dose and the chronic daily dose that would cause poisoning. In sublethal dosages dichlorvos is rapidly degraded by mammals to O,O-dimethyl phosphate and other nontoxic metabolites (22). It can be incorporated into resin strips from which a sustained release of vapor takes place. Such preparations are used in buildings to control flying insects and on animals as collars to control ectoparasites.

Dichlorvos is easily hydrolyzed yielding dimethylphosphoric acid and dichloroacetaldehyde:

$$
\begin{array}{c}
CH_3O \\
CH_3O
\end{array}
P
\begin{array}{c}
O \\
O-CH=CCl_2
\end{array}
\longrightarrow
\begin{array}{c}
CH_3O \\
CH_3O
\end{array}
P
\begin{array}{c}
O \\
OH
\end{array}
+ CHCl_2CHO
$$

The former compound is practically nontoxic and the latter readily decomposes and evaporates; these qualities of the degradation products make dichlorovos a "safe" chemical to use where no pesticide residues are desirable. In living organisms it undergoes similar hydrolysis and, in addition, demethylation takes place to yield desmethyl dichlorvos (22, 23). To prevent hydrolysis of technical grade dichlorvos in storage, epichlorhydrin (2–4%) is added.

Chlorine and bromine can easily add at the double bond of the dichlorovinyl group of this chemical. Both chlorine and bromine derivatives thus formed are good insecticides; the latter is discussed below.

Naled or Dibrom®, 1,2-dibromo-2,2-dichloroethyl dimethyl phosphate,

$$
\begin{array}{c}
CH_3O \\
CH_3O
\end{array}
P
\begin{array}{c}
O \\
O-CHCBrCl_2 \\
| \\
Br
\end{array}
$$

is closely related to dichlorvos and has both insecticidal and acaricidal properties. Naled is a crystalline substance, mp 25°C; it is soluble in most organic
solvents but insoluble in water. Naled is moderately toxic to mammals, birds,
and fish; its acute oral $LD_{50}$ for rat is 430 mg/kg.

*Mevinphos* or *Phosdrin®*, methyl 3-hydroxy-α-crotonate, dimethyl
phosphate,

$$
\begin{array}{c}
CH_3O \diagdown \\
CH_3O \diagup P \diagup \diagup O \qquad \overset{\displaystyle CH_3}{\underset{}{|}} \quad \overset{\displaystyle O}{\overset{\|}{}} \\
\qquad O-C=CHC-OCH_3
\end{array}
$$

is an insecticide with systemic properties. It is a colorless liquid miscible with
water, acetone, and benzene, bp 99–103°C at 0.03 mm Hg. The technical
product is a mixture of *cis* and *trans* isomers in the approximate ratio of 60:40
respectively. The *cis* isomer is by far the most active insecticide of the two
isomers. Mevinphos is very toxic to warm blooded animals (acute oral $LD_{50}$
for rat is 6–7 mg/kg). It is easily hydrolyzed and is effective for a short period
only. In mammals the metabolism of the two isomers follows different routes:
the *cis* isomer undergoes extensive *O*-demethylation, whereas the *trans* isomer
suffers hydrolysis (24).

Two related insecticidal phosphates containing nitrogen in the molecule are:
*phosphamidon*, *Dimecron®* 2-chloro-*N,N*-diethyl-3-hydroxycrotonamide
dimethyl phosphate,

$$
\begin{array}{c}
CH_3O \diagdown \quad \overset{\displaystyle O}{\overset{\|}{}} \qquad \overset{\displaystyle CH_3}{\underset{}{|}} \quad \overset{\displaystyle O}{\overset{\|}{}} \\
CH_3O \diagup P \diagdown O-C=CC-N \diagdown \overset{C_2H_5}{\underset{C_2H_5}{}} \\
\qquad\qquad\qquad \underset{Cl}{|}
\end{array}
$$

which possesses both contact and systemic action; and *monocrotophos*,
*Azodrin®*, 3-hydroxy-*N*-methyl-*cis*-crotonamide dimethyl phosphate,

$$
\begin{array}{c}
CH_3O \diagdown \quad \overset{\displaystyle O}{\overset{\|}{}} \qquad \overset{\displaystyle CH_3}{\underset{}{|}} \quad \overset{\displaystyle O}{\overset{\|}{}} \\
CH_3O \diagup P \diagdown O-C=CHCN \diagdown \overset{H}{\underset{CH_3}{}}
\end{array}
$$

which is an acaricide in addition to being an insecticide. It is very effective
against lepidopterous insect pests of cotton that have acquired resistance to
other chemicals. Azodrin is soluble in water and acts both as a contact and a
systemic insecticide.

The vinyl and aryl derivatives of phosphoric acid have yielded some useful insecticides; some of the interesting chemicals in this group are discussed below.

*Gardona*® or *Rabon*®, 2-chloro-1-(2,4,5-trichlorophenyl)vinyl dimethyl phosphate, which has been given the common name *tetrachlorvinphos*

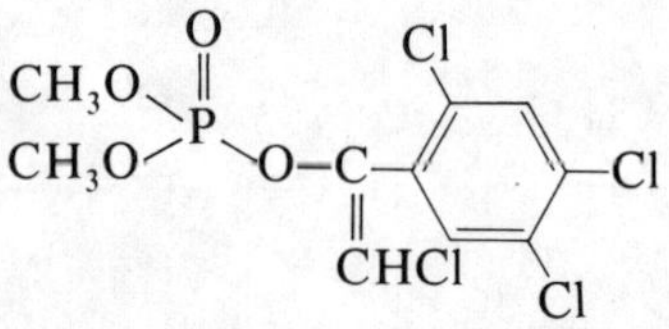

is a solid, mp 97–98°C, very slightly soluble in water (11 ppm) but soluble in organic solvents. It has low mammalian toxicity, the acute oral $LD_{50}$ for rat being 4000–5000 mg/kg. Gardona is relatively nonhazardous to fish and wildlife. It is used against some agricultural pests and as a residual wall spray for fly control.

*Chlorfenvinphos* or *Birlane*®, 2-chloro-1-(2,4-dichlorophenyl)vinyl diethyl phosphate,

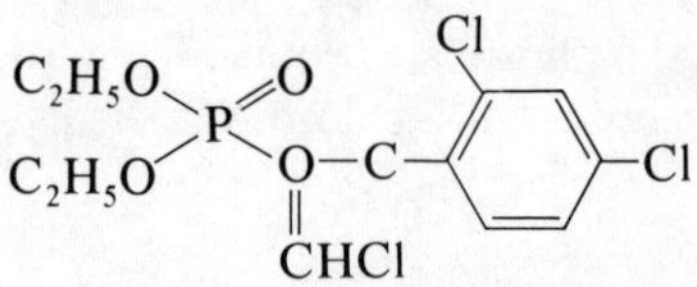

is an amber colored liquid, bp 167–170°C at 0.05 mm Hg. It is sparingly soluble in water and soluble in some organic solvents. Birlane is highly toxic to mammals, its acute oral $LD_{50}$ for rat being 12–30 mg/kg. It is particularly effective for the control of rootworms, root maggots, and cutworms. Chlorfenvinphos suffers an unusual form of *O*-dealkylation in mammals, cleaving at the ethyl group to yield acetaldehyde (25). This indicates that the mechanism of cleavage is not by hydrolysis (p. 61).

*Ciodrin*® or *crotoxyphos*, α-methylbenzyl 3-hydroxycrotonate dimethyl phosphate,

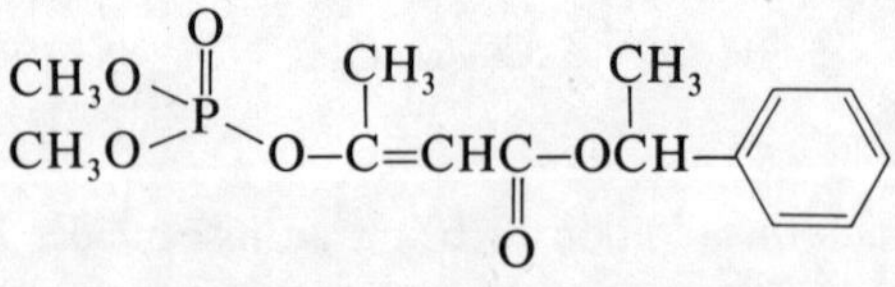

is a light-straw-colored liquid, bp 135°C at 0.03 mm Hg. It is slightly soluble in water (0.1%) and soluble in some organic solvents. Ciodrin is moderately toxic to mammals, its acute oral $LD_{50}$ for rat being 125 mg/kg. It can be used for controlling ectoparasites on dairy and beef cattle.

*Cruformate, Ruelene®*, 4-*tert*-butyl-2-chlorophenyl methyl methylphosphoramidate,

$$
\begin{array}{c}
CH_3O \\
\phantom{} \searrow \\
H_3CHN
\end{array}
P
\begin{array}{c}
\nearrow O \\
\searrow O
\end{array}
-\!\!\!\!\!
\begin{array}{c}
Cl \\
\end{array}
-C(CH_3)_3
$$

is a white crystalline solid, mp 57.4–59.8°C. It is slightly soluble in water and highly soluble in polar organic solvents. Ruelene has very low mammalian toxicity, its acute oral $LD_{50}$ for rat being 770–950 mg/kg; it is therefore used as an animal systemic for the control of intestinal worms and warble flies in cattle. Ruelene can be administered with cattle feed.

### Derivatives of Thiophosphoric and Dithiophosphoric Acids

Replacement of oxygen in P=O in organophosphates by sulfur results in thiono derivatives,

$$
\begin{array}{c}
RO \\
\phantom{} \searrow \\
RO
\end{array}
P
\begin{array}{c}
\nearrow S \\
\searrow OR
\end{array}
$$

and the replacement of oxygen in P–O–R by sulfur results in thiolo derivatives.

$$
\begin{array}{c}
RO \\
\phantom{} \searrow \\
RO
\end{array}
P
\begin{array}{c}
\nearrow O \\
\searrow SR
\end{array}
$$

When both oxygens are replaced by sulfur, the resulting compounds are called phosphorodithioates or dithiophosphates.

The thiono derivatives easily rearrange to form thiolo isomers; this tautomeric conversion results in dual reactive capabilities of these compounds and is discussed later p. 48. Thiono and thiolo derivatives are of additional interest because the replacement of one of the oxygen atoms by sulfur in a phosphate, in general, decreases its mammalian toxicity without a corresponding reduction in its insecticidal or acaricidal activity. These compounds have therefore found wide use for pest control in agriculture. The thiono derivatives, however, are converted into potent cholinesterase inhibitors in living organisms. This change is brought about by the replacement of sulfur in P=S by oxygen.

*Parathion, O,O*-diethyl *O-p*-nitrophenyl phosphorothioate,

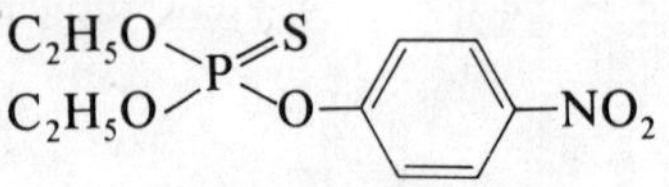

$$\begin{array}{c} C_2H_5O \\ C_2H_5O \end{array} P \!\!\! \underset{O}{\overset{S}{<}} \!\!\!- \!\!\! \bigcirc \!\!\!-NO_2$$

is one of the best known and one of the most widely used insecticides of this group. It is highly toxic to man because in living organisms it is readily converted to paraoxon, which is a potent cholinesterase inhibitor. Oxidizing agents bring about similar conversion *in vitro*; reducing agents convert parathion to aminoparathion

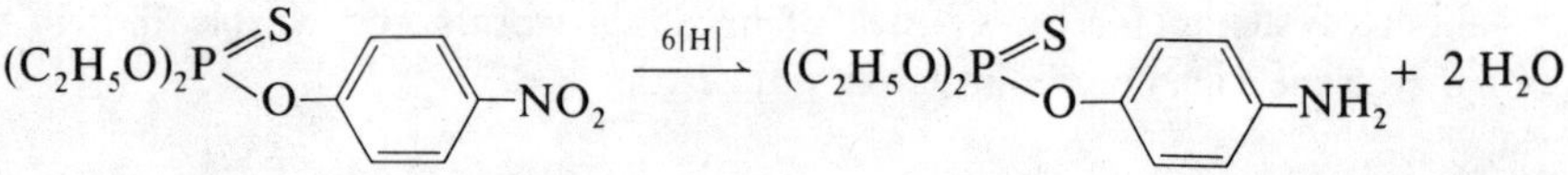

$$(C_2H_5O)_2P\!\!\overset{S}{\underset{O}{<}}\!\!-\!\!\bigcirc\!\!-NO_2 \xrightarrow{\;6|H|\;} (C_2H_5O)_2P\!\!\overset{S}{\underset{O}{<}}\!\!-\!\!\bigcirc\!\!-NH_2 + 2\,H_2O$$

which is nontoxic to insects and mammals. This reduction is one of the detoxifying pathways for parathion in the bovine rumen (26) and, to a limited extent, in other mammals. In the bovine rumen parathion is further converted to aminophenol, which forms aminophenyl glucuronide before being excreted. In other mammals and in man parathion cleaves at the aryl phosphate bond to yield *p*-nitrophenol and other degradation products (see Fig. 3.5).

*Methyl parathion* is the methyl analog of parathion; it is less toxic than parathion to mammals.

Another compound related to methyl parathion is *Sumithion*® or *fenitrothion, O,O*-dimethyl *O*-(4-nitro-*m*-tolyl) phosphorothioate,

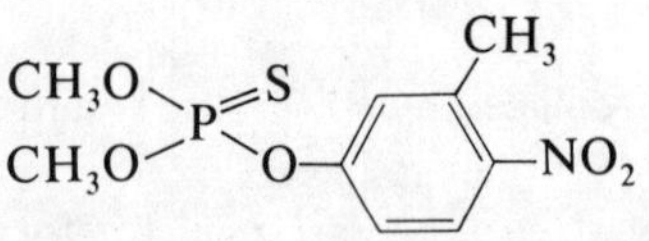

$$\begin{array}{c} CH_3O \\ CH_3O \end{array} P \!\!\! \underset{O}{\overset{S}{<}} \!\!\!- \!\!\! \bigcirc\!\!\!\overset{CH_3}{-}\!\!NO_2$$

which has low mammalian toxicity because it is rapidly degraded in mammals.

Other compounds that have similar structures include *dicapthon*

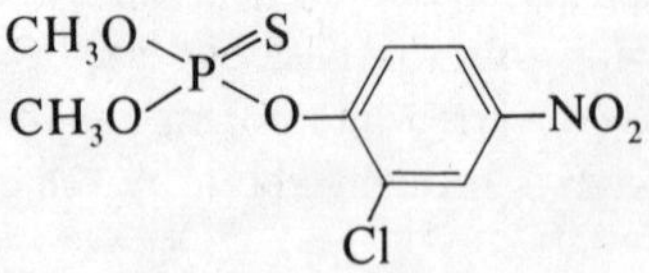

$$\begin{array}{c} CH_3O \\ CH_3O \end{array} P \!\!\! \underset{O}{\overset{S}{<}} \!\!\!- \!\!\! \bigcirc\!\!\!-NO_2$$
$$Cl$$

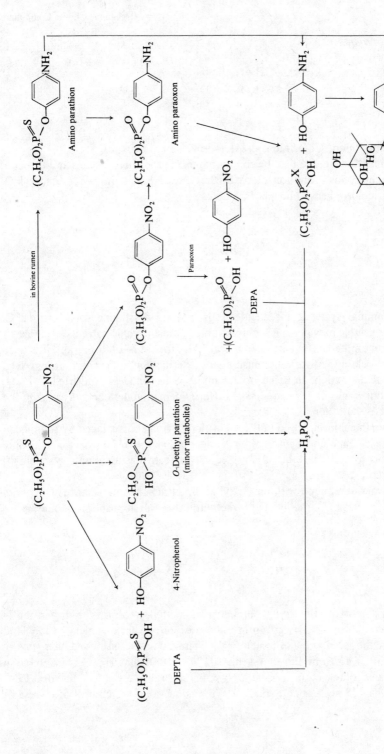

**Fig. 3.5** Metabolic scheme of parathion. DEPA = Diethyl phosphoric acid; DEPTA = $O,O$-Diethyl phosphorothioic acid; X = O or S.

45

*and Chlorthion*®:

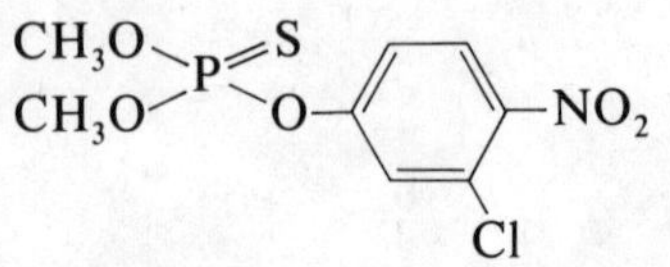

Like Sumithion, Chlorthion also has low mammalian toxicity.

*Ronnel*® (ronnel is also the common name in the United States and Canada) or fenchlorphos (common name in the United Kingdom). *O,O*-dimethyl *O*-2,4,5-trichlorophenyl phosphorothioate,

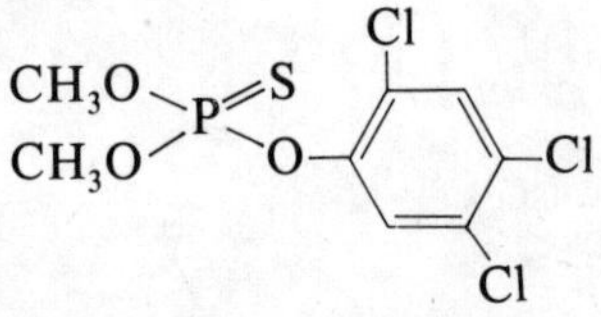

is an animal systemic insecticide. It is a white crystalline solid, mp 41°C, slightly soluble in water but highly soluble in most organic solvents. Because of its low mammalian toxicity (acute oral $LD_{50}$ for rat is 1740 mg/kg), it is used on livestock in the form of emulsions for external application and through feed for systemic action. In living organisms P=S in ronnel is converted to P=O. Both ronnel and its oxygen analog suffer *O*-demethylation and cleavage at the aryl phosphate bond (27, 28).

When the chlorine atom in the 4-position in ronnel is replaced by bromine, a white crystalline chemical, *bromophos*, is obtained. It is an insecticide as safe as ronnel to mammals, but it is not equally effective as an animal systemic insecticide.

Another compound with an *O*-aryl group attached to phosphorus is *fenthion*, or *Baytex*®, *O,O*-dimethyl *O*-[4-(methylthio)-*m*-tolyl]phosphorothioate:

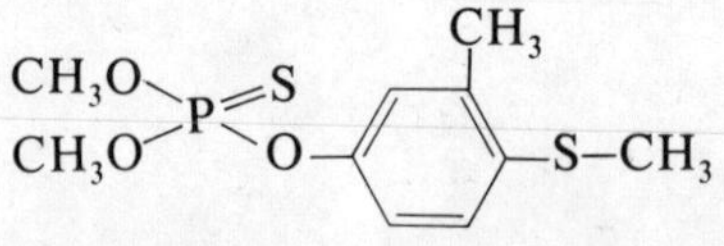

In this compound the aryl group carries a methyl group on the ring in the 3-position and an —$SCH_3$ group in the 4-position; an arrangement of this kind often reduces mammalian toxicity of a chemical and is discussed later (p. 47). Fenthion is a yellow tan liquid, bp 105°C at 0.01 mm Hg; it is insoluble in water, but soluble in most organic solvents. It is relatively safe (acute oral $LD_{50}$ for rat is 313 mg/kg) and stable. Baytex is useful in the control of insects of public health importance that have acquired resistance to chlorinated hydrocarbons.

A related methyl sulfinyl compound that is both an insecticide and a nematicide is *fensulfothion* or *Dasanit*®, *O,O*-diethyl *O-p*-[(methylsulfinyl)-phenyl] phosphorothioate:

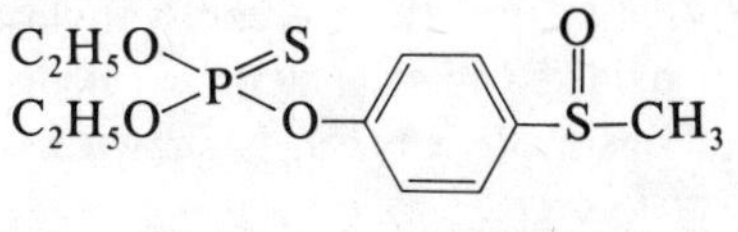

$$\begin{array}{c} C_2H_5O \\ C_2H_5O \end{array} P \overset{S}{\diagup} O \overset{}{-}\!\!\left\langle \bigcirc \right\rangle\!\!\overset{}{-}\overset{O}{\underset{}{S}}\!-CH_3$$

It is highly toxic to mammals (acute oral $LD_{50}$ for rat is 2–11 mg/kg). Fensulfothion is used to control soil insects and nematodes.

Another insecticide with low mammalian toxicity (acute oral $LD_{50}$ for rat is 2000 mg/kg) is *Abate*® or *temephos*, *O,O*-dimethyl phosphorothioate *O,O*-diester with 4,4'-thiodiphenol:

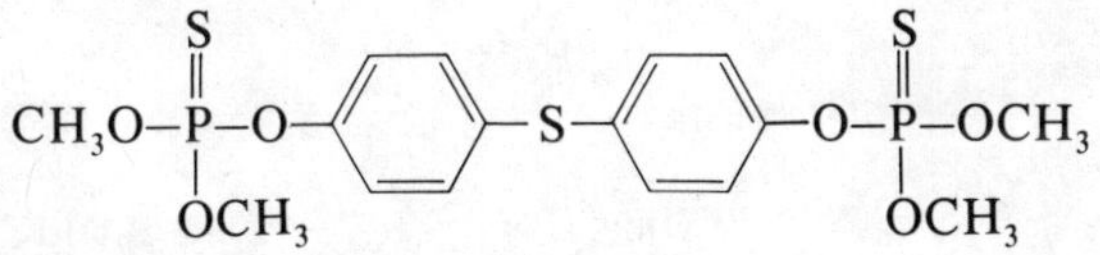

$$CH_3O-\overset{S}{\underset{OCH_3}{\overset{\|}{P}}}-O-\!\!\left\langle \bigcirc \right\rangle\!\!-S-\!\!\left\langle \bigcirc \right\rangle\!\!-O-\overset{S}{\underset{OCH_3}{\overset{\|}{P}}}-OCH_3$$

Pure Abate is a white crystalline substance, mp 30–35°C; however the technical grade material is a viscous brown liquid that is fairly stable. It is finding use in mosquito control.

A close inspection of the structures of parathion, fenitrothion (Sumithion)®, Chlorthion®, ronnel, bromophos, and fenthion indicates that, in general, in many mixed esters of thiophosphoric acid, the introduction in the 3-position or *meta* position to the ester group of a halogen (as in the case of ronnel) or methyl group (as in the case of fenthion) lowers their mammalian toxicity without appreciably changing their insecticidal properties.

Compounds containing a thioether or a sulfoxyl group in one of the ester radicals have yielded useful systemic insecticides; two of these are discussed below.

*Demeton* or *Systox*® or *mercaptophos* is a mixture of two isomers (thiono and thiolo) in the approximate ratio of 2 : 1. The chemical name, *O,O*-diethyl S-(and *O*)-2-[(ethylthio) ethyl] phosphorothioates, indicates the presence of these isomers.

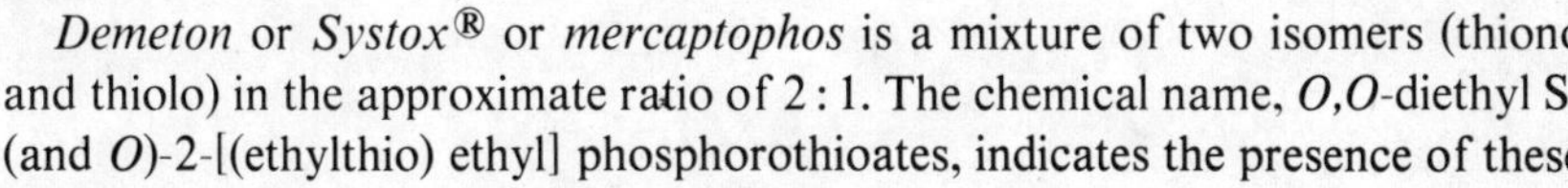

$$(C_2H_5O)_2P\overset{O}{\underset{S-C_2H_4-S-C_2H_5}{\diagup}} \quad\rightleftharpoons\quad (C_2H_5O)_2P\overset{S}{\underset{O-C_2H_4-S-C_2H_5}{\diagup}}$$

Thiono isomer                                                                    Thiolo isomer

Demeton is a long-lasting systemic insecticide and acaricide that is readily absorbed by and distributed in plants. It is highly toxic to mammals (acute oral $LD_{50}$ for rat is 4–12 mg/kg). Increase in temperature and polar solvents favor the conversion of the thiono isomer to the thiolo isomer by rearrangements of the positions of oxygen and sulfur. This rearrangement also takes place readily in plant tissues. In addition, in plants the thiono isomer is desulfurated to yield the oxygen analog. Sulfur in the ethylmercaptoethyl group of the thiolo and thiono isomers is also oxidized in two steps to form sulfoxide and sulfone, respectively. The metabolism of demeton is shown in Fig. 3.6. The dithioate analog of demeton is *disulfoton*.

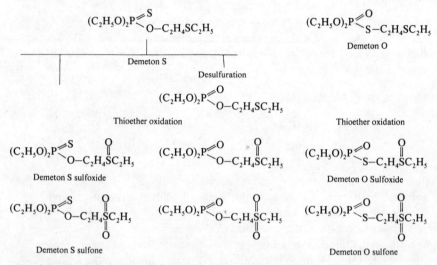

Fig. 3.6 Metabolic scheme of demeton. These metabolites are further fragmented by hydrolysis and oxidation. The decomposition products cannot be distinguished from natural animal and plant materials.

A related systemic insecticide and acaricide is *oxydemetonmethyl* or *Meta-Systox-R®*, *S*-[2-(ethylsulfinyl)ethyl] *O,O*-dimethyl phosphorothioate:

$$
\begin{array}{c}
CH_3O \\
CH_3O
\end{array}
P {\Large \substack{\diagup O \\ \diagdown}} S-CH_2CH_2-\overset{\displaystyle O}{\overset{\|}{S}}-C_2H_5
$$

It has an acute oral toxicity of 65–75 mg/kg for rat.

The derivatives of dithiophosphoric acid, in many respects, have more desirable properties than those of the thiophosphoric acid discussed previously. Some of these compounds, such as malathion, follow different metabolic routes in insects and in vertebrates and thus are selectively toxic to insects.

They are in general more stable and show insecticidal activity toward a wide range of insects.

*Malathion, Cythion®*, diethyl mercaptosuccinate, *S*-ester with *O,O*-dimethyl phosphorodithioate,

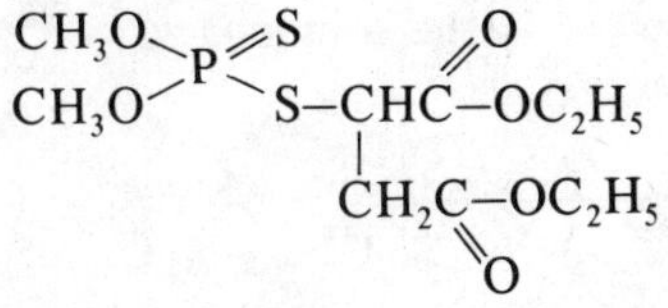

$$\begin{array}{c} CH_3O \\ \diagdown \\ CH_3O \diagup \end{array} P \diagup\!\!\!\diagdown \begin{array}{c} S \\ \diagdown \\ S-CHC-OC_2H_5 \\ | \\ CH_2C-OC_2H_5 \\ \diagdown\!\!\!\diagdown O \end{array} \begin{array}{c} O \\ \diagup\!\!\!\diagup \end{array}$$

is one of the best known insecticides of this group. It is a colorless liquid bp 120°C at 0.2 mm Hg. Malathion is slightly soluble in water but highly soluble in most organic solvents except saturated hydrocarbons. It has low mammalian toxicity; its $LD_{50}$ for white rat is 1000 mg/kg for males and 1375 mg/kg for females and its dermal $LD_{50}$ is $> 4444$ mg/kg for both sexes. As indicated earlier, malathion is selective in its toxicity to insects (compare schradan). This is due to different degrees and routes of metabolism followed by malathion in insects and in vertebrates as first reported by March et al. (29). Later O'Brien (30) published similar results for malaoxon and malathion.

Malathion is degraded in living organisms by two principal metabolic pathways: one involves carboxylesterase which vigorously attacks one of the carbethoxy groups, and the other involves phosphatases which attack the P—O— and P—S— bonds.* The former is the predominant and rapid route in mammalian liver and the latter the predominant but slow route in susceptible insects.

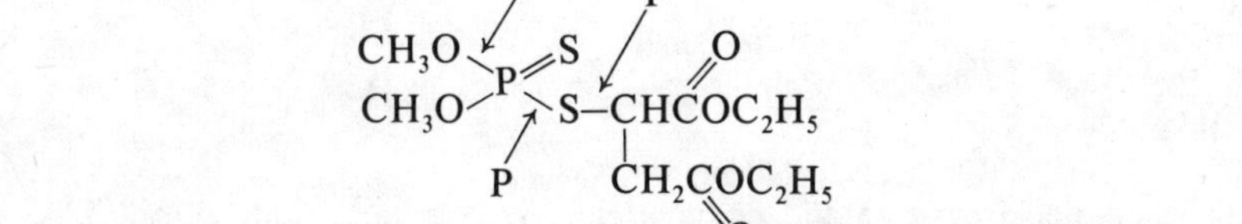

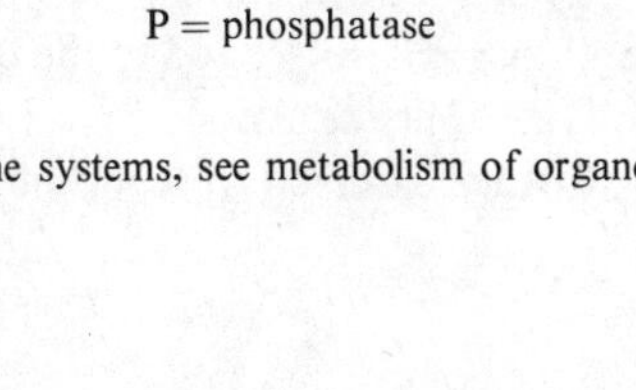

$$(CH_3O)_2P\diagup\!\!\!\diagdown\begin{array}{c}S\\S-CHCOC_2H_5\\|\\CH_2COC_2H_5\\\diagdown\!\!\!\diagdown O\end{array}\begin{array}{c}O\\\diagup\!\!\!\diagup\end{array}\xrightarrow{\text{carboxylesterase}}(CH_3O)_2P\diagup\!\!\!\diagdown\begin{array}{c}S\\S-CHCOOH\\|\\CH_2COOC_2H_5\end{array}$$

$$\begin{array}{c}P\\\diagup\\CH_3O\diagup\\CH_3O\diagup\end{array}P\diagup\!\!\!\diagdown\begin{array}{c}S\\S-CHCOC_2H_5\\|\\CH_2COC_2H_5\\\diagdown\!\!\!\diagdown O\end{array}\begin{array}{c}P\\\diagup\\O\\\diagup\!\!\!\diagup\end{array}$$

P = phosphatase

* For discussion of enzyme systems, see metabolism of organophosphorus insecticides, p. 54.

The acid released on the hydrolysis of malathion by rat liver homogenate was first identified by Cook and Yip (31) as the half acid of malathion. These authors named the catalyzing enzyme malathionase. The enzyme was partially purified by Main and Braid (32), who characterized it as a carboxylesterase. Chen et al. (33) identified the half acid produced as a result of hydrolysis of malathion by carboxylesterase as the $\alpha$-monoacid.

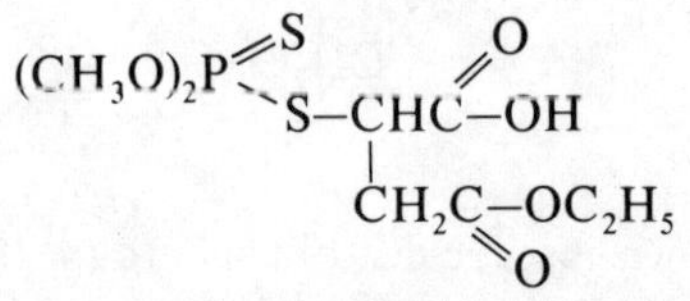

$$(CH_3O)_2P \overset{S}{\underset{S-CHC-OH}{\diagup}} \overset{O}{\diagup}$$
$$\underset{\overset{|}{CH_2C-OC_2H_5}}{\overset{}{}}$$
$$\overset{}{O}$$

The carboxylesterase hydrolyses only one of the carbethoxy groups of malathion and is unable to catalyze the hydrolysis of the monoacid. The mechanism by which malathion diacid, found in rat urine, is produced is not fully known. Like other P=S compounds, malathion is converted to its oxygen analog, malaoxon, in living organisms.

*Dimethoate*, *O,O*-dimethyl *S*-(*N*-methylcarbamoylmethyl) phosphorodithioate,

$$\overset{CH_3O}{\underset{CH_3O}{\diagdown}} P \overset{S}{\underset{S-CH_2CN}{\diagup}} \overset{O}{\underset{CH_3}{\diagup}} \overset{H}{\diagdown}$$

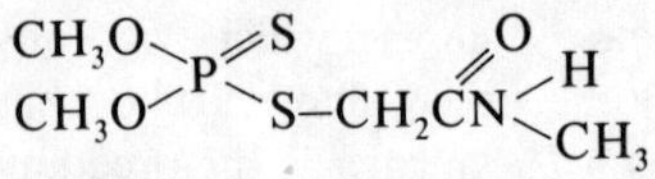

is a snow-white crystalline chemical, mp 51–52°C. It is highly soluble in water and in most organic solvents. Its acute oral $LD_{50}$ for rat is 250 mg/kg for males. Dimethoate is not very toxic to fish. It is rapidly absorbed and metabolized by plants. It is also converted to its oxygen analog. In the liver of mammals, dimethoate is rapidly degraded to nontoxic metabolites by the dimethoate–amidase system; the C–N bond is cleaved and the metabolites are excreted.

*Phorate* or *Thimet*®, *O,O*-diethyl-*S*-(ethylthio)methyl phosphorodithioate,

$$\overset{C_2H_5O}{\underset{C_2H_5O}{\diagdown}} P \overset{S}{\underset{S-CH_2-S-C_2H_5}{\diagup}}$$

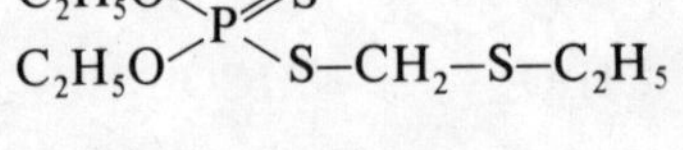

is a systemic insecticide and an acaricide. It is a clear liquid, bp 100°C at 0.4 mm Hg. Phorate is highly toxic to mammals and fish; its acute oral $LD_{50}$ for white rat is 1–5 mg/kg. Phorate is converted to the corresponding sulfoxide which is resistant to hydrolysis. This conversion easily takes place in treated plants, and the sulfoxide is further oxidized to form the sulfone. Conversion of

P=S to P=O in phorate and its oxidation products also takes place in plants. Finally, the metabolites are degraded by hydrolysis. In mammals the metabolism of phorate is qualitatively similar. In insects Menn (34) found only sulfoxide and sulfone of phorate as metabolites. No evidence for the conversion of P=S to P=O was found.

Two closely related insecticides are thiometon, *S*-[2-(ethylthio)ethyl] *O,O*-dimethyl phosphorodithioate,

$$CH_3O \diagdown \underset{\diagup}{P} \diagup\!\!\!\diagup S$$
$$CH_3O \diagup \phantom{P} \diagdown S-CH_2CH_2-S-C_2H_5$$

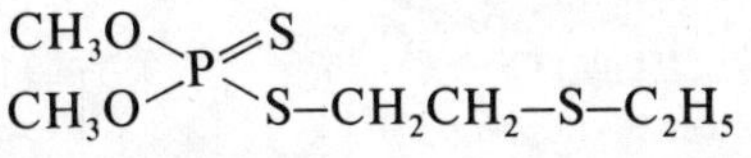

and disulfoton or Di-Syston®, *O,O*-diethyl *S*-[2-(ethylthio)ethyl] phosphorodithioate,

$$C_2H_5O \diagdown \underset{\diagup}{P} \diagup\!\!\!\diagup S$$
$$C_2H_5O \diagup \phantom{P} \diagdown S-CH_2CH_2-S-C_2H_5$$

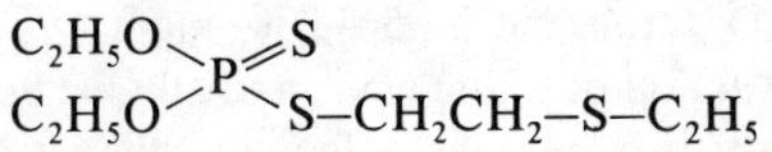

Thioether systemics like demeton, disulfoton, and phorate are very effective against sucking insects, but they are not very toxic to chewing insects.

### Heterocyclic Derivatives of Phosphorus Compounds

Some mixed esters of thiophosphoric acid of the heterocyclic series have shown desirable insecticidal properties and some of these insecticides are discussed on the following pages.

*Diazinon*® (*diazinon* is also the common name), *O,O*-diethyl *O*-(2-isopropyl-4-methyl-6-pyrimidyl) phosphorothioate,

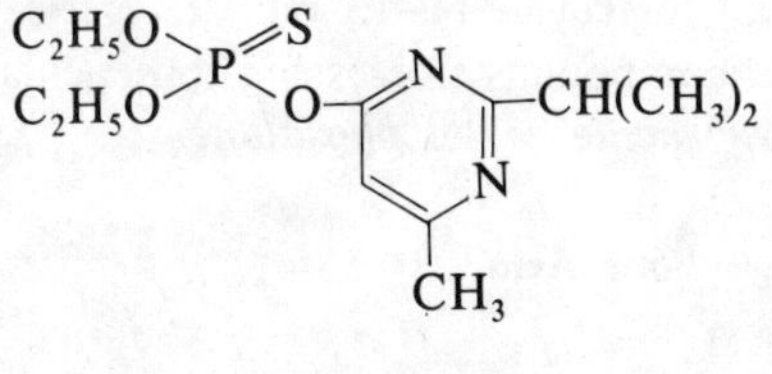

is an insecticide and an acaricide. It is a colorless oil, bp 89°C at 0.1 mm Hg. Diazinon is toxic to mammals, its acute oral $LD_{50}$ for rat being 66–600 mg/kg.

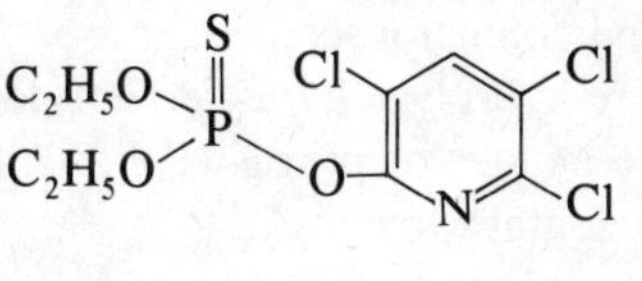

*Dursban*® or *chlorpyriphos*, *O,O*-diethyl *O*-(3,5,6-trichloro-2-pyridyl) phosphorothioate, has shown promise as an insecticide against a number of insects. It appears to be of special interest in mosquito control and in the control of household pests. Its acute oral $LD_{50}$ for rat is 97–267 mg/kg.

A useful insecticide for the control of ectoparasites of domestic animals is *coumaphos* or *Co-RAL*®, *O*-(3-chloro-4-methyl-2-oxo-2*H*-1-benzopyran-7-yl) *O*,*O*-diethyl phosphorothioate:

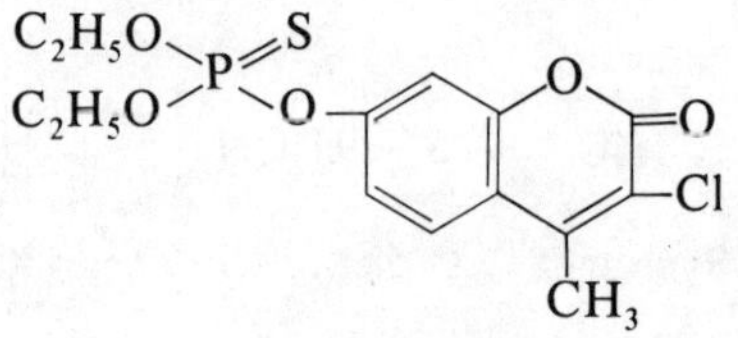

Its acute oral $LD_{50}$ for rat is 56–230 mg/kg. Coumphos is not easily hydrolyzed by the action of dilute acids and alkalis; however it is rapidly metabolized in mammals. It is one of the few insecticides that have low toxicity to fish.

*Azinphosmethyl* or *Guthion*®, *O*,*O*-dimethyl *S*-[4-oxo-1,2,3-benzotriazin-3-(4*H*)-ylmethyl] phosphorodithioate,

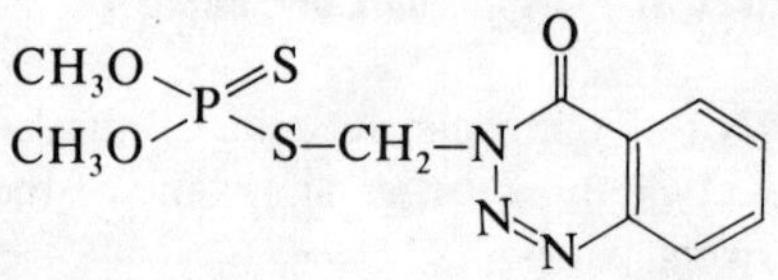

is an insecticide and an acaricide. It is a white crystalline solid, mp 73–74°C. Azinphosmethyl is slightly soluble in water. It is highly toxic to fish and mammals (acute oral $LD_{50}$ for rat is 11–13 mg/kg).

Other heterocyclic compounds possessing insecticidal properties are *Imidan*® or *phosmet*, *Supracide*®, and *phosalone*.

### Derivatives of Pyrophosphoric Acid

Although the derivatives of pyrophosphoric acid are losing their importance as commercial insecticides, they still retain their historical significance. Tetraethyl pyrophosphate or tepp was the first organophosphorus insecticide manufactured in Germany in 1943. This discovery resulted in a systematic investigation of the possible insecticidal properties of organophosphates. Most pyrophosphates have strong contact insecticidal properties, but because of the rapidity with which they are hydrolyzed, they do not possess a systemic insecticidal effect. However the amides of pyrophosphoric acid show increasingly systemic action and little contact toxicity.

Tetraethyl pyrophosphate or tepp

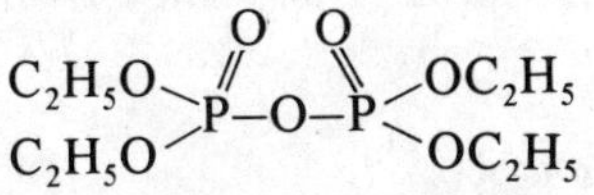

$$\begin{array}{c} C_2H_5O \\ C_2H_5O \end{array} \overset{\overset{O}{\underset{\parallel}{}}}{P} - O - \overset{\overset{O}{\underset{\parallel}{}}}{P} \begin{array}{c} OC_2H_5 \\ OC_2H_5 \end{array}$$

is a colorless liquid with an unpleasant odor, bp 82°C at 0.05 mm of Hg. It is miscible with water, but water rapidly hydrolyzes tepp. It is highly toxic, the acute $LD_{50}$ for white rats being 0.5–1.05 mg/kg.

*Schradan*, octamethylpyrophosphoramide,

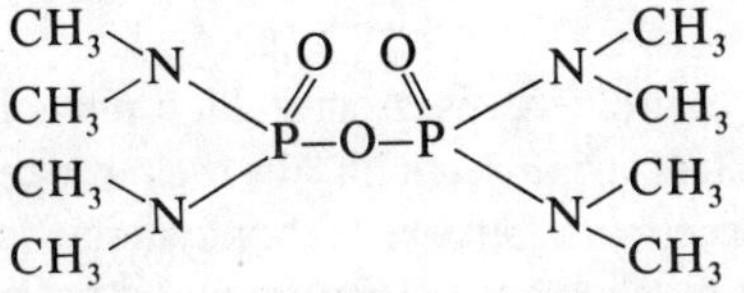

$$\begin{array}{c} CH_3 \\ CH_3 \end{array} N \qquad \begin{array}{c} O \quad O \\ \end{array} \qquad N \begin{array}{c} CH_3 \\ CH_3 \end{array}$$
$$\begin{array}{c} CH_3 \\ CH_3 \end{array} N \qquad \overset{\parallel}{P} - O - \overset{\parallel}{P} \qquad N \begin{array}{c} CH_3 \\ CH_3 \end{array}$$

is no longer used as an insecticide in the United States, but it is of historical importance because the conversion of schradan to the N-hydroxymethyl derivative is the first example of activation among organophosphorus insecticides. Schradan is more toxic to sucking insects than to chewing insects. According to Saito (15, 35) this selective toxicity is due to the structural differences between the nerve sheaths of the two types of insects. Schradan is also a selective mammalicide because the mammalian nerve lacks a protective sheath found in insects. This sheath is impermeable to sodium, potassium, acetylcholine, and *Tetram*®, S-[2-(Diethylamino)ethyl] *O,O*-diethyl phosporothioate hydrogen oxalate, (15).

## O,S-Dimethyl Phosphoramidothioates

Two newly introduced organophosphorus insecticides belonging to this group are discussed here.

*Monitor*® or *methamidophos*, *O,S*-dimethyl phosphoramidothioate, introduced by Chevron Chemical Corporation (36) is known as *Tamaron*® in Europe and was commercially introduced by Farbenfabriken Bayer in 1970.

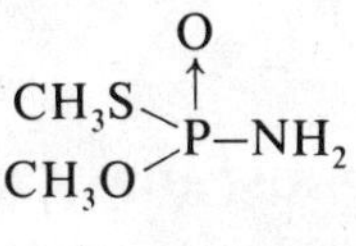

$$\begin{array}{c} CH_3S \\ CH_3O \end{array} \overset{\overset{O}{\uparrow}}{P} - NH_2$$

it is a broad-spectrum insecticide and an acaricide with considerable residual properties. Its oral $LD_{50}$ for rat is 20 mg/kg.

*Orthene*® or *acephate O,S*-dimethyl acetyl phosphoramidothioate, is obtained by the *N*-acetylation of Monitor and was developed for commercialization

$$\begin{array}{c} \text{CH}_3\text{S} \\ \text{CH}_3\text{O} \end{array}\!\!\!\!\nearrow\!\!\overset{\overset{O}{\uparrow}}{\text{P}}\!\!-\!\!\text{NH}\overset{\overset{O}{\|}}{\text{C}}\!\!-\!\!\text{CH}_3$$

tion in 1972 (37). Its acute oral $LD_{50}$ is 45 mg/kg for male rat. Acephate is mildly persistent and has a residual systemic activity of about 2 weeks.

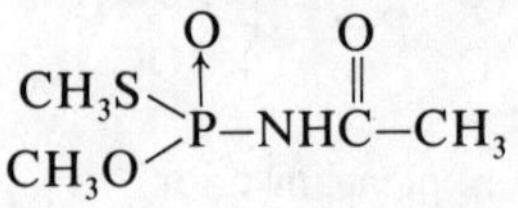

### Saligenin Phosphorus Esters

Saligenin, $\alpha$,2-dihydroxytoluene, is a crystalline phenolic alcohol whose alkyl cyclic phosphorus esters are insecticidal in their properties. The homologous aryl esters are not insecticidal but are highly synergistic with malathion. Some cause damage to the myelin sheath surrounding most nerves in mammals and birds. The esters can be prepared by the reaction shown below:

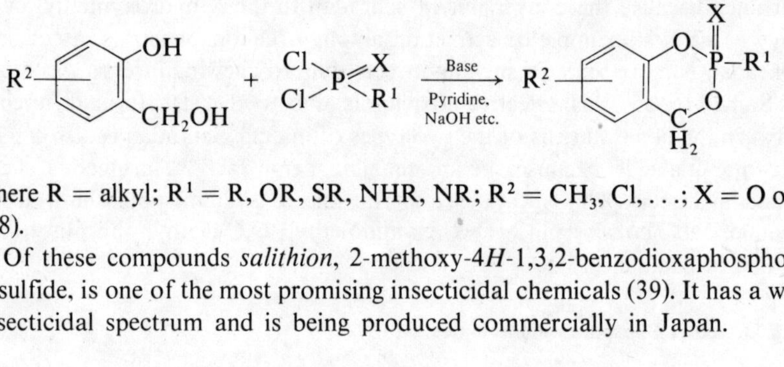

where R = alkyl; $R^1$ = R, OR, SR, NHR, NR; $R^2$ = $CH_3$, Cl, . . .; X = O or S (38).

Of these compounds *salithion*, 2-methoxy-4*H*-1,3,2-benzodioxaphosphorin 2-sulfide, is one of the most promising insecticidal chemicals (39). It has a wide insecticidal spectrum and is being produced commercially in Japan.

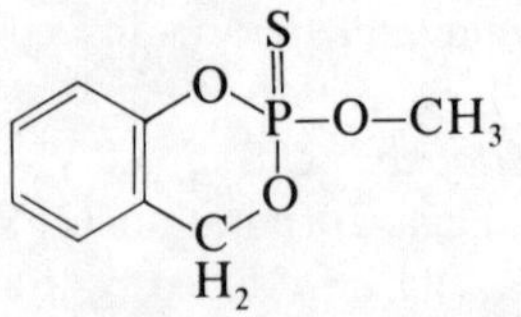

Salithion

## METABOLISM OF ORGANOPHOSPHORUS INSECTICIDES

All organophosphorus compounds are metabolized in living organisms, but the extent and nature of metabolism depends on both the chemistry of the compound and the biological makeup of the species concerned. Metabolism of organophosphorus compounds, or for that matter of any foreign chemical (xenobiotic), is brought about by enzyme systems and chemical reactions. In order to understand the interactions involved, a brief discussion of enzyme systems is necessary. Organophosphorus esters interact with two major groups of enzyme systems: the microsomal oxidative system and nonoxidative enzymes, such as esterases. For recent reviews see Boyland and Chasseaud (40), Furst (41), and Terriere (42).

The microsomal oxidative system in mammals is frequently localized in the microsomal fraction of the liver, while in insects the comparable enzymes reside mostly in the fat body. By centrifugation it is possible to separate discrete fractions of microsomes and use them in *in vitro* biochemical studies. Such experiments enable us to study specific enzymatic reaction. *In vitro*, however, the microsomal oxidative reactions require the presence of a cofactor, such as nicotinamide-adenine dinucleotide phosphate, in its reduced form (NADPH) and oxygen. Possibly the microsomal enzymes catalyze the conversion of molecular oxygen from air to an "active" form of oxygen. This oxygen forms a complex with cytochrome P-450 and thus brings about oxidative reactions.

The nonoxidative enzymes are of several types, for example, an enzyme that degrades diethyl phosphate and phosphorothioates, a malathion carboxylesterase, and a dealkylation enzyme. All those enzymes that hydrolyze any phosphorus ester or "anhydride" bond are called phosphatases. Aldridge (43, 44) introduced the terms A- and B-esterases. They are both serum enzymes that hydrolyze *p*-nitrophenyl acetate; But, whereas A-esterase can also hydrolyze paraoxon and therefore is a phosphatase, B-esterase is inhibited by paraoxon. With this background in mind, the metabolism of organophosphorus insecticides appears especially interesting because some of them can be both activated and detoxified during metabolism by enzymes. Some recent reviews on the subject include those by Dauterman (45), Hollingworth (46), and Menzer and Dauterman (47).

## Activation Reactions

These reactions are mostly catalyzed by mixed-function oxidases that are important in the metabolism of xenobiotics in mammals. Four types of reactions catalyzed by these enzymes generally increase the toxicity of organophosphorus compounds before their degradation to nontoxic metabolites.

### Desulfuration

One of the commonest reactions is the conversion of P=S to P=O in phosphorothionate or phosphorodithioate esters. This change generally results in an increase in their toxicity to the organism, because the oxygen analogs are better inhibitors of target enzymes than the parent compounds (3, 10). The *in vivo* conversion of parathion to paraoxon has been demonstrated in both insects and mammals (48–50). *In vitro* studies with microsomal preparations from insect and mammalian tissues have shown that desulfuration of parathion is catalyzed by a pyridine-nucleotide-linked mixed-function oxidase system (51–56). Using $^{35}S$-labeled parathion, Nakatsugawa et al. (57) also investigated the fate of the sulfur atom detached in the reaction. They found that an $^{35}S$ metabolite "was bound onto microsomes, probably as a result of desulfuration in the activation reaction" (18). Evidence for desulfuration has been produced for a variety of insecticides, including dimethoate (58), Dyfonate (fonofos) (21), fenitrothion (59) and malathion (30).

### *Hydroxylation of N-methyl group(s) followed by N-dealkylation*

Many of the nitrogen-containing organophosphorus insecticides undergo activation by oxidative *N*-dealkylation catalyzed by mixed-function oxidases. An example of this reaction is the *N*-demethylation of dicrotophos (Bidrin ®) and monocrotophos (Azodrin®), which are *N*-alkylamide containing vinyl phosphates. Initial investigations on their metabolism were conducted by Menzer and Casida (60). They used $P^{32}$- and *N*-methyl-$C^{14}$-labeled compounds and showed the formation of some metabolites as a result of hydroxylation and demethylation (Fig. 3.7). The toxicities of the metabolites are shown in Table 3.1. Similarly, phosphamidon (61) and dimethoate (62) are oxidatively dealkylated *in vivo* to yield more toxic products. Toxicities of the metabolites of dimethoate are given in Table 3.2. All toxic metabolites in the living organism are ultimately degraded to nontoxic metabolites.

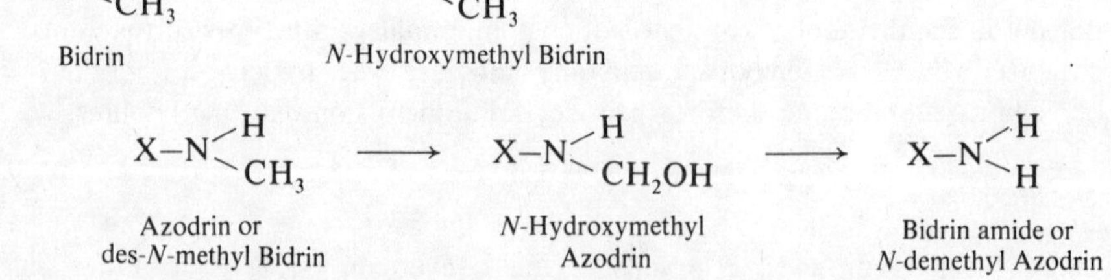

**Fig.3.7**  *N*-demethylation of dicrotophos (Bidrin), and monocrotophos (Azodrin).

**Table 3.1 Biological Activity in Bidrin and Its Analogs**

| *cis*-$(CH_3O)_2P(O)OC(CH_3)=CHC(O)R$ | LD$_{50}$ (mg/kg) | | |
|---|---|---|---|
| | | Housefly, ♀, topical[a] | |
| | Female mouse, i.p | −Sesamex | +Sesamex |
| R = $N(CH_3)_2$ (Bidrin) | 14 | 38 | 1.0 |
| R = $N(CH_2OH)CH_3$[b] | 18 | 14 | 1.2 |
| R = $NHCH_3$ (Azodrin) | 8 | 6.4 | 0.8 |
| R = $NHCH_2OH$[b] | 12 | 30 | 3.4 |
| R = $NH_2$ | 3 | 1.0 | 0.9 |

*Source:* Menzer and Casida (60). Reprinted with permission from *J. Ag. Food Chem.* **13:** 102 (1965). Copyright by the American Chemical Society.

[a]Fly LD$_{50}$ values for organophosphates determined with and without simultaneous treatment with 10 $\mu$g sesamex per fly. Slopes for log dose-probit mortality lines were similar for all compounds with and without synergist.

[b]Impure reaction products.

**Table 3.2 Biological Activity of Dimethoate and Its Metabolites**

| Compound | | LD$_{50}$ (mg/kg) | | |
|---|---|---|---|---|
| $(CH_3O)P(X)SCH_2C(O)NHR$ | | Mouse, ♂, i.p. | Housefly, ♀, topical[a] −Sesamex | Housefly, ♀, topical[a] +Sesamex |
| X = S | R = $CH_3$ | 151 (137–166) | 0.83 (0.72–0.95) | 0.71 (0.59–0.86) |
| X = S | R = $CH_2OH$ | — | — | — |
| X = S | R = H | 190 (168–215)[b] | 0.69 (0.54–0.88) | 0.84 (0.54–1.28) |
| X = O | R = $CH_3$ | 13 (11–16) | 0.21 (0.17–0.25) | 0.05 (0.03–0.07) |
| X = O | R = $CH_2OH$ | — | — | — |
| X = O | R = H | 10 (8–13) | 0.09 (0.07–0.12)[b] | 0.06 (0.04–0.08) |

*Source:* Lucier and Menzer (62). Reprinted with permission from *J. Agr. Food Chem.* **18:** 698 (1970). Copyright by the American Chemical Society.

[a]Numbers in parentheses are 95% confidence limits as calculated by method of Litchfield and Wilcoxon (1949).

[b]Value is significantly different from value immediately above at 5% level.

In the case of phosphoramidates there is hydroxylation of one of the *N*-methyl groups. Thus the insecticide schradan is activated to the *N*-hydroxymethyl derivative, possibly through the formation of schradan *N*-oxide. Formerly the active metabolite itself was suspected to be *N*-oxide (63), but Spencer et al. (64) synthesized schradan *N*-oxide and found that it had different properties than the metabolite. The controversy over the role of *N*-oxide formation has been discussed by Furst (41). Dimefox, bis(dimethyl-amino) fluorophosphine oxide, probably suffers comparable hydroxylation (11, 65).

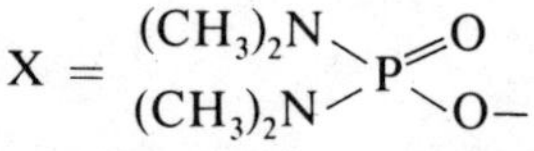

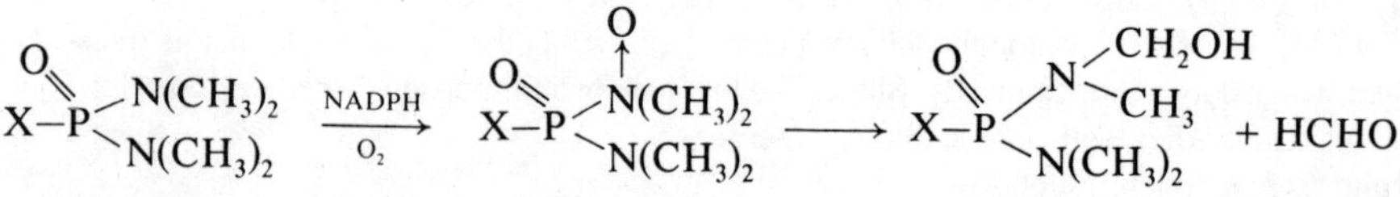

$$X = \begin{array}{c} (CH_3)_2N \\ (CH_3)_2N \end{array} P \begin{array}{c} O \\ O- \end{array}$$

Schradan                          Schradan *N*-oxide                *N*-hydroxymethyl schradan

## Thioether Oxidation

The divalent sulfur in thioether-containing compounds provides an additional site for oxidation to sulfoxide and sulfone derivatives. These reactions take place in the case of demeton, disulfoton, fenthion, phorate, and other sulfides (see metabolism of demeton). There is a rapid conversion to sulfoxide (which is probably the main toxicant) followed by a slow conversion to sulfone. No *in vitro* data are available for enzyme systems catalyzing these reactions in the case of organophosphorus compounds. However investigations with carbamates (66–68) indicate that these reactions are catalyzed by microsomal oxidases *in vitro*. Sulfoxides like oxydemetonmethyl are oxidized to sulfone derivatives. In thioether compounds conversion of P=S to P=O usually occurs. Metcalf et al. (68a) found that the rate of conversion of P=S to P=O *in vivo* is comparable to that of the oxidation of S to S → O to O ← S → O.

## Side-Group Oxidation

Casida and co-workers (69) showed an interesting activation reaction for tri-*o*-tolyl phosphate (TOCP). This chemical per se is not an insecticide, but it potentiates the toxicity of many compounds like malathion and dimethoate.

The activation reaction in this case takes place in two steps. The first step, which is catalyzed by mixed-function oxidases, involves the hydroxylation of a ring methyl group. The second step involves the cyclization of this intermediate product through intramolecular transphosphorylation. The latter reaction is catalyzed by plasma albumin and the resulting product is seligenin cyclic phosphate (38, 70).

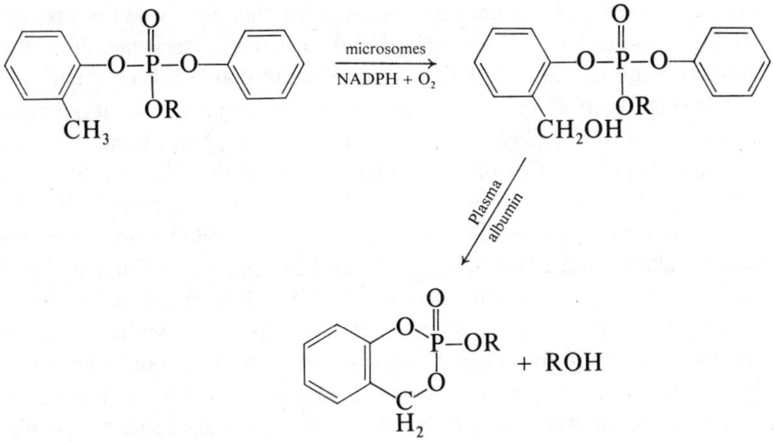

In the case of TOCP the final product is *o*-tolyl seligenin phosphate, which inhibits carboxylic esterases and amidases involved in the detoxication of compounds that TOCP potentiates.

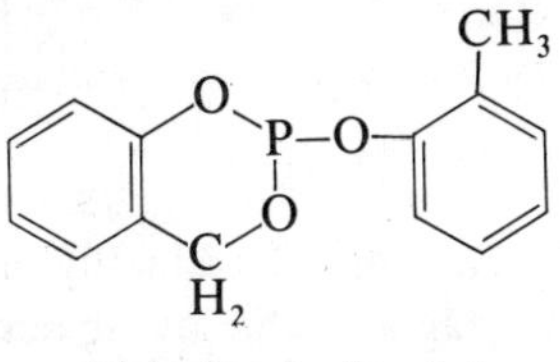

*o*-tolyl seligenin phosphate

**Degradation reactions**

Enzymatic hydrolysis of an organophosphorus compound, which is by far the most important detoxication reaction, was first demonstrated for DFP, diisopropyl phosphorofluoridate, by Mazur (71). Later investigators found evidence for the enzymatic hydrolysis of the anhydride bond in the case of tabun, ethyl dimethylphosphoramidocyanidate (72); paraoxon and analogs (73, 74); diazinon (75); and other organophosphates. These observations, coupled with the fact that some organophosphates, such as tepp, are easily hydrolyzed lead to the surmise that the degradation of almost all organophosphorus compounds occurs primarily by hydrolysis and the action of phosphatases on the anhydride linkage is of prime importance. Some dealkylating routes were discovered later, but they were considered to be of minor importance. The phosphatases accordingly were divided into "the more common ones," which catalyze the hydrolysis of leaving group, and "the less common ones," which dealkylate (10). However details of the reactions were not fully examined and mechanistic evidence for this view was not provided. Recent investigations have provided us with a better understanding of these reactions and the identity of the enzyme systems that catalyze them.

The degradation of an organophosphorus triester, ronnel, both at the methyl phosphate bond ($O$-dealkylation) and phenyl phosphate bond was first reported by Plapp and Casida (27). They surmised that the degradation at either bond took place as a result of hydrolysis (reactions II and III, Fig. 3.8). Recently, however, various mechanisms for $O$-dealkylation have been proposed. Fukami and Shishido (76, 77) and Shishido and Fukami (78) investigated the enzyme system responsible for $O$-dealkylation of methyl parathion, paraoxon, and Sumithion (reaction II, Fig. 3.8). They found that the enzyme was located in the supernatant fraction of rat liver homogenates and that glutathione was a required cofactor for $O$-dealkylation by both insect and mammalian tissue supernatants. It has since been established that the products of this enzyme system are $S$-methylglutathione and the monomethyl derivative of the organophosphate, which may be degraded further (18, 79, 80).

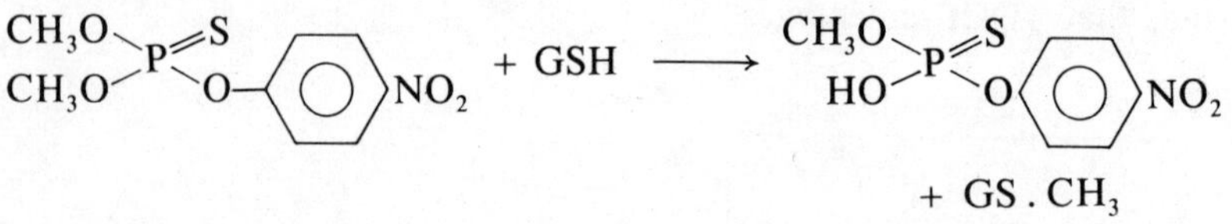

It seems that this enzyme system attacks dimethyl analogs more strongly than diethyl analogs; this in part explains the low toxicity to mammals of methyl

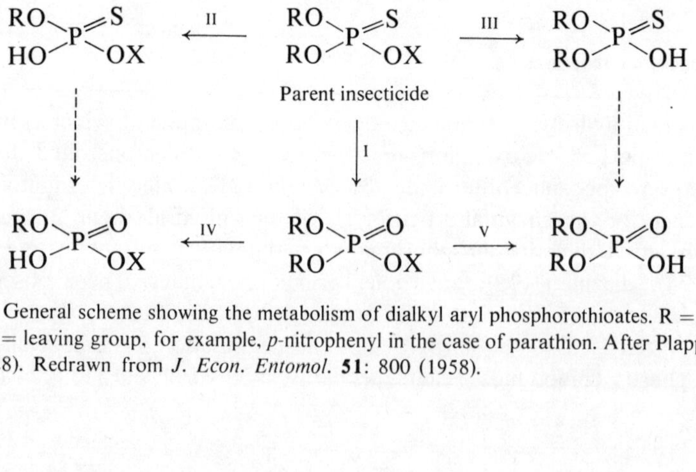

Fig. 3.8  General scheme showing the metabolism of dialkyl aryl phosphorothioates. R = alkyl group; X = leaving group, for example, $p$-nitrophenyl in the case of parathion. After Plapp and Casida (28). Redrawn from *J. Econ. Entomol.* 51: 800 (1958).

parathion and fenitrothion (Sumithion) as compared to parathion. In the last few years several investigators have presented evidence to suggest that in living organisms a significant contribution toward detoxification of organophosphorus triesters may be made by the cleavage of a single esterified alkyl group (24, 25, 79–82). In some instances, for example, in the case of the *cis* isomer of mevinphos (Phosdrin), *O*-dealkylation may become the predominant detoxication mechanism (24).

An entirely different mechanism for the oxidative *O*-dealkylation of chlorfenvinphos has been described by Donninger et al. (25). The enzyme that catalyzes this reaction is located in the microsomal fraction of liver homogenate and is dependent on molecular oxygen and NADPH for its activity. It therefore possesses the characteristics of a typical microsomal mixed-function oxidase. The ethyl group in the deethylation of chlorfenvinphos is removed as acetaldehyde. The dealkylation mechanism is therefore not hydrolysis but involves hydroxylation at the $\alpha$-carbon atom of the alkyl group that is removed as acetyldehyde. This reaction is analogous to reactions catalyzed by mixed-function oxidases (see pp. 39 and 195). If the cleavage were by hydrolysis, the ethyl group would have yielded ethyl alcohol instead of acetyldehyde. The authors have given the following mechanism for the dealkylation of chlorfenvinphos:

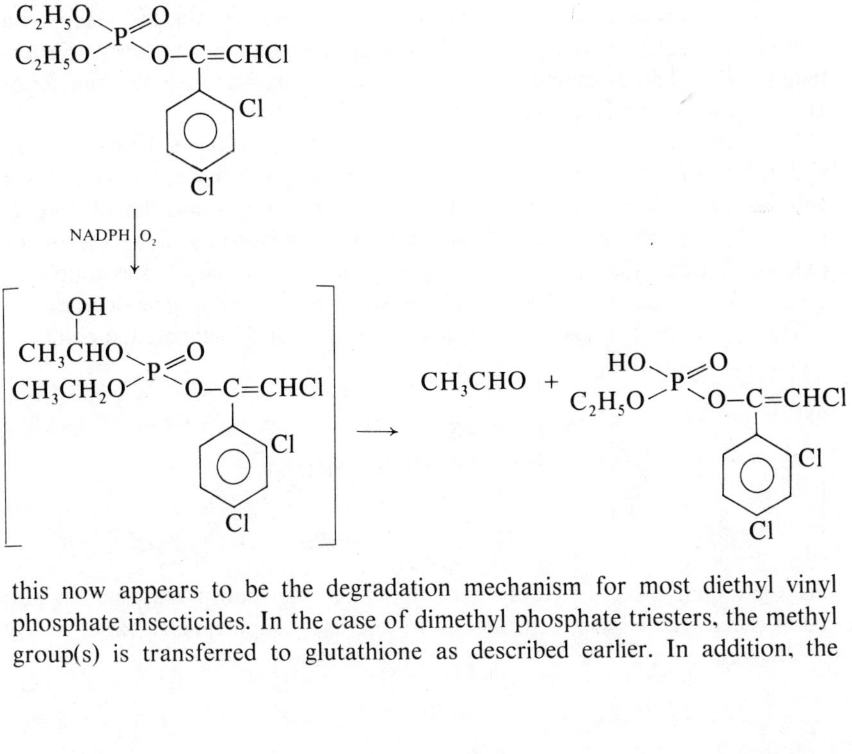

this now appears to be the degradation mechanism for most diethyl vinyl phosphate insecticides. In the case of dimethyl phosphate triesters, the methyl group(s) is transferred to glutathione as described earlier. In addition, the breakage of the P—O—vinyl bonds takes place in these pesticides; the P—O—C bonds also may be attacked, especially in the case of hydrophylic compounds (82a).

## Degradation of Organophosphorothionates and Related Compounds

The organophosphorothionates and related compounds are important because of two reasons: first, they have yielded some of the best known insecticides, and second, their metabolism has received considerable attention. In order to understand their degradation it would be useful to discuss the basic metabolic scheme that Plapp and Casida (27) first proposed for ronnel. Reaction I, as stated earlier, is an activation reaction and the remaining reactions result in detoxicatiòn. For several years many investigators felt that reaction V (Fig. 3.8) was the major detoxication mechanism. This view was based mostly on the *in vitro* studies by Aldridge (73), Augustinsson (83), Mounter (84), and others and on the assumption that thionophosphorus compounds were more stable to cleavage by phosphatases than the corresponding phosphates 85–87). However, *in vivo* both in mammals (27, 28, 88) and in insects (75, 89), *O,O*-dialkyl phosphorothioates have been reported as major metabolites (reaction III, Fig. 3.8). It was hypothesized that these metabolites were produced as a result of hydrolytic degradation by the action of phosphatases; but recent *in vitro* investigations indicate that microsomal and soluble liver enzymes play a role in the desulfuration as well as detoxication of organophosphorus triesters (54, 90–95). There is some evidence that they may play a role in the transfer of aryl groups as well (18, 76, 77, 94, 95).

Other detoxication pathways involve the hydrolysis of the carbethoxy group and carboxyamide group. These reactions are catalyzed by carboxylesterases and carboxyamidases, as in the cases of malathion and dimethoate, respectively (29, 31, 96, 97). In several mammals this appears to be the major pathway for detoxication; there also is some indication that the susceptible insects lack this mechanism and the resistant ones possess it. The degradation scheme of malathion has already been discussed; for dimethoate the points of attack are shown below.

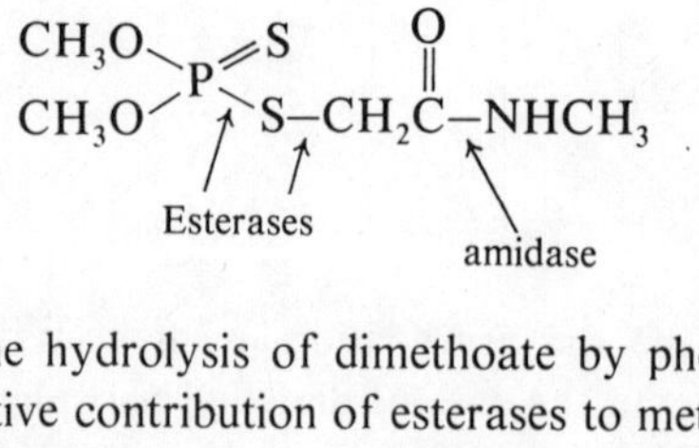

$$
\begin{array}{c}
CH_3O \\
\quad\quad\;\; \searrow \\
CH_3O \;\nearrow
\end{array}
P
\begin{array}{c}
\nwarrow S \\
\\
\searrow S-CH_2\overset{\displaystyle O}{\overset{\|}{C}}-NHCH_3
\end{array}
$$

Esterases                amidase

As for malathion, the hydrolysis of dimethoate by phosphatase action also takes place. The relative contribution of esterases to metabolites varies widely in different species.

# REFERENCES

1. J. A. Cohen and R. A. Oosterbaan. In *Cholinesterase and Anticholinesterase Agents*. Handbuch der Experimentellen Pharmakologie (G. B. Koelle, ed.), Vol. 15, Springer, Berlin, 1963, Chapter 7.
2. T. R. Fukuto. In *Residue Rev.* **25**: 327 (1969).
3. D. F. Heath. *Organophosphorus Poisons*, Macmillan (Pergamon), New York, 1961, 403 pp.
4. R. M. Krupka. *Can. J. Biochem.* **42**: 677 (1964).
5. R. D. O'Brien. *Toxic Phosphorus Esters*. Academic Press, New York, 1960, 434 pp.
6. T. Narahashi. In *The Physiology of Insect Central Nervous System* (J. E. Treherne and J. W. L. Beament, eds.), Academic Press, London, 1965, p. 1.
7. W. N. Aldridge. *Biochem. J.* **46**: 451 (1950).
8. A. R. Main. *Science* **144**: 992 (1964).
9. W. N. Aldridge and A. N. Davison. *Biochem. J.* **51**: 62 (1952).
10. R. D. O'Brien. *Insecticides: Action and Metabolism*, Academic Press, New York, 1967, 332 pp.
11. B. W. Arthur and J. E. Casida. *J. Agr. Food Chem.* **5**: 186 (1957).
12. A. Hassan, S. M. A. D. Zayed, and F. M. Abdel-Hamid. *Can. J. Biochem.* **43**: 1263 (1965).
13. R. L. Metcalf, R. B. March, and T. R. Fukuto. *J. Econ. Entomol.* **52**: 44 (1959).
14. W. E. Robbins, T. L. Hopkins, D. I. Darrow, and G. W. Eddy. *J. Econ. Entomol.* **49**: 801 (1956).
15. T. Saito. In *Residue Rev.* **25**: 175 (1969).
16. S. M. A. D. Zayed, I. Y. Mostafa, and A. Hassan. *Arch. Mikrobiol. (Berlin)*, **51**: 118 (1965).
17. R. A. Neal and K. P. DuBois. *J. Pharm. Exp. Ther.* **148**: 185 (1965).
18. P. A. Dahm. In *Biochemical Toxicology of Insecticides* (R. D. O'Brien and I. Yamamoto, eds.), Academic Press, New York, 1970, p. 51.
19. L. J. Hoffman, I. M. Ford, and J. Menn. *Pestic. Biochem. Physiol.* **1**: 349 (1971).
20. J. B. McBain, L. J. Hoffman, and J. J. Menn. *Pestic. Biochem. Physiol.* **1**: 356 (1971).
21. J. B. McBain, I. Yamamoto, and J. E. Casida. *Life Sci.* **10**: 947 (1971).
22. E. Hodgson and J. E. Casida. *J. Agr. Food Chem.* **10**: 208 (1962).
23. L. Dicowsky and A. Morello. *Life Sci.* **10**: 1031 (1971).
24. A. Morello, A. Vardanis, and E. Y. Spencer. *Can. J. Biochem.* **46**: 885 (1968).
25. C. Donninger, D. H. Hutson, and B. A. Pickering. *Biochem. J.* **126**: 701 (1972).
26. M. K. Ahmed and J. E. Casida. *J. Econ. Entomol.* **51**: 59 (1958).
27. F. W. Plapp and J. E. Casida. *J. Agr. Food Chem.* **6**: 662 (1958).
28. F. W. Plapp and J. E. Casida. *J. Econ. Entomol.* **51**: 800 (1958).
29. R. B. March, R. L. Metcalf, T. R. Fukuto, and F. A. Gunther. *J. Econ. Entomol.* **49**: 679 (1956).
30. R. D. O'Brien. *J. Econ. Entomol.* **50**: 159 (1957).
31. J. W. Cook and G. Yip. *J. Assoc. Off. Agr. Chem.* **41**: 399 (1958).
32. A. R. Main and P. E. Braid. *Biochem. J.* **84**: 255 (1962).
33. P. R. Chen, W. P. Tucker, and W.,C. Dauterman. *J. Agr. Food Chem.* **17**: 86 (1969).
34. J. J. Menn. *J. Econ. Entomol.* **55**: 90 (1962).
35. T. Saito. In *Radiation and Radioisotopes Applied to Insects of Agricultural Importance*, International Atomic Energy Agency, Vienna, 1963, p. 255.
36. Chevron Research Corp. Netherlands Patent appl. 6,602,588, Jan. 1967; *Chem. Abstr.* **67**: 10691Y (1967).
37. S. Magee. In *Residue Rev.* **53**: 3 (1974).
38. M. Eto. *op. cit.* **25**: 187 (1969).

39. M. Eto, Y. Kinoshita, T. Kato, and Y. Oshima. *Nature.* **223**: 210 (1963).
40. E. Boyland and L. F. Chasseaud. *Adv. Enzymol. Relat. Areas Mol. Biol.* **32**: 173 (1969).
41. C. I. Furst. *Ann. Rep. Prog. Chem. (Chem. Soc. Lond.)* **61**: 465 (1964).
42. L. C. Terriere. *Annu. Rev. Entomol.* **13**: 375 (1968).
43. W. N. Aldridge. *Biochem. J.* **53**: 110 (1953).
44. W. N. Aldridge. *op. cit.* **53**: 117 (1953).
45. W. C. Dauterman. *Bull. WHO* **44**: 133 (1971).
46. R. M. Hollingworth. *op. cit.* **44**: 155 (1971).
47. R. E. Menzer and W. C. Dauterman. *J. Agr. Food Chem.* **18**: 1031 (1970).
48. W. A. Brindley and P. A. Dahm. *J. Econ. Entomol.* **57**: 47 (1964).
49. J. C. Gage. *Biochem. J.* **54**: 426 (1953).
50. R. L. Metcalf and N. *Ann. Entomol. Soc. Amer.* **46**: 63 (1953).
51. A. N. Davison. *Biochem. J.* **61**: 203 (1955).
52. S. El Bashir and F. J. Oppenoorth. *Nature* **233**: 210 (1969).
53. S. D. Murphy and K. P. DuBois. *Proc. Soc. Exp. Biol. NY.* **96**: 813 (1957).
54. T. Nakatsugawa and P. A. Dahm. *Biochem. Pharmacol.* **16**: 25 (1967).
55. R. A. Neal. *Biochem. J.* **103**: 183 (1967).
56. R. A. Neal. *op. cit.* **105**: 289 (1967).
57. T. Nakatsugawa, N. M. Tolman, and P. A. Dahm. *Biochem. Pharmacol.* **10**: 1103 (1969).
58. W. C. Dauterman, G. B. Viado, J. E. Casida, and R. D. O'Brien. *J. Agr. Food Chem.* **8**: 115 (1960).
59. R. M. Hollingworth, R. L. Metcalf, and T. R. Fukuto. *J. Agr. Food Chem.* **15**: 250 (1967).
60. R. E. Menzer and J. E. Casida. *J. Agr. Food Chem.* **13**: 102 (1965).
61. G. P. Clemons and R. E. Menzer. *J. Agr. Food Chem.* **16**: 312 (1968).
62. G. W. Lucier and R. E. Menzer. *op. cit.* **18**: 698 (1970).
63. C. S. Hartley. Address to 15th International Chemistry Congress, New York, 1951. Cited by C. S. Hartley. *Chem. Ind.* No. 19. May 8, 1954. p. 529.
64. E. Y. Spencer, R. D. O'Brien, and R. W. White. *J. Agr. Food Chem.* **5**: 123 (1957).
65. M. L. Fenwick, J. R. Barron, and W. A. Watson. *Biochem. J.* **65**: 58 (1957).
66. M. Tsukamoto and J. E. Casida. *J. Econ. Entomol.* **60**: 617 (1967).
67. M. Tsukamoto and J. E. Casida. *Nature* **213**: 49 (1967).
68. N. R. Andrawes, H. W. Dorough, and R. A. Lundquist. *J. Econ. Entomol.* **60**: 979 (1967).
68a. R. L. Metcalf, T. R. Fukuto, and R. B. March. *J. Econ. Entomol.* **50**: 338 (1957).
69. J. E. Casida, M. Eto, and R. L. Baron. *Nature* **191**: 1396 (1961).
70. M. Eto, J. E. Casida, and T. Eto. *Biochem. Pharmacol.* **11**: 337 (1962).
71. A. Mazur. *J. Biol. Chem.* **164**: 271 (1946).
72. K. B. Augustinsson and G. Heimburger. *Acta Chem. Scan.* **8**: 1533 (1954).
73. W. N. Aldridge. *Biochem. J.* **54**: 442 (1953).
74. A. R. Main. *Biochem. J.* **75**: 188 (1960).
75. F. Matsumura and C. J. Hogendijk. *J. Agr. Food Chem.* **12**: 447 (1964).
76. J. Fukami and T. Shishido. *Botyu-Kagaku* **28**: 77 (1963).
77. J. Fukami and T. Shishido. *J. Econ. Entomol.* **59**: 1338 (1966).
78. T. Shishido and J. Fukami. *Botyu-Kagaku* **28**: 69 (1963).
79. R. M. Hollingworth. *op. cit.* **17**: 987 (1969).
80. R. M. Hollingworth. In *Biochemical Toxicology of Insecticides* (R. D. O'Brien and I. Yamamoto, eds.), Academic Press, New York, 1970, p. 75.
81. D. H. Hutson, B. A. Pickering, and C. Donninger. *Biochem. J.* **127**: 285 (1972).
82. D. H. Hutson, B. A. Pickering, and C. Donninger. *Biochem. J.* **106**: 20P (1968).
82a. K. I. Benyon, D. H. Hutson, and A. N. Wright. *Residue Rev.* **47**: 55 (1973).
83. K. B. Augustinsson. *Biochem. Biophys. Acta* **13**: 303 (1954).
84. L. A. Mounter. *J. Biol. Chem.* **209**: 813 (1954).

85. L. Lykken and J. E. Casida. *Can. Med. Assoc. J.* **100**: 145 (1969).
86. S. D. Murphy. *In Residue Rev.* **25**: 201 (1969).
87. J. H. Vinopal and T. R. Fukuto. *Pestic. Biochem. Physiol.* **1**: 44 (1971).
88. H. R. Krueger, J. E. Casida, and R. P. Niedermeyer. *J. Agr. Food Chem.* **7**: 182 (1959).
89. T. R. Fukuto, I. P. Wolf III, R. L. Metcalf, and M. Y. Minton. *J. Econ. Entomol.* **49**: 149 (1956).
90. T. Nakatsugawa, N. M. Tolman, and P. A. Dahm. *J. Econ. Entomol.* **62**: 408 (1969).
91. T. Nakatsugawa, N. M. Tolman, and P. A. Dahm. *Biochem. Pharmacol.* **17**: 1517 (1969).
92. T. Nakatsugawa, N. M. Tolman, and P. A. Dahm. *op. cit.* **18**: 685 (1969).
93. W. Welling, P. Blaakmeer, G. J. Vink, and S. Voerman. *Pesti. Biochem. Physiol.* **1**: 161 (1971).
94. R. S. H. Yang, E. Hodgson, and W. C. Dauterman. *J. Agr. Food Chem.* **19**: 10 (1971).
95. R. S. H. Yang, E. Hodgson, and W. C. Dauterman. *op. cit.* **19**: 14 (1971).
96. W. C. Dauterman, J. E. Casida, J. B. Knaak, and T. Kowalczyk. *J. Agr. Food Chem.* **7**: 188 (1959).
97. T. Uchida, W. C. Dauterman, and R. D. O'Brien. *J. Agr. Food Chem.* **12**: 48 (1964).

# SYNTHETIC ORGANIC INSECTICIDES

# CHAPTER 4 | Carbamates

Though free carbamic acid is not known to occur in nature, some of its esters, the carbamates, are potent pharmacological agents. A naturally occurring indole derivative of methyl carbamic acid, physostigmine or eserine,

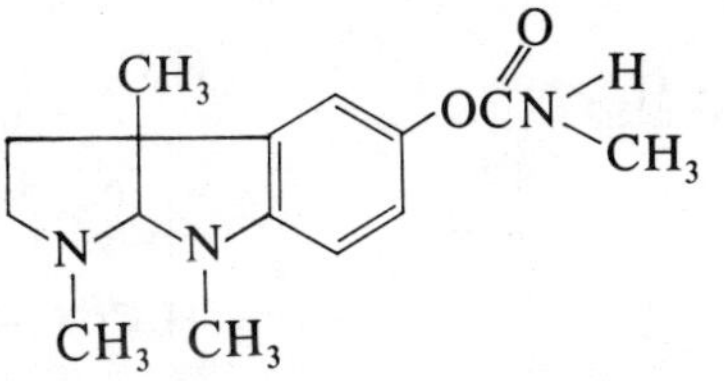

is the active pharmacological principle of Calabar beans, the seeds of *Physostigma venenosum* (Balfour), which have been used in West African witchcraft trials. Miotic and cholinergic properties of physostigmine were discovered in 1862 and its use in ophthalmology was suggested by Robertson (1).

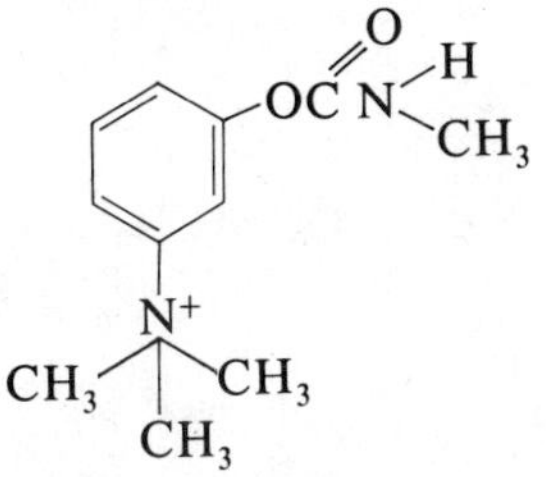

The structure of physostigmine was elucidated by Stedman (2) who also prepared a number of its synthetic analogs. One of these, *m*-trimethyl-ammoniumphenyl methylcarbamate, was pharmacologically active, but its instability in aqueous solution limited its effective use. Aeschlimann and Reinert (3) synthesized the corresponding dimethylcarbamate, neostigmine or prostigmine,

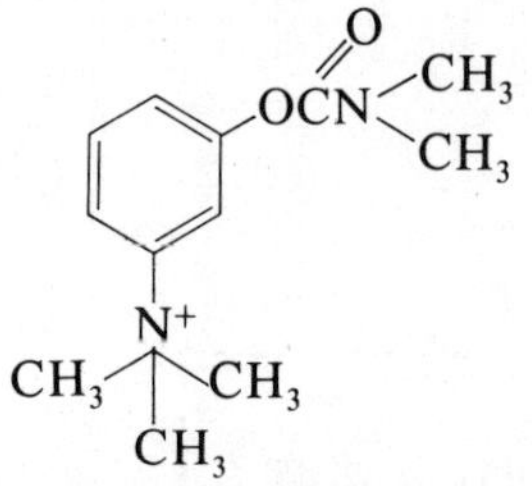

which possessed pharmacological activity as well as stability. For details of the development of synthetic analogs of physostigmine for use in clinical medicine, see Stempel and Aeschlimann (4).

In the late 1940s Hans Gysin and co-workers of the Geigy Company, discovered the insecticidal properties of *N*-dimethylcarbamates (5). Partly because of the success of neostigmine and partly because of the synthetic route followed by the Geigy Company of Switzerland, the earlier insectical compounds were all *N*-dimethylcarbamates. Some of the insecticidal carbamates synthesized by Gysin and co-workers are:

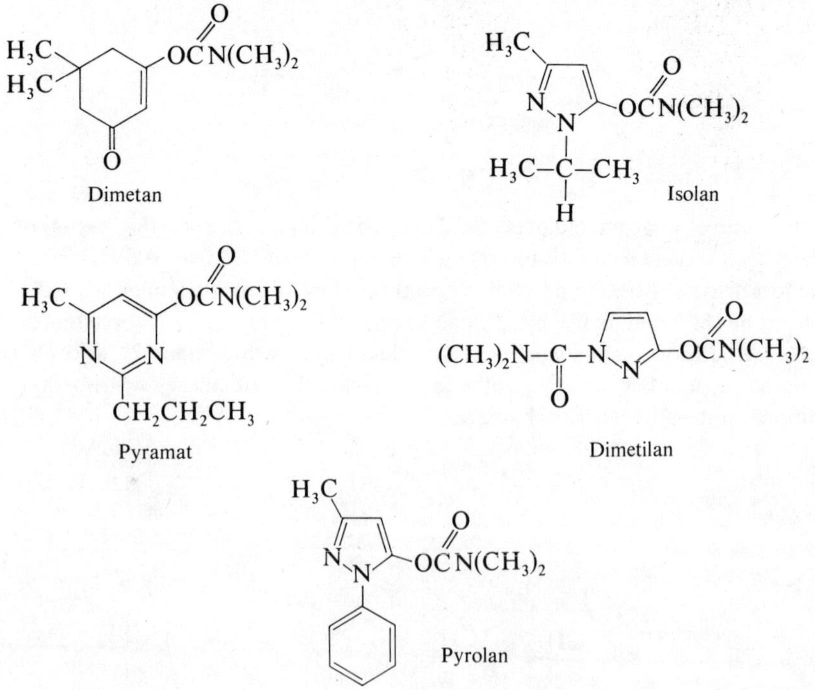

Dimetan

Isolan

Pyramat

Dimetilan

Pyrolan

These compounds were highly toxic to aphids and house flies but possessed limited toxicity toward other insects. This narrow spectrum of toxicity resulted in restricted commerical interest in these insecticides.

At about the same time Metcalf and co-workers at the University of California were studying the properties of anticholinesterase agents. They started work on various carbamates in an effort to resolve why charged carbamates like physostigmine, which were good inhibitors of house fly cholinesterase, were totally ineffective as contact insecticides. These investigators modified the structures of simple phenylmethyl carbamates to increase their lipid solubility and discovered that substitution in the *ortho* or *meta* position with an alkyl group or a halogen in a phenyl methylcarbonate increased its insecticidal activity. They further found that *m-tert*-butylphenyl methylcarbamate, the uncharged isostere of *m*-trimethylammoniumphenyl methylcarbamate, was highly insecticidal (6).

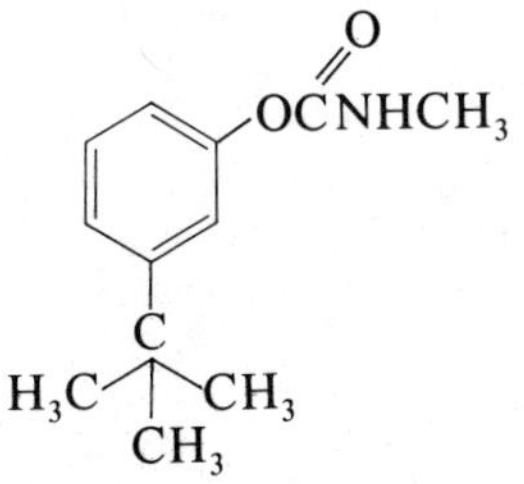

$$\underset{\text{C}}{\overset{\text{O}}{\parallel}} \text{OCNHCH}_3$$

*m-tert*-butylphenyl methylcarbamate

Their careful investigations led Metcalf and co-workers to surmise that there were three major requirements for optimum contact toxicity and insecticidal activity: first, to be able to penetrate to the site of action in the insect nerve, the molecule should be nonpolar; second, to effectively inhibit the enzyme, the carbamate molecule should be structurally complementary to the active surface of acetylcholinesterase, both in stereochemistry and reactivity; and third, it should possess sufficient stability to multifunction-oxidase detoxication (6, 7). Though the University of California group synthesized a series of nonpolar substituted phenyl methylcarbamates with insecticidal activity (8), the first commercially successful carbamate, Sevin®, carbaryl, was synthesized by scientists at Union Carbide Corporation (9, 10).

The *N*-dimethylcarbamates and the *N*-methylcarbamates discussed until now have one structural feature in common: the carbamic acid was esterified with a weakly acidic (enolic or phenolic) hydroxyl compound represented by

$$\text{>C}\overset{\cdot}{=}\text{C}-\text{OH} \qquad \text{or} \qquad \text{>C}=\text{C}-\text{OH}$$

respectively. However in the early 1960s scientists at Union Carbide Corporation (11) discovered that carbamoyl oximes of cyclic ketones possessed, in addition to insecticidal properties, acaricidal and nematicidal properties. This discovery led to the synthesis of trisubstituted acetaldehyde O-(methylcarbamoyl) oximes bearing structural similarities to acetylcholine (12), and the open-chain carbamoyloxime, Temik®, aldicarb, was subsequently synthesized. Though initially the activity of aldicarb was assumed to be due to its being isosteric with acetylcholine, it now appears that its toxicity is due more to the hydrophilic nature of its metabolite, aldicarb sulfoxide, than to its conformational similarity to acetylcholine (13, 14).

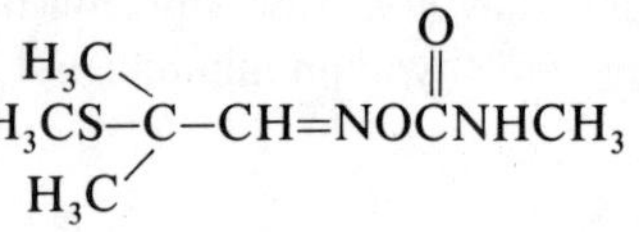

$$\begin{array}{c} H_3C \\ \phantom{H_3C}\diagdown \\ H_3CS-\underset{\underset{\displaystyle H_3C}{\diagup}}{C}-CH=NO\overset{\displaystyle O}{\overset{\|}{C}}NHCH_3 \end{array}$$

Aldicarb

$$\begin{array}{c} H_3C \\ \phantom{H_3C}\diagdown \\ H_3C-\underset{\underset{\displaystyle H_3C}{\diagup}}{N}-CH_2-CH_2O\overset{\displaystyle O}{\overset{\|}{C}}CH_3 \end{array}$$

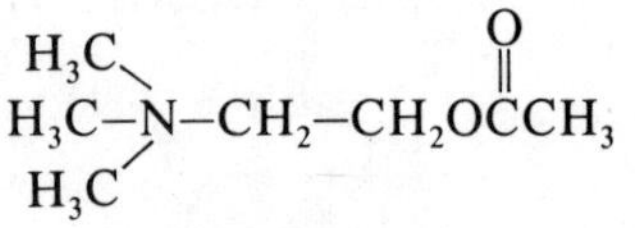

Acetylcholine

## MODE OF ACTION

Carbamate insecticides, like the organophosphorus compounds, discussed earlier, inhibit AChE by acting as a substrate for the enzyme. The reaction between a carbamate inhibitor (CI) and AChE (HE) may be represented as follows (15–17):

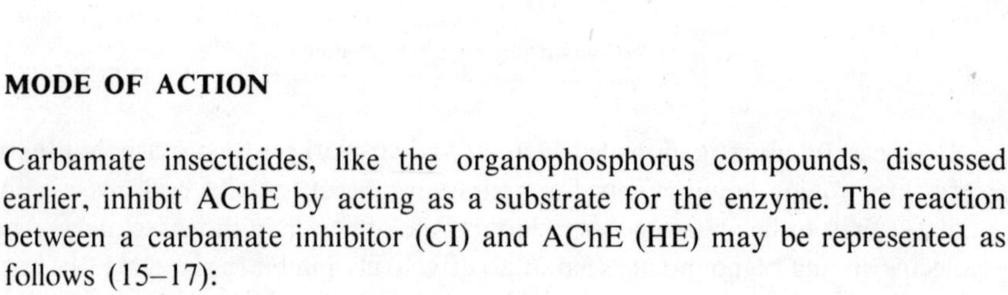

$$HE + CI \left\{ \overset{k_1}{\underset{k_{-1}}{\rightleftharpoons}} \right\} \underset{\substack{\text{Enzyme} \\ \text{inhibitor complex}}}{HE.CI} \xrightarrow{k_2} EC \xrightarrow[HOH]{k_3} HE + \text{Carbamic acid}$$

(with $KI$ and $k_i$ bracket above, and leaving group $HI$ below)

$$\tag{4.1}$$

where HE·CI is a reversible complex whose formation is controlled by the equilibrium constant $KI = (k_{-1}/k_{-2})$, and EC is the carbamylated active site of the enzyme whose formation is governed by the carbamylation rate constant $k_2$

(compare $k_p$ Fig. 3.3). The overall bimolecular rate constant and the regeneration or decarbamylation constant are represented by $k_i$ and $k_3$, respectively (compare reaction between ACh and AChE Chapter 3 pp. 30–33).

By applying the kinetic treatment and experimental procedures developed by Main (18) and Main and Iverson (19) for organophosphate inhibitors, Main and Hastings (20) successfully evaluated the binding constant $KI$ and the carbamylation rate constant $k_2$ for the inhibition of acetylcholinesterase by eserine. Yu et al. (17) used the same method with some modifications to obtain acetylcholinesterase inhibition constants (binding, carbamylation, and overall bimolecular rate constants, $KI$, $k_2$, and $k_i$, respectively) for selected substituted phenyl $N$-alkyl carbamates. Similarly, the decarbamylation rate constant $k_3$ can be determined by the method described by Wilson et al. (15) for organophosphates and used by Yu et al. (17) for bis carbamates.

The carbamylated enzyme EC is more stable by far (as much as $10^6$ or more) than the transitory acetylated enzyme produced during the hydrolysis of acetylcholine under natural conditions. Stage 3 therefore becomes rate limiting and results in the accumulation of acetylcholine at the synaptic junctions (21). This, at the organismic level, results in typical cholinergic symptoms involving irritability, myosis, tremors, lachrymation, salivation, incoordinated movements, convulsions, paralysis, and ultimately death.

Metcalf and Fukuto (22) and Metcalf (6) have synthesized and evaluated more than 600 carbamates in order to investigate their physiological interactions with insects. These authors hypothesize that an active insecticidal carbamate molecule should be uncharged lipid partioning in order to be able to penetrate to the site of action in the insect nerve. It should have some structural resemblance to acetylcholine, both in stereochemistry and reactivity, and also its structure should be complementary to the surface features of acetylcholinesterase. It should possess sufficient stability to multifunction oxidases. These features would thus facilitate the formation of the enzyme inhibitor complex by providing a close fit of the carbamate molecule to the acetylcholinesterase molecule and subsequent carbamylation of the serine hydroxyl at the active site in the enzyme. Thus, in the case of insecticidally active carbamates, according to Metcalf and Fukuto, steric factors predominate over others. In this respect these compounds differ from the organophosphates where electronic factors play an important role in the inhibition of acetylcholinesterase. For details see Kolbezen et al. (8), Metcalf (6), Metcalf and March (23), and Metcalf and Fukuto (22, 24).

## STRUCTURE–ACTIVITY CORRELATIONS

If the insecticidal activity of carbamates is due to the inhibition of acetylcholinesterase, a direct correlation between toxicity and anticholinesterase activity would be expected. This activity is usually measured in terms of $I_{50}$ value as determined by the Warburg manometric method using house fly head acetylcholinesterase as the substrate. $I_{50}$ is defined as the molar concentration required to give 50% inhibition of the enzyme under specific conditions. The conversion of the $I_{50}$ values to the overall bimolecular rate constant $k$ values can be achieved by the relationship (25, 26):

$$k = \frac{0.695}{I_{50}\ t}$$

where $t$ is the time of reaction of the carbamate with acetylcholinesterase. The $I_{50}$ is strictly applicable only to reversible inhibitors, but, as indicated earlier (Equation 4.1), carbamate inhibition of acetylcholinesterase involves a carbamylation step. Nevertheless, $I_{50}$ is a useful parameter and provides a satisfactory basis for structure–activity correlations. However, when the inhibition of acetylcholinesterase by a carbamate is compared with its toxicity to insects, a direct correlation between the two is not obtained because of the detoxification of the chemical by the microsomal enzyme system. A reasonably good correlation can be obtained for many substituted phenyl methylcarbamates by coadministration of a synergist like piperonyl butoxide (27) which counteracts detoxification by microsomal enzymes. Fukuto et al. (28) and Jones et al. (29) have subjected such data to multiple regression analysis in order to obtain useful information from correlations between synergized toxicity and anticholinesterase activity.

Certain compounds, such as neostigmine and carbamates, with substituents such as the thiocyano group do not show toxicity even when used with piperonyl butoxide because of the inability of the former compounds to penetrate insect tissue (8) and the extramicrosomal detoxification of the latter (30).

### Alkylphenyl and alkoxyphenyl Methylcarbamates

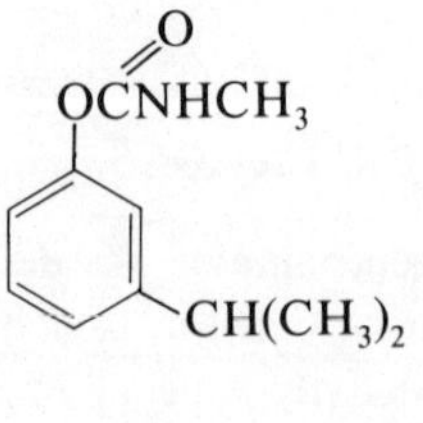

In alkyl-substituted carbamates the anticholinesterase activity is usually correlated with ring position (*meta* > *ortho* > *para*) and "activity increases progressively with the addition of methyl groups: $CH_3 < C_2H_5 < i\text{-}C_3H_7 = tert\text{-}C_4H_9 < sec\text{-}C_4H_9$" (6). However, in the case of isopropylphenyl methylcarbamate, UC 10854, the difference in insecticidal activity between the *meta* and *ortho* isomers is small, whereas the *para* isomer is much less insecticidal (14). The *meta* isomer of propoxur is almost inactive. Meltzer and Welle (31) tested the activities of a number of substituted phenyl *N*-methylcarbamates and concluded that the surmise that meta isomers of alkylphenyl *N*-methylcarbamates are most active has to be restricted to certain insects like flies and caterpillars. "In the case of Colorado potato beetle, *Leptinotarsa decemlineata* (Say) *o*- and *m*-isomers were equally active, for aphids the *o*-isomer was the most toxic one." In alkoxyphenyl *N*-methyl carbamates the order of activity is *ortho* > *meta* > *para* because of the spatial effect of an extra O atom (6).

Most aryl methylcarbamates show a limited toxicity to the house fly, but their activity can be synergized by piperonyl butoxide; however the compound

$$
\begin{array}{c}
\overset{\displaystyle O}{\underset{\displaystyle }{\parallel}} \\[-2pt]
\text{OC\,NHCH}_3 \\
\end{array}
$$

O–CH$_2$C≡CH

which contains a propargyloxy group, is highly active against the house fly and is not synergized by piperonyl butoxide. It is possible that this chemical provides for its own synergism (14). Wilkinson (32) considers the proparglyoxy group as an essential feature of certain synergists. A similar explanation for the unusually high insecticidal activity of 3,4-methylenedioxyphenyl methylcarbamate was given by Fukuto et al. (27).

The propynyloxy group is known to be a synergophore. Kooy (32a) discovered the synergistic activity of various propynyl aryl ethers and Sacher et al. (32b) synthesized and tested 30 derivatives of propynyl naphthyl ether as synergists for carbaryl against the house fly. 1-Naphthyl-3-butynyl ether was found to give a synergistic ratio of 176.5 against the insect but did not synergize the insecticide against the white mouse. These authors hypothesized that the propynyloxyaryl synergists complex *in vivo* with an essential metal cofactor of the mixed function oxidases. This retards the oxidative degradation of xenobiotics.

Incorporation of two alkyl groups into the proper position substantially enhances the anticholinesterase activity of phenyl methylcarbamates. The greatest activity is shown by di-*meta*, that is, 3,5-substituted, compounds (6). Successful insecticides of the dialkyl series include Butacarb®, 3,5-di-*t*-butylphenyl methylcarbamate; (meobal®; and promecarb, 3-methyl-5-isopropylphenyl-*N*-methylcarbamate.

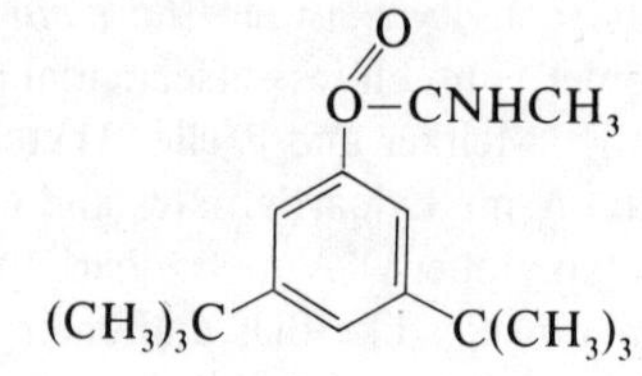

$$\underset{\text{Butacarb}}{(CH_3)_3C} \overset{\displaystyle \overset{O}{\parallel}}{\underset{}{\diagdown}} O-CNHCH_3 \quad C(CH_3)_3$$

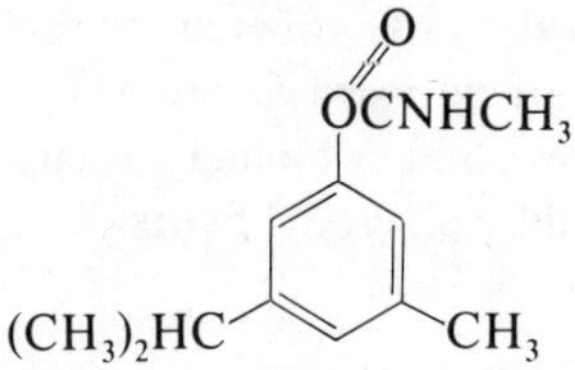

Butacarb

$$\underset{\text{Promecarb}}{(CH_3)_2HC} \quad \overset{\displaystyle \overset{O}{\parallel}}{OCNHCH_3} \quad CH_3$$

Promecarb

$$O-\overset{\displaystyle \overset{O}{\parallel}}{C}NHCH_3$$

$$CH_3$$
$$CH_3$$

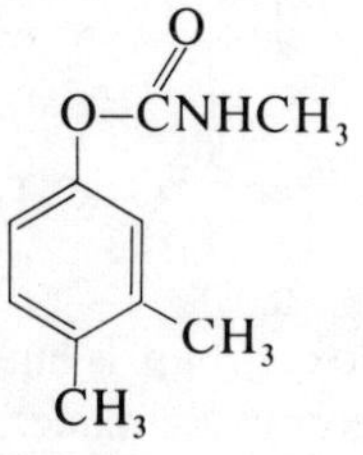

Meobal

Landrin is a mixture of two trisubstituted isomers, namely, 3,4,5-trimethyl-phenyl methylcarbamate, 75%, and 2,3,5-trimethylphenyl methylcarbamate, 18%.

### Other Substituted Phenyl Methylcarbamates

Substitution of the phenyl nucleus with an amino group does not confer appreciable insecticidal activity; p-dimethylaminophenyl methylcarbamate

$$\overset{\displaystyle \overset{O}{\parallel}}{OCNHCH_3}$$

$$CH_3 \overset{\diagup}{\underset{\diagdown}{N}} CH_3$$

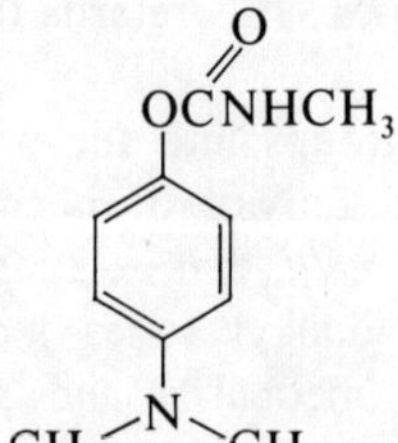

is almost inactive as an insecticide (31, 33). However, when one methyl group is added in position 3, an insecticidal compound, aminocarb, 4-dimethyl amino-3,5-xylyl methylcarbamate, is obtained. Addition of two methyl groups in positions 3 and 5 yields an even more potent insecticide, Zectran[®], 4-dimethylamino-3,5-xylyl methylcarbamate, which has found commercial use.

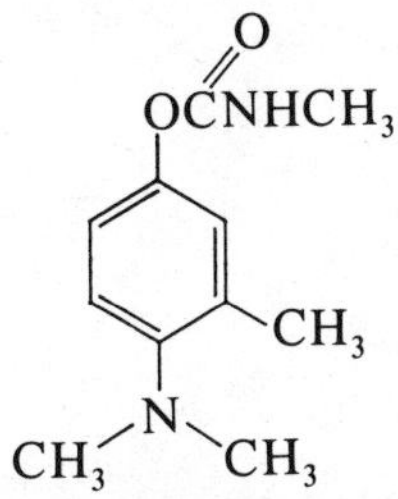

Aminocarb

Substitution of the phenyl nucleus with a thioether moiety markedly enhances the insecticidal activity of the carbamate (6, 22, 34). For details on the effects of a sulfur atom, see Metcalf (6).

**Carbaryl and Other Multiring Carbamates**

Because of the commercial success of carbaryl, the structure–activity relationships of multiring carbamates have received considerable interest. Carbaryl itself can be considered as a 2,3-disubstituted phenyl methylcarbamate; the closely related 2-naphthyl compound, which can be considered as 3,4-disubstituted phenyl methylcarbamate is much less active as an insecticide. 5,6,7,8-Tetrahydrocarbaryl

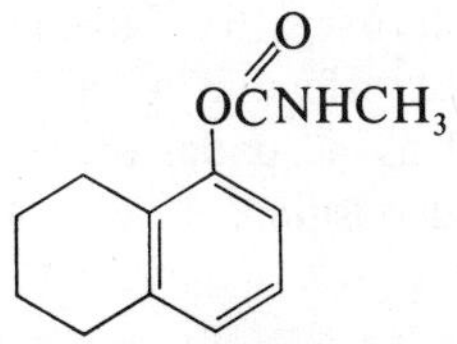

5,6,7,8-Tetrahydrocarbaryl

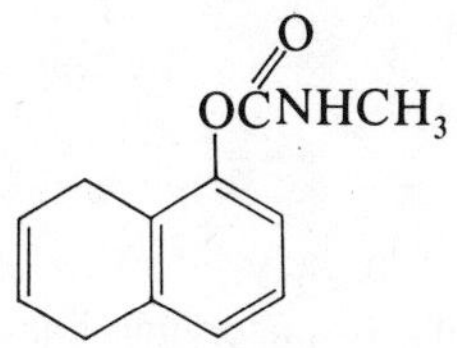

is slightly less toxic than carbaryl to insects. The 5,8-dihydro analog of carbaryl, however, shows increased toxicity to house fly and *Culex fatigans* larvae, suggesting that the detoxifying OH moiety does not readily attack non-aromatic rings (6).

Demethyl carbaryl is unstable even at neutral pH and shows little anticholinesterase activity. 5-Hydroxy carbaryl and methylol of carbaryl show very little insecticidal activity.

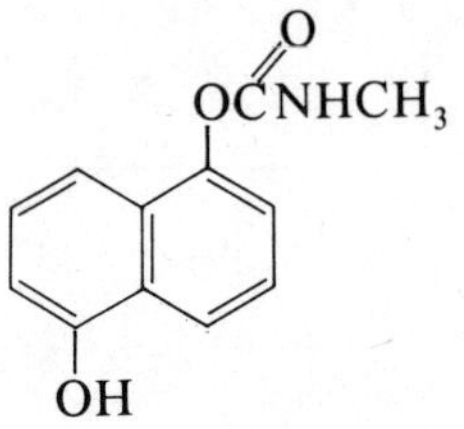

5-Hydroxycarbaryl 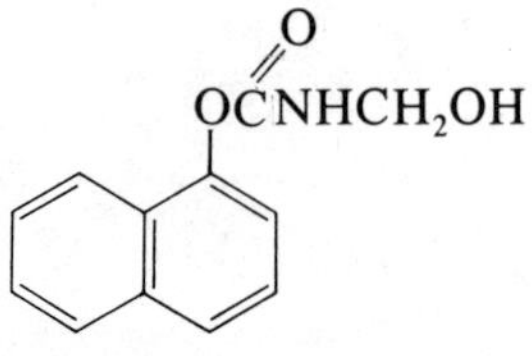                    Methylol of carbaryl

Mobam®, 4-benzothienyl methylcarbamate,

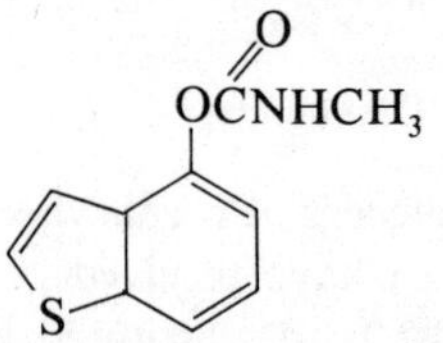

is insecticidally about as active as carbaryl against many insects. It is isosteric with carbaryl because of the aromatic character of the S-containing five-membered ring. However, despite this similarity, Mobam is unexpectedly much more toxic to house fly than carbaryl (6). Carbofuran has low mammalian toxicity but is one of the most effective of all carbamate insecticides. This chemical can be considered as a spatial analog of propoxur in which the isopropyl group is in a fixed position.

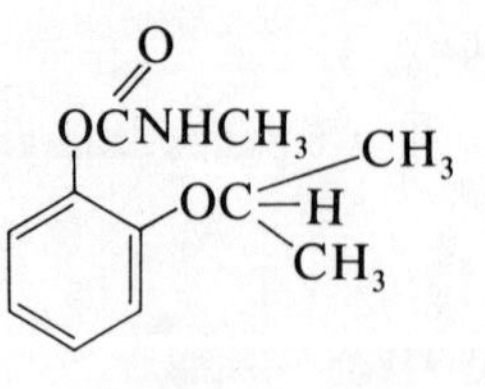

Propoxur 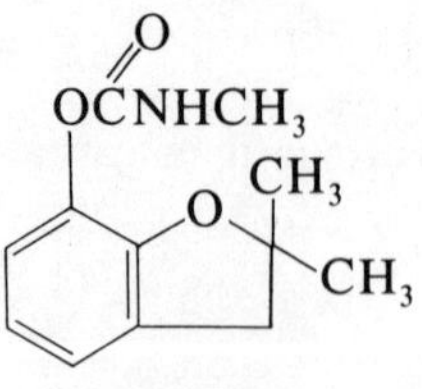                    Carbofuran

This fusion of the isopropoxy moiety into the benzofuran ring, according to Metcalf (6), enhances affinity for acetylcholinesterase and decreases the *in vitro* detoxification in insects. The related methylcarbamate, 2,3-dihydro-2-methyl-7-benzofuranyl methylcarbamate, though as insecticidal as carbofuran, possesses lower affinity for acetylcholinesterase and is therefore one-tenth as toxic to mammals as carbofuran. This is probably due to lower van der Waals interaction (6). Larger ring compounds like fluorenyl methylcarbamates have proved to be nontoxic to insects, though they interact with acetylcholinesterase. Metcalf (6) hypothesizes that the presence of two aromatic rings provides a greatly enhanced opportunity for detoxification.

### N-Acyl and Other Substitutions at the Carbamate Nitrogen

Fraser et al. (35) discovered that the acyl derivatives of insecticidal carbamates, formed by the substitution of the proton of the N-methylcarbamates by various acyl groups, were generally less toxic to mammals than the parent carbamates. Such compounds show substantially less anticholinesterase activity than the parent N-methyl compounds (36), but their insecticidal activity is highly variable. For instance, *m*-isopropylphenyl methylcarbamate shows considerable insecticidal activity, but its N-acetyl derivative is practically inactive as an insecticide (14). However the N-acetyl derivative of Zectran is about half as toxic to the western budworm, *Choristoneura occidentalis* Freeman, as the parent compound (37).

Fahmy and Fukuto (38) reported that the acyl derivatives of carbaryl, when tested against the carbaryl-susceptible Egypian cotton leafworm, *Spodoptera littoralis* (Boisd.), were generally less toxic than the parent insecticide. The order of toxicity was, however, reversed in the case of carbaryl-resistant strain of the same insect. This, the authors surmised, was due to the increased presence of deacylation enzyme in the resistant strain. The N-formyl derivative was an exception; it showed three times the activity of carbaryl against the susceptible strain. The N-acetyl derivative of carbaryl, when tested against the susceptible strain, showed only one-fifteenth of the activity of the parent compound. In the cases of carbaryl and propoxur, the butyryl derivatives were considerably more toxic than the corresponding acetyl derivatives (38, 39). Barlow and Hadaway (40) tested propoxur and three closely related N-methylcarbamates, all of which were toxic to mosquitoes, and their corresponding acetyl derivatives. They found that N-acetylation was not accompanied by excessive loss of toxicity to mosquitoes, but this change in chemical structure increased the volatility of the derivatives to the extent than none of them could be considered as potential residual insecticide against adult mosquitoes. Hadaway et al. (41) obtained similar results with several carbamates. In some cases they found contact toxicity to be maximal in N-acetyl derivatives.

Initially there was a feeling that a suitable acylation might further enhance the insecticidal action and selectivity pattern of an N-methylcarbamate. Later metabolic investigations, however, suggested that the toxic principle was still the parent N-methylcarbamate that was produced as a result of deacylation *in vivo* (35, 39). But it has been shown by Fraser et al. (35) that in some cases acyl substitution imparts insecticidal activity better than that of the parent carbamate. If one assumes that toxicity is still due to the *in vivo* deacylation of the acyl derivative, then acylation possibly results in better penetration to the site of action (35).

There is also evidence of different metabolic routes followed by the N-acyl derivative and its parent carbamate in insects and in mammals. Thus N-acetyl Zectran® is metabolized by the spruce budworm, *Choristoneura fumiferana* (Clemens), to yield the parent carbamate, whereas in mice it is detoxified by hydrolysis to 4-dimethylamino-3,5-xylenol (37, 42). The instability of the N-acyl derivatives in general has interfered with their practical use.

Bearing in mind the different routes of metabolism of Zectran and malathion, Fahmy et al. (38) hypothesized that the substitution of a proton on the N-methyl moiety of carbamate insecticides possessing mammalian toxicity by a substituted phosphorothioyl group might yield compounds that are more likely to be detoxified by alternative routes in mammals. Derivatization of insecticidal carbamate esters with the *O,O*-dimethylphosphorothioyl group by these scientists "produced compounds which were remarkably less toxic to the white mice than the original carbamate, but were of equal or increased toxicity to susceptible and carbamate-resistant strains of house flies."

Another substituent, the phenylthio group, for the proton on the N-methyl moiety of a carbamate ester has yielded an unusually potent mosquito larvicide. The new carbamate, RE 11775, *m-sec*-butylphenyl[methyl(phenylthio)-carbamate], is highly effective against adults as well as larvae of susceptible and organophosphorus-resistant strains of *Aedes nigromaculis* (Ludlow) (43).

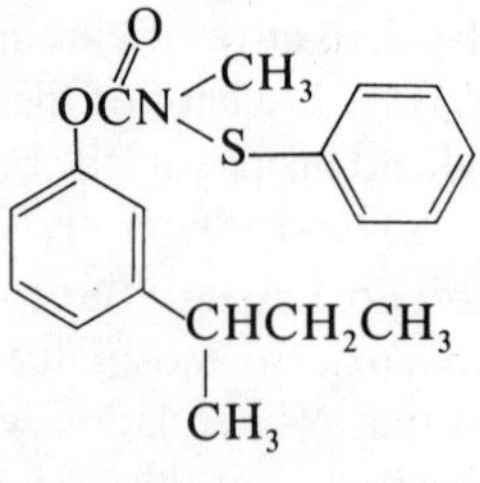

## N-METHYLCARBAMOYL OXIMES

Methylcarbamates of a variety of oximes have given the newest category of insecticidal and acetylcholinesterase-inhibiting carbamates. The structure–activity relationships of this group of compounds has been discussed by Fenton (44), Fukuto et al. (28), Metcalf (6), and Weiden (13, 14). The possibilities of synthesizing chemicals belonging to this group is enormous because the structural variety of oximes is virtually limitless. In their biological activity, carbamoyl oximes show marked resemblance to the carbamates of cyclic enols and phenols; moreover, the electronic effect of substituents on acetylcholinesterase-inhibiting activity is more obvious in the case of the former than in the latter. Increasing electronegativity of substituents in the 2-position generally results in a corresponding increase in anticholinesterase activity. Alkylthio substituents affect the biological activity in two ways: first, by providing potential for sulfoxidation, thereby increasing polarity, and secondly, by increasing affinity for acetylcholinesterase (13, 14, 45). Generally an alkyl sulfide substituent attached directly to the oxime carbon, as in the case of methomyl, confers outstanding activity against lepidopterous larvae (14).

Compounds belonging to this group show marked changes in insecticidal spectrum even with a relatively minor modification of structure. Possibly species-specific detoxification of the carbamoyl oximes is a major cause of this selectivity (13). When the $CH_3S$ group in aldicarb is replaced by an *S*-phenyl group,

$$\langle\ \rangle - S - \overset{\overset{\displaystyle CH_3}{|}}{\underset{\underset{\displaystyle CH_3}{|}}{C}} CH = N - O - \overset{\overset{\displaystyle O}{\parallel}}{C} NHCH_3$$

there is a remarkable reduction in toxicity that is probably associated with a high detoxification rate of the compound thus formed, because its toxicity is considerably increased when it is used with a synergist (14).

Carbamates of aldoximes and unsymmetrical ketoximes exhibit *syn–anti* isomerism in which oximino oxygen is either *cis* or *trans* to the aldehyde hydrogen. These isomers are difficult to separate and published literature on the comparison of isomeric pairs is limited. Felton (44) found that the *syn* isomer of methomyl was more active by far than the *anti* isomer. Temik aldicarb, is supposed to be the *syn* isomer (13).

Aldicarb and methomyl are two of the successful pesticides belonging to this category. Neither of these is either synergized or significantly antagonized by piperonyl butoxide.

Oxime methylcarbamates are generally superior in pesticidal activity to oxime dimethylcarbamates. *N*-Phenylcarbamates of oximes are inactive as insecticides.

## SELECTIVE TOXICITY OF CARBAMATES

Unlike the organophosphorus compounds, the carbamate esters do not show a broad spectrum of insecticidal activity; some of these compounds are among the most specific of all insecticidal chemicals. Though a few carbamates like aldicarb and isolan are highly toxic to both insects and mammals, others like carbaryl show differential toxicity to insects and vertebrates. Even among insects and other arthropods, some of the carbamates show surprising species specificity, for example, dimetan is highly selective for aphids and Dimetilan is an effective stomach poison for flies; but both these chemicals show little activity against other insect pests. Carbaryl is a relatively broad-spectrum carbamate, but it is weak against aphids and house flies. Carbaryl is also toxic to many arthropods belonging to the class Acarina. Included here are ticks, parasitic mites, predacious mites, and rust mites; however it is inactive against tetranychid mites (46).

The phenyl N-methylcarbamates, in general, show good activity against the larva of the Mexican bean beetle *Epilachna varivestis* Mulsant, and against the bean aphid, *Aphis fabae* Scopoli, but poor activity (with some exceptions) toward the tetranychid mites, *Tetranychus* spp., and the southern armyworm, *Spodoptera eridania* (Cramer). Zectran, a substituted aminophenyl methylcarbamate is effective against aphids, the southern armyworm, and the Mexican bean beetle, but it is ineffective against mites and the house fly. *p*-Dimethylaminomethylphenyl N-methylcarbamates show a high level of activity toward several insects but are nontoxic to the house fly (31).

Carbamoyloximes as a class are generally effective against a broad spectrum of plant pests, but individual compounds, like those belonging to other classes of carbamate esters, tend to be selective. For example, aldicarb, which is a potent insecticide, nematicide, and miticide, is ineffective against the southern armyworm. Carbamoyloximes with 1-thio substituents, such as methomyl, Lannate®, are toxic to this pest. The activity of carbamoyloximes toward the house fly is variable, but ineffective compounds can be synergized by piperonyl butoxide (13).

The reasons for the selective toxicity of carbamates is probably the species-specific detoxification of the compounds, because ineffective compounds can often be synergized more effectively by methylene dioxyphenyl type synergists; however those carbamates that are intrinsically most toxic are synergized the least. It has also been suggested (47) that the cause for this differential toxicity may lie in the gross differences between the target acetylcholinesterase, but the absence of satisfactory assay methods for this enzyme limits the test of this hypothesis. Suggestion has been made by Hellenbrand and Krupka (48) that the fly-head acetylcholinesterase differs from bovine erthrocyte enzyme in that it possesses a second anionic site, this makes it possible for the former to bind two substrate molecules simultaneously.

In addition to the chemical structure, physicochemical and biophysical factors also influence the toxicity of carbamates. Strongly hydrophilic carbamates are slow to penetrate the cuticle and do not show contact toxicity. However, once they gain access to the inside, knockdown is quick and there is usually no recovery. Strongly lipophilic carbamates, on the other hand, penetrate easily and give a rapid knockdown that is followed by a degree of recovery dependent on the dose (13, 14). This recuperation is probably due to the oxidative attack by microsomes on lipid-soluble material; a positive association between good lipid solubility and uptake by microsomes has been reported by Gaudette and Brodie (49). The final detoxification by the microsomes naturally depends on the chemistry of the molecule. A balance between lipophylic and hydrophilic properties and chemical structure is necessary to insure penetration and to prevent degradation and excretion. This balance possibly makes aldicarb a successful pesticide.

If a specific group of compounds possesses a clearly defined mode of action, and a relatively large amount of quantitative data relating to biological activity, structure, and physicochemical parameters are available for a series of compounds in the group, then the prediction of structure versus activity relationships appears within the realm of possibility. For substituted phenyl diethyl phosphates, a reasonable correlation between electrophilic properties of the phosphorus atom as quantitatively determined by Hammett's $\sigma$ constant and the $I_{50}$ values for the inhibition of the house-fly-head acetylcholinesterase was found (33, 50). Using data from Metcalf and co-workers, Hansch and Deutsch (51) computed Equation 4.2 for the substituted phenyl diethyl phosphates.

$$\log \frac{1}{I_{50}} = 3.451\sigma + 4.461 \qquad (r = 0.957) \qquad (4.2)$$

Inclusion of a term for $\pi$ constant as a measure of hydrophobic bonding (52) did not significantly improve the correlation.

Jones and coworkers (29) used multiple regression analyses to correlate topical $LD_{50}$ values for the house fly and $I_{50}$ values for house-fly-head acetylcholinesterase with linear free-energy parameters. For 13 *para*-substituted phenyl $N$-methylcarbamates correlation was improved by replacing Hammett's $\sigma$ (53) values by field ($F$) and resonance ($R$) constants that Swain and Lupton (54) elaborated to separate these parameters from the overall effects of Hammett's $\sigma$. The computed equation was as follows:

$$\log \frac{1}{I_{50}} = 0.457\pi - 1.180R - 0.853F + 3.784 \qquad (r = 0.91) \quad (4.3)$$

However, in the case of *meta*-substituted phenyl methylcarbamates where the computed equation was

$$\log \frac{1}{I_{50}} = 1.031\pi - 1.015F + 4.394 \qquad (r = 0.855) \qquad (4.4)$$

inclusion of $R$ values raised $r$ to 0.88 only; the additional correlation was not significant at the 5% level. This was expected because *meta* substituents do not produce substantial resonance effects as compared with *para* substituents (29).

Hansch and Deutsch (51) drew attention to the fact that in the case of phenyl diethyl phosphates, lipophylic bonding plays almost no role in the inhibition of acetylcholinesterase because $\pi$ values do not improve the correlation, but in the case of carbamates as much as 60% of the variation in acetylcholinesterase inhibition is accounted for by lipophylic bonding (positive $\pi$ term). The inhibition of acetylcholinesterase by the two types of compounds is therefore distinctly different. In the case of phosphorus esters the reversible step $k_1/k_{-1}$ plays no appreciable role in inhibition, indicating direct phosphorylation of the esteratic site of cholinesterase. On the contrary, in the case of carbamates the lipophylic character of the interaction with the anionic site of the enzyme determines to a large extent the degree of complex formation between acetylcholinesterase and the carbamate (29). Fukuto et al. (28) also used the same types of multiple regression analyses to correlate reactivity and biological data with free-energy parameters in the cases of $O$-(diethyl phosphoryl) and $O$-(methylcarbamoyl) oximes of substituted acetophenones and $\alpha$-substituted benzaldehydes.

From kinetic studies on the reaction of acetylcholinesterase with $O$-dimethylmethylcarbamyl esters of quaternary quinolinium compounds, Kitz et al. (55) concluded, "the structure of the leaving group is very important in determining the activity of the carbamate anticholinesterase agents and the p$K_a$ value of the leaving group is relatively unimportant. This conclusion is just the opposite to that reached for diethylphosphoryl anticholinesterases." Hastings et al. (56) examined the carbamylation and affinity constants of some anticholinesterase carbamates and arrived at similar conclusions. They found that there is a "fairly close analogy between the substrate and the carbamate reaction, but the analogy did not hold for the comparable nitrophenyl phosphates," suggesting that "the phosphorylation reaction occur at a different region of the active site than did acylation."

Resistance to carbamate insecticides develops more rapidly in insects against which the compounds is inherently not very potent. For instance, the house fly, toward which carbaryl is relatively ineffective at the onset, developed a high degree of resistance after a few generations of laboratory selection at the $LD_{70}$ level; the Mexican bean beetle, which is highly susceptible to carbaryl, showed no development of resistance when subjected to the same pressure for 14 generations (57).

## SYMPTOMS OF INTOXICATION

In mammals carbamate insecticides produce typical symptoms of anticholinesterase poisoning. Compounds that induce intoxication symptoms of short duration, for example, propoxur and promecarb, have short persistence in blood (1–2 hours) while those that induce prolonged symptoms, such as carbaryl and Landrin, persist for a considerably longer period in blood (6 hours or more). This inhibition of cholinesterase produced by most carbamates of low toxicity is of short duration and recovery of cholinesterase activity, accompanied by disappearance of symptoms, is rapid after cessation of exposure; this is in contrast to poisoning by organophosphorus compounds.

In addition to typical symptoms of anticholinesterase poisoning, Wilhelm and Vandekar (58) discovered that intravenous administration of near lethal or lethal doses of certain monomethylcarbamates caused deep anesthesia with dyspnea (difficult breathing), and in the cases of lethal doses, respiration ceased within a few minutes. Spontaneous respiration resumed if artificial respiration was given within the first minute of apnea.

In insects the typical symptoms of carbamate intoxication are initial hyperactivity, followed by incoordinated movements and convulsions. Later, paralysis ensues, during the early stages of which there may be erratic tremors. When house flies are injected with toxic carbamates, symptoms appear within a few seconds after treatment, indicating that carbamates do not require activation before exerting their toxic action. Unlike mammals, treated insects are able to recover after prolonged periods of paralysis. Apnea, which is a decisive factor in mammalian toxicology, is not critical in insects because of their diffusion-type respiratory system. Because of this difference, synergists greatly influence the toxicity of carbamates to insects (14).

## CARBAMATE INSECTICIDES

*Sevin*®, *carbaryl*, 1-naphthyl *N*-methylcarbamate,

is a white crystalline material, mp 142°C, that is soluble in most polar organic solvents but sparingly soluble ($< 0.1\%$) in water. Carbaryl is moderately toxic to mammals; its acute oral $LD_{50}$ for the rat is 500 mg/kg, and its vapor pressure is $< 0.005$ mg Hg at 26°C. Sevin is a broad spectrum insecticide that has been used to control over 150 major pests on fruits, vegetables, cotton, and other crops of economic importance. It is also used as fruit thinner for apples. Carbaryl is not effective against the house fly and the tetranychid mites; it is weak against aphids and dipterous insects. However it is highly effective against arachnid ectoparasites and certain phytophagous eriophyid mites and against some adult mosquitoes. It is especially effective against cotton pests.

*Temik®, aldicarb*, 2-methyl-2-(methylthio)-propionaldehyde *O*-(methylcarbamoyl)oxime,

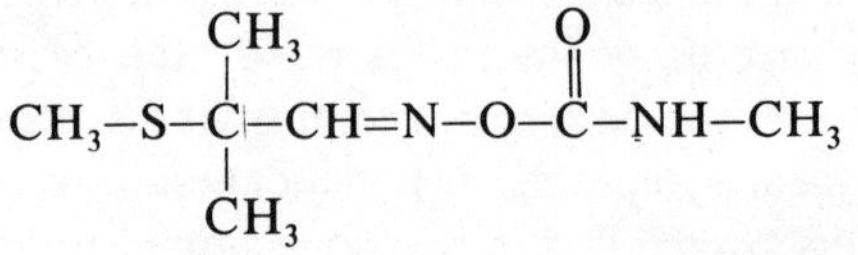

$$CH_3-S-\underset{\underset{CH_3}{|}}{\overset{\overset{CH_3}{|}}{C}}-CH=N-O-\overset{\overset{O}{\|}}{C}-NH-CH_3$$

is a white crystalline material; it is a systemic insecticide for soil use. Aldicarb is highly toxic, $LD_{50}$ for rat being 0.93 mg/kg and acute dermal $LD_{50}$ for rabbit being 5.0 mg/kg. Because it is also highly toxic by inhalation only 10% granules are offered for sale. Aldicarb is easily leached from soil and is fairly persistent in water (59).

*Furadan®, carbofuran*, 2,3-dihydro-2,2-dimethyl-7-benzofuranyl methylcarbamate

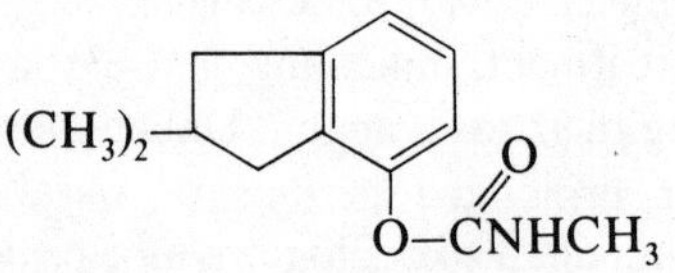

is a white crystalline solid that is slightly soluble in water; it is unstable in alkaline media. Carbofuran is highly toxic, its oral $LD_{50}$ for rat being 5 mg/kg. It is one of the most persistent and active carbamates and for this reason is widely used in soil applications for the control of insects attacking corn. It is effective against a broad spectrum of insects and mites and also against nematodes. It is both a soil and systemic insecticide.

*Zectran®*, 4-dimethylamino-3,5-xylyl-*N*-methylcarbamate,

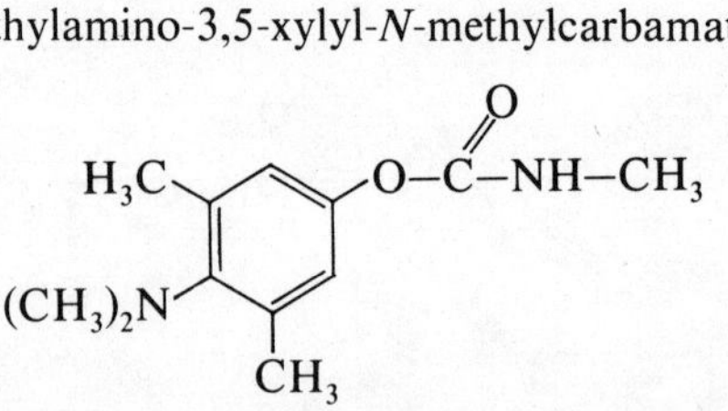

is highly toxic to mammals; the acute oral $LD_{50}$ for male rat is 19 mg/kg. It controls a wide range of pests, including thrips, leafhoppers, mites, snails, and slugs.

*Meobal®*, 3,4-dimethyl phenyl-*N*-methyl carbamate, is widely used in Japan for the control of various plant hoppers on rice and scales on fruits.

*Landrin®*, contains approximately 75% 3,4,5-trimethylphenyl methylcarbamate and 18% of the 2,3,5-isomer. It is useful in the control of certain soil insects, including corn rootworm. Landrin is also effective against black cutworm.

*Bux*® is a mixture of *m*-(1-methylbutyl)phenyl methylcarbamate and *m*-(1-ethylpropyl)phenyl methylcarbamate in the ratio of 3:1, respectively. It is applied to the soil in a band at the time of cultivation for the control of corn rootworm.

*Methomyl, Lannate*®, *S*-methyl *N*-([methylcarbamoyl]oxy)thioacetamide, is insecticidal against beetles, aphids, thrips, leafhoppers, and caterpillars, especially loopers.

A carbamate that is effective against insects of medical and veterinary importance is *Baygon*®, *propoxur*, *o*-isopropoxyphenylmethylcarbamate;

$$
\begin{array}{c}
\text{O} \\
\parallel \\
\text{O-C-NH-CH}_3 \\
\text{O-CH(CH}_3)_2
\end{array}
$$

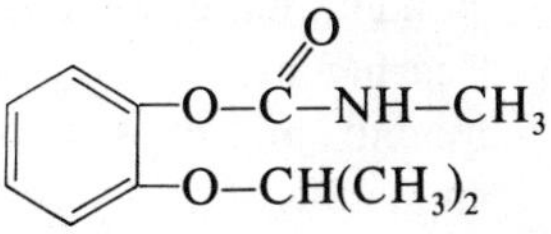

It is a white to tan crystalline solid, mp 91°C. Propoxur is slightly (0.2%) soluble in water and soluble in all polar organic solvents. It is toxic to mammals; acute oral $LD_{50}$ for rat is 128 mg/kg. Baygon is unstable in alkaline media. It is used in the control of cockroaches, spiders, crickets, wasps, hornets, flies, and mosquitoes. It possesses rapid knockdown and residual properties and flushes pests from their hiding places. It is highly toxic to fish and wildlife.

## METABOLISM OF CARBAMATES

The carbamate insecticides are absorbed and metabolized rapidly in living systems. Some recent reviews on the metabolism of carbamates are available (60–64). Carbamate metabolism, in general, is very complicated and the number of metabolites is considerable. Carbaryl is degraded to at least 13 ether-soluble metabolites by rat liver microsomes (65). Fifteen metabolites of carbaryl were found in chicken urine (66), and Zectran was metabolized to 17 products by human liver microsomes (67).

The metabolic attack on methylcarbamate insecticides usually involves oxidation or hydroxylation of an *N*-methyl group as in carbaryl (1), of a ring substituent as in Meobal (2), or of the ring itself as in carbaryl (3) or carbofuran (4); sulfoxidation as in aldicarb (5); formation of dihydrodihydroxy derivatives as in carbaryl (6); and hydrolysis of the carbamate ester bond, which is an important pathway for all *N*-methylcarbamates. In the case of carbamates like Zectran (7) and Metacil, which have a dimethylamino group attached to the ring, *N*-dealkylation of this group also takes place. There is subsequent hydrolysis and/or conjugation of the degradation products. The water-soluble products in mammals and birds are glucuronides and sulfates, probably a combination of sulfate, phosphate, and sugar conjugates in insects, and glucosides in plants.

Carbamates are rapidly and more or less completely degraded by animals and there is a prompt excretion of the metabolites. Usually within 24 hours almost all the administered dose is eliminated, about 80% by way of the urine, and the major portions of the remainder through exhalation and feces. In the case of lactating animals between 0.1 and 1% is eliminated in milk. Residues in tissues, milk, and eggs of treated animals 24–48 hours after treatment are very small, and continued feeding of small amounts of carbamate insecticides does not cause a continued corresponding increase of residues in milk (60, 66, 68, 69). Since the metabolism of carbamates is complicated, it is not possible to discuss all the important compounds in a book of this sort. Interested persons should refer to individual compounds in the literature. Some compounds are discussed below.

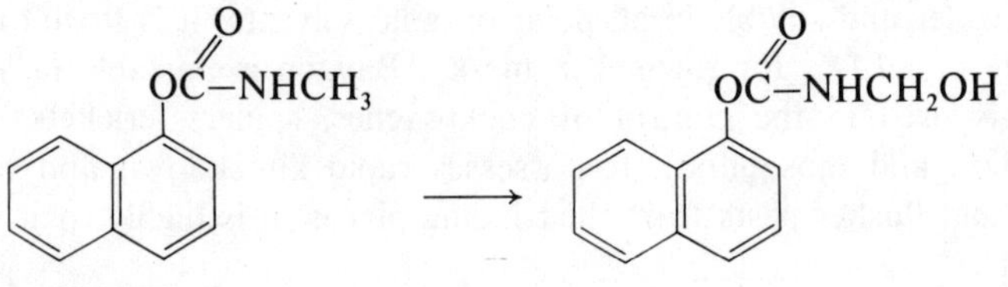

Carbaryl **I**

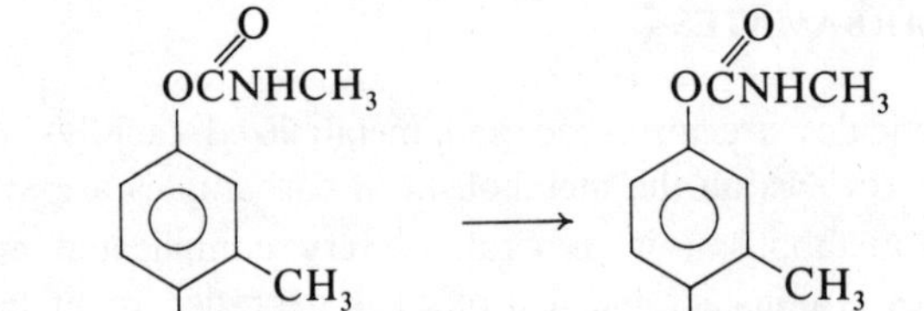

Meobal **2**

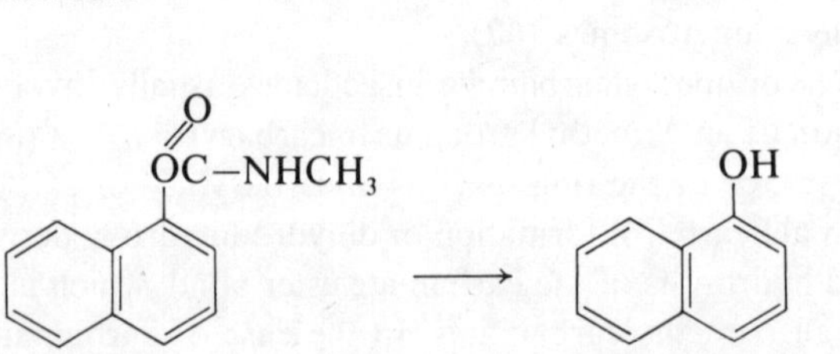

Carbaryl **3**

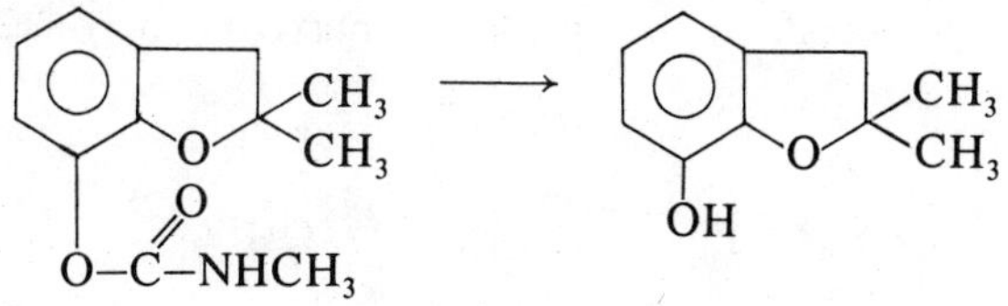

Carbofuran **4**

Aldicarb **5**

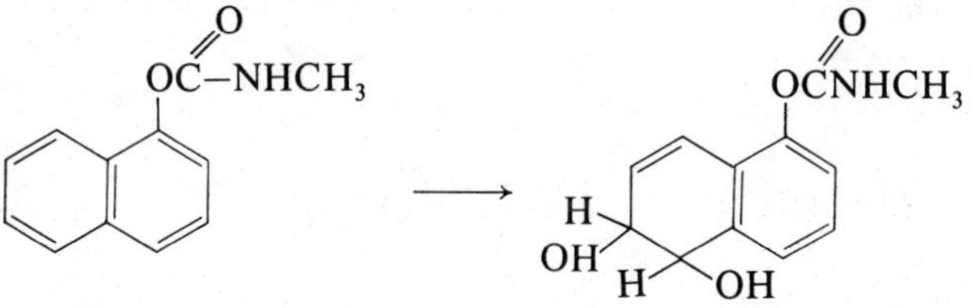

Carbaryl **6**

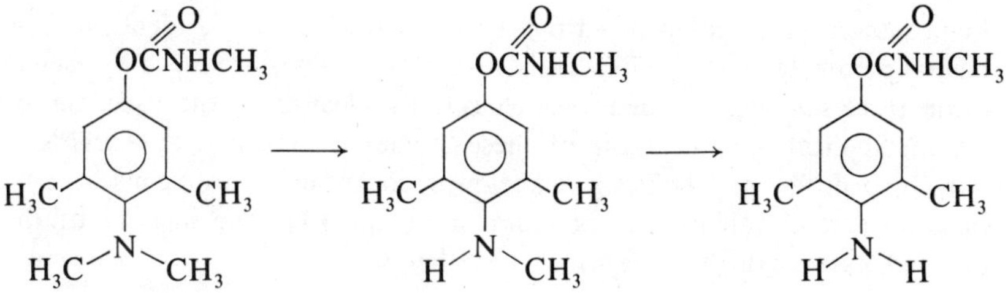

Zectran **7**

## METABOLISM OF CARBARYL

Metabolism of carbaryl has been studied by, among others, Abdel-Wahab et al. (70), Carpenter et al. (71), Dorough and Casida (72), Eldefrawi and Hoskins (73), Knaak et al. (74), Ku and Bishop (75), Kuhr (63), Leeling and Casida (79), and Dorough and Casida (77). Earlier investigations had indicated that the metabolism of carbamate insecticides involved the hydrolysis of the carbamate ester bond. In the case of carbaryl this may be represented as follows:

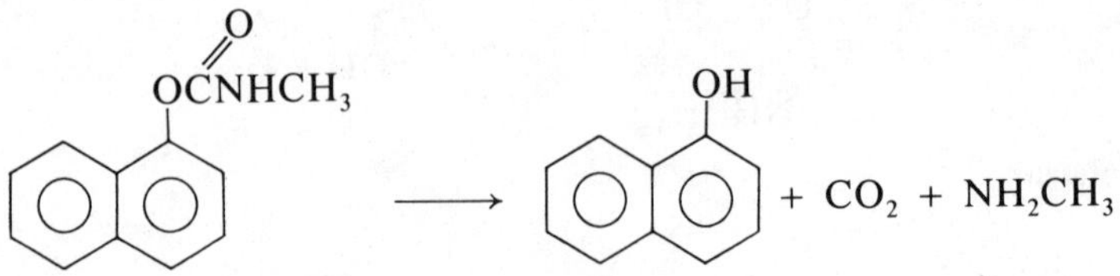

Hodgson and Casida (78, 79) produced evidence to suggest that oxidative pathways were also involved in the metabolism of carbamate compounds. Further investigations with the aid of isotopically labeled carbaryl were carried out by Casida and co-workers (76, 77, 80). These investigators used three radioactive samples of carbaryl with a different carbon atom ($^{14}C$) radioactive in each. The positions of the radioactive carbon atoms in the three samples are indicated below by an asterisk.

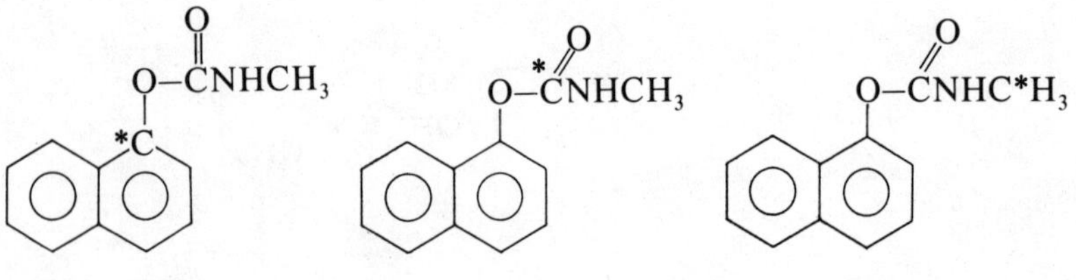

Each labeled preparation of carbaryl was then subjected to identical experimetnal parameters. The radioactivity of the metabolites produced by each one of the three samples provided some clues to the cleavage of the insecticide during metabolism. On the basis of these studies Casida and co-workers tentatively identified metabolites given below in cockroaches among insects and in rats, mice, and rabbits among mammals (Fig. 4.1). The identity of these metabolites was further confirmed by other workers.

> 1-Naphthyl-N-hydroxymethyl carbamate (N-hydroxymethyl carbaryl)
> 4-Hydroxy-1-naphthyl methylcarbamate (4-hydroxy carbaryl)
> 5-Hydroxy-1-naphthyl methylcarbamate (5-hydroxy carbaryl)
> 5,6-Dihydro-5,6-dihydroxy-1-naphthylmethyl carbamate, formed through
>   an epoxide intermediate (5,6-dihydro-5,6-dihydroxy carbaryl)
> 1-Hydroxy-5,6-dihydro-5,6-dihydroxynaphthalene
> 1-Naphthol

Paulson et al. (81) also report 1,4-dihydroxy- and 1,5-dihydroxy-naphthalene and 1,5,6-trihydroxynaphthalene from chicken urine, in addition to other metabolites reported above.

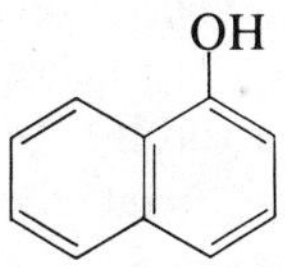

1-naphthol

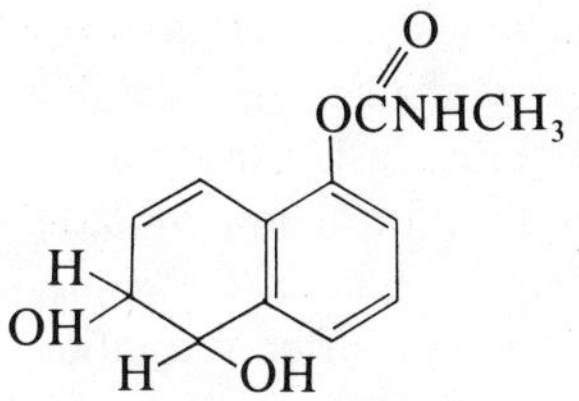

5,6-dihydro,5,6-dihydroxy carbaryl

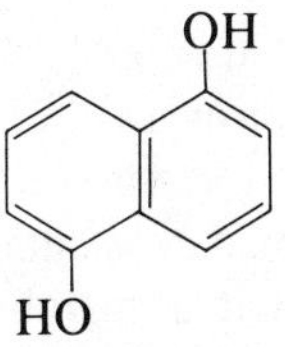

1,5-dihydroxy naphthalene

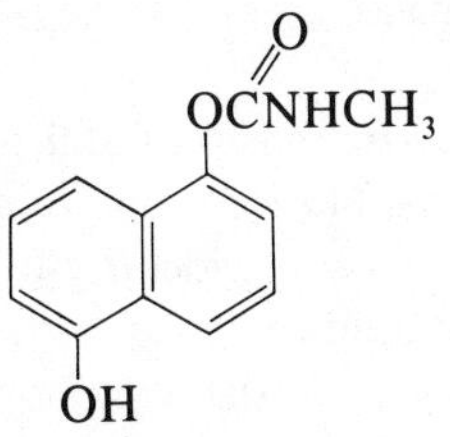

5-hydroxycarbaryl

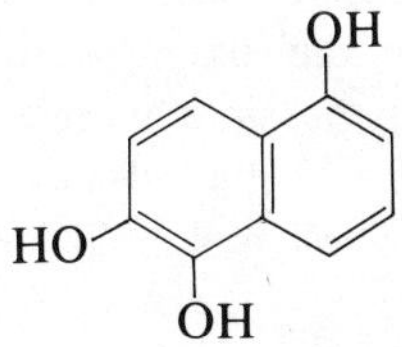

1,5,6-trihydroxy naphthalene

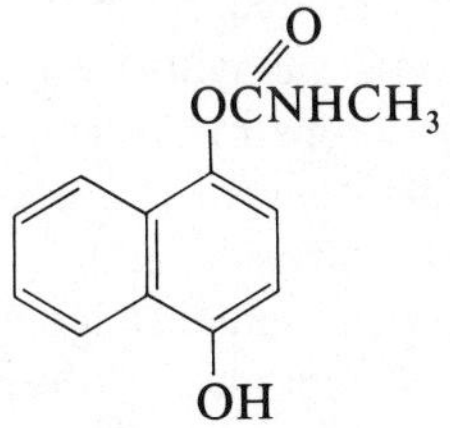

4-hydroxycarbaryl

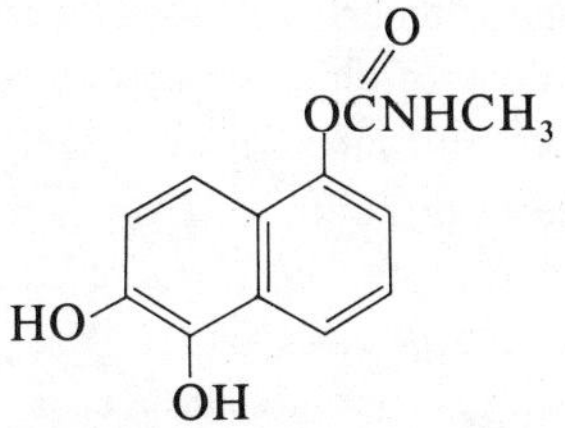

5,6-dihydroxy carbaryl

**Fig. 4.1** Some of the metabolites of carbaryl.

In the German cockroach carbaryl is largely metabolized to 1-naphthol, and 1-naphthol conjugates (75). Among mammals, the metabolism of carbaryl is similar in most species, but the dog presents a different picture; the liberation of 1-naphthol or the hydroxylation of carbaryl does not appear to take place. Comparative *in vitro* metabolism of carbaryl and other methylcarbamates by human and rat liver fractions has been studied by Strother (67).

In plants the metabolism of carbaryl has been studied by, among others, Abdel-Wahab et al. (70), Mostafa et al. (82), Dorough and Casida (77), and Kuhr and Casida (83). Hydroxylation of the aromatic rings in the 4- and 5-positions, formation of 5,6-dihydroxy carbaryl, and hydrolysis of the carbamate ester bond are some of the common reactions. Metabolites are conjugated soon after formation with glucose and other sugars.

## METABOLISM OF ALDICARB

The metabolism of aldicarb (Fig. 4.2) has been studied in animals (68, 84–86), insects (45, 47), and plants (87–89). Investigators prior to 1970 reported mainly on the identity of less polar organoextractable metamolites. Based on these studies it was shown that both hydrolytic and oxidative pathways were involved in the metabolism of this chemical. The initial metabolic attack resulted in the rapid oxidative conversion of aldicarb to its sulfoxide [2-methyl-2(methylsulfinyl) propionaldehyde-O-(methylcarbamoyl) oxime], and a minor attack resulted in the conversion of the parent insecticide to aldicarb oxime (2-methyl-2-methylthiopropionaldoxime), which was further oxidized to its sulfoxide (2-methyl-2-methylsulfinylpropionaldoxime) or the oxime sulfoxide. Further conversions of aldicarb sulfoxide resulted in its possible hydrolysis to oxime sulfoxide and its oxidation to aldicarb sulfone [2-methyl-2-(methylsulfonyl)propionaldehyde-O-(methylcarbomyl)oxime]. These metabolites were further degraded to their corresponding nitriles.

Both aldicarb sulfoxide and sulfone were more potent *in vitro* anticholinesterase agents than the parent insecticide (45), but quantitatively sulfoxide was much more important than aldicarb sulfone. Knaak et al. (86) reported that 40% of the administered dose of aldicarb was extreted in the rat urine as aldicarb sulfoxide and none as sulfone. Andrawes et al. (84), in the same animal, found 20% of the dose in urine as sulfoxide and 1% as sulfone in 24 hours. These findings indicate that the conversion of aldicarb sulfoxide to aldicarb sulfone is a slow process *in vivo*.

Another metabolite with the carbamate moiety intact, namely, the N-hydroxymethyl analog of aldicarb sulfone, has been reported in laying hens by Hicks et al. (85). It is suspected that this chemical is formed rather slowly.

## METABOLISM OF CARBOFURAN

Metabolism of carbofuran has been studied in animals (Fig. 4.3) (90–93), plants (Fig. 4.4) (94–96), and soil (97). In animals carbofuran is metabolized by hydrolytic and oxidative mechanisms, and, within 24 hours, more than 70%

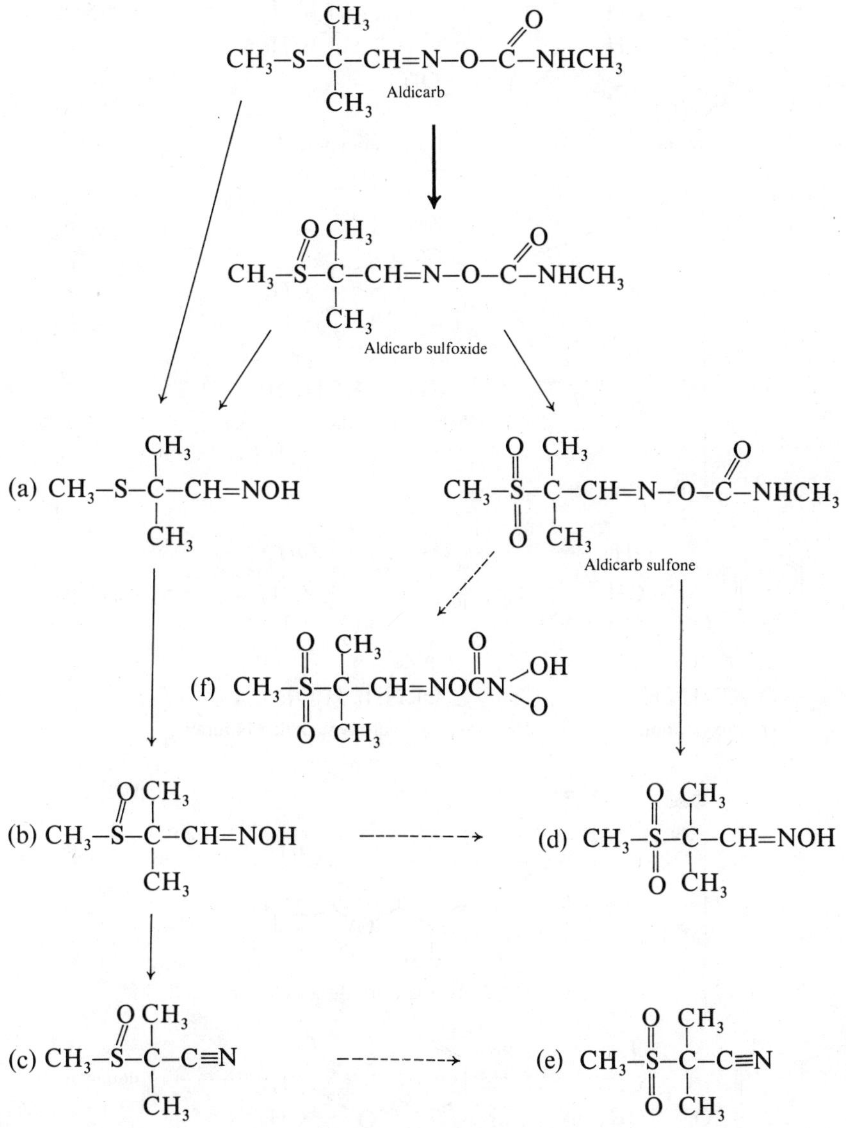

**Fig. 4.2** Some of the metabolites of aldicarb in animals; the thickness of arrows indicates the rate of formation. (*a*) 2-Methyl-2-methylthiopropionaldoxime; (*b*) 2-methyl-2-methylsulfinyl-propionaldoxime; (*c*) 2-methyl-2-methylsulfonylpropionaldoxime; (*d*) and (*e*) corresponding nitriles of *b* and *c*, respectively; (*f*) *N*-hydroxymethyl analog of aldicarb sulfone.

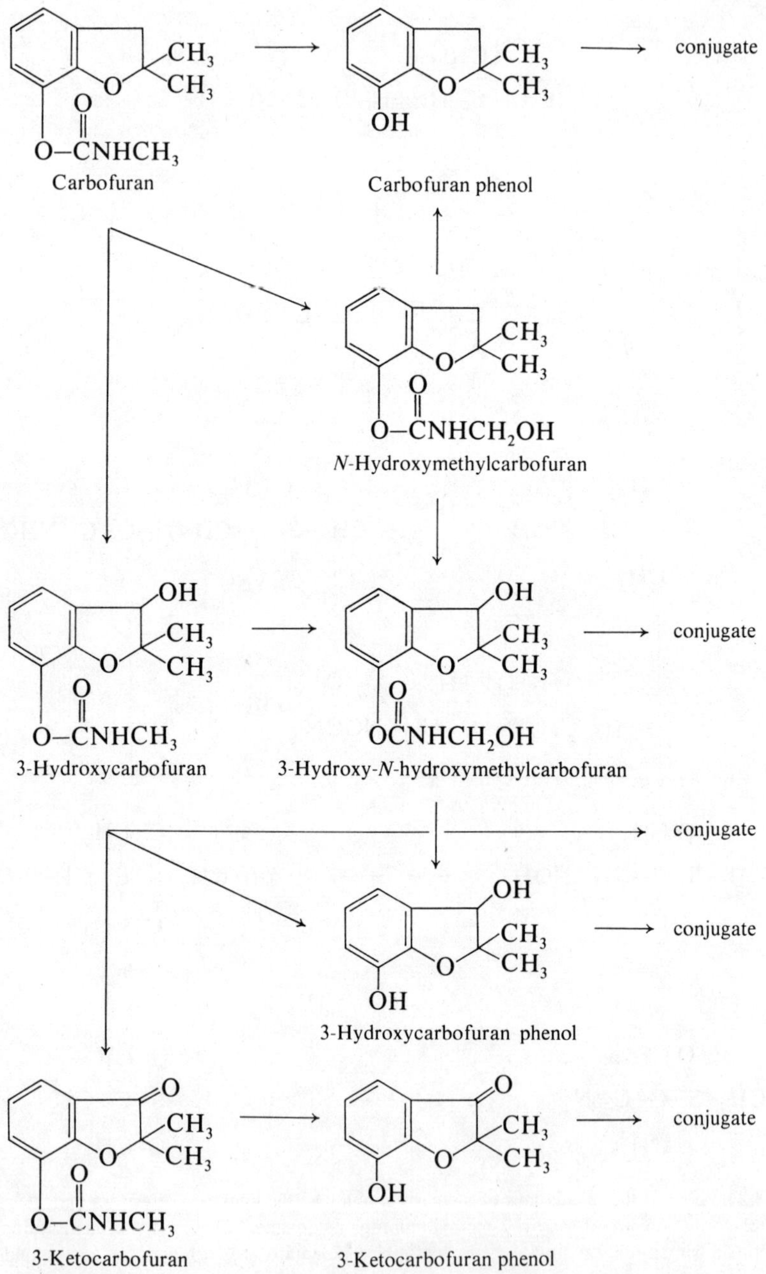

**Fig. 4.3** Proposed metabolic pathway of carbofuran in animals. Reprinted with permission from Knaak et al. *J. Agr. Food Chem.* **18**: 1018 (1970). Copyright by the American Chemical Society.

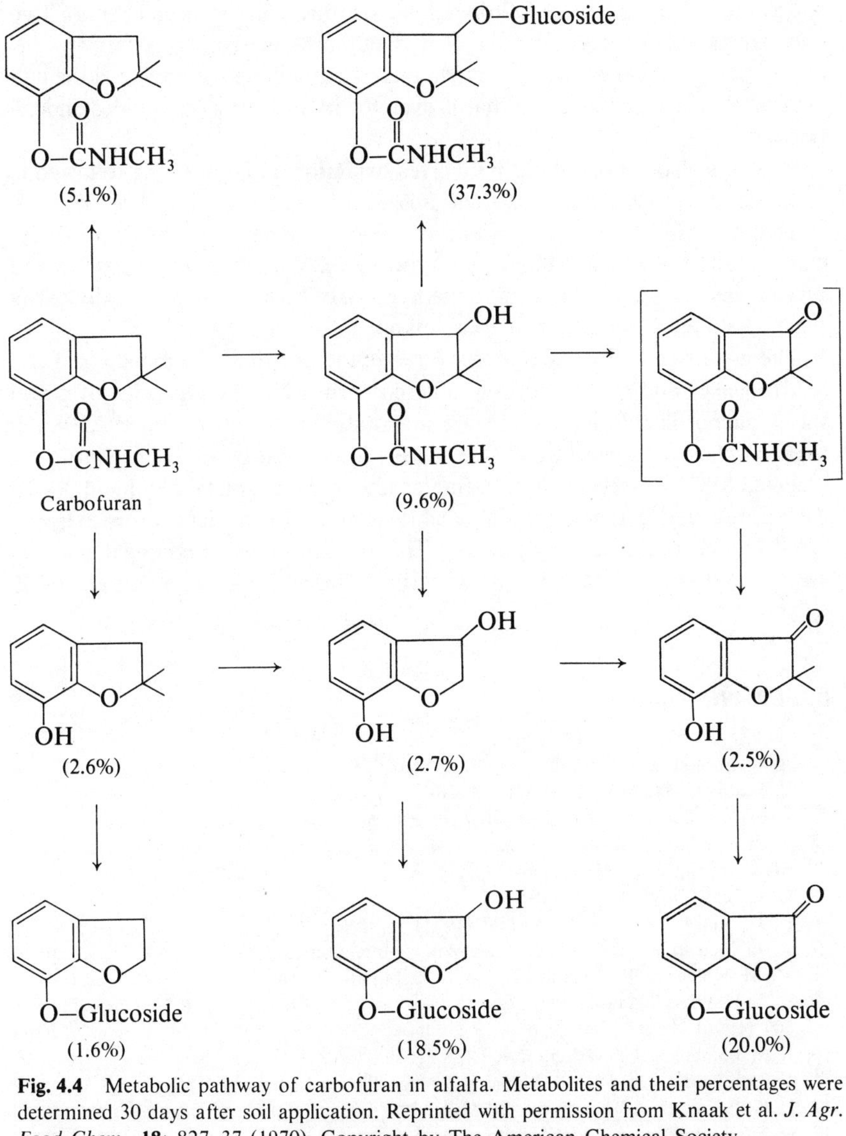

**Fig. 4.4** Metabolic pathway of carbofuran in alfalfa. Metabolites and their percentages were determined 30 days after soil application. Reprinted with permission from Knaak et al. *J. Agr. Food Chem.*, **18**: 827–37 (1970). Copyright by The American Chemical Society.

of the administered dose is eliminated by way of the urine; of this 95% is mostly in the form of conjugated metabolites. The principal metabolic product in urine is 2,3-dihydro-7-hydroxy-2,2-dimethyl-3-oxobenzofuran (3-keto-carbofuran phenol), which is conjugated as glucuronides and sulfates. Other minor metabolites, which are present both in free and conjugated form are 2,3-dihydro-3,7-dihydroxy-2,2-dimethylbenzofuran (3-hydroxycarbofuran phenol) and 2,3-dihydro-7-hydroxy-2,2-dimethylbenzofuran (carbofuran phenol). Two minor metabolites found unconjugated in animals are 3-ketocarbofuran and *N*-hydroxymethylcarbofuran. The predominant carbamate metabolite is the conjugated 3-hydroxycarbofuran that is excreted in urine as glucuronides and sulfates.

In plants also the metabolism involves oxidation and hydrolysis followed by conjugation of the metabolites. The following metabolites have been reported as glycosides (92): 3-hydroxycarbofuran, 3-hydroxycarbofuranphenol, 3-keto-carbofuran phenol, 2,3-dihydro-7-hydroxy-2,2-dimethylbenzofuran (carbofuran phenol), 2,3-dihydro-2,2-dimethyl-3-oxo-7-benzofuranyl methylcarbamate (3-ketocarbofuran), and four unknowns.

The glycoside of 3-hydroxycarbofuran is the principal metabolic product in alfalfa leaves and stems and in weathered corn, while the glycoside of carbofuran phenol is the main glycoside in potato roots and tubers (92, 98). In tobacco plants, Ashworth and Sheets (94) found that in 4 days carbofuran is oxidized to 3-hydroxycarbofuran (major unconjugated metabolite) and 3-ketocarbofuran. The parent carbamate and their oxidation products are also hydrolyzed to their corresponding phenols. The three phenols are conjugated as glycosides and so also is the 3-hydroxycarbofuran, which is the major glycosidic aglycone.*

## REFERENCES

1. A. Robertson. *Edinb. Med. J.* **8**: 815 (1863).
2. E. Stedman. *Biochem. J.* **20**: 719 (1926).
3. J. A. Aeschlimann and M. Reinert. *J. Pharm. Exp. Ther.* **43**: 413 (1931).
4. A. Stempel and J. A. Aeschlimann. In *Medicinal Chemistry* (F. F. Blicke and R. H. Cox, eds.), Vol. 3, Wiley New York, 1956, p. 238.
5. H. Gysin. *Chimia* **8**: 205 (1954).
6. R. L. Metcalf. *Bull. WHO.* **44**: 43 (1971).
7. R. L. Metcalf. *Organic Insecticides*, Wiley—Interscience, New York, 1955, 392 pp.
8. M. M. Kolbezen, R. L. Metcalf, and T. R. Fukuto. *J. Agr. Food Chem.* **2**: 864 (1954).
9. H. L. Haynes, J. A. Lambrech, and H. H. Moorefield. *Contrib. Boyce Thompson Inst.* **18**: 507 (1958).
10. J. A. Lambrech. U.S. Patent 2,903,478 (1959).
11. J. R. Kilsheimer and D. T. Manning. French Patent 1,343, 654 (1963).
12. L. K. Payne, H. A. Stansbury, and M. H. J. Weiden. *J. Agr. Food Chem.* **14**: 356 (1966).
13. M. H. J. Weiden. *J. Sci. Food Agr.* Suppl. p. 19 (1968).
14. M. H. J. Weiden. *Bull. WHO* **44**: 203 (1971).
15. I. B. Wilson, M. A. Hatch, and S. Ginsburg. *J. Biol. Chem.* **235**: 2312 (1960).

* Chemically, it is not proper to employ the common name of the parent chemical (carbofuran in this case) in naming the derivatives in which the basic molecular structure is not essentially intact; however the nomenclature employed in this chapter has been commonly used in designating the hydrolytic products in literature.

16. R. D. O'Brien, B. D. Hilton, and L. Gilmour. *Mol. Pharm.* **2**: 593 (1966).

17. C.-C. Yu, C. W. Kearns, R. L. Metcalf, and J. D. Davies. *Pestic. Biochem. Physiol.* **1**: 241 (1971).

18. A. R. Main. *Science* **144**: 992 (1964).

19. A. R. Main and F. Iverson. *Biochem. J.* **100**: 525 (1966).

20. A. R. Main and F. L. Hastings. *Biochem. J.* **101**: 584 (1966).

21. G. Booth and R. L. Metcalf. *Ann. Entomol. Soc. Amer.* **63**: 197 (1970).

22. R. L. Metcalf and T. R. Fukuto. *J. Agr. Food Chem.* **12**: 220 (1965).

23. R. L. Metcalf and R. B. March. *J. Econ. Entomol.* **43**: 670 (1950).

24. R. L. Metcalf and T. R. Fukuto. *J. Econ. Entomol.* **58**: 1151 (1965).

25. R. D. O'Brien. *Toxic Phosphorus Esters*, Academic Press, New York, 1960, 434 pp.

26. R. D. O'Brien. *Insecticides: Action and Metabolism.* Academic Press, New York, 1967, 332 pp.

27. T. R. Fukuto, R. L. Metcalf, M. Y. Winton, and P. A. Roberts. *J. Econ. Entomol.* **55**: 341 (1962).

28. T. R. Fukuto, R. L. Metcalf, R. J. Jones, and R. O. Myers. *J. Agr. Food Chem.* **17**: 923 (1969).

29. R. L. Jones, R. L. Metcalf, and T. R. Fukuto. *J. Econ. Entomol.* **62**: 801 (1969).

30. H. Okhawa, R. Okhawa, I. Yamamoto, and J. E. Casida. *Pestic. Biochem. Physiol.* **2**: 95 (1972).

31. J. Meltzer and H. B. A. Welle. *Entomol. Exp. Appl.* **12**: 169 (1969).

32. C. F. Wilkinson. *Bull. WHO* **44**: 171 (1971).

32a H. J. Kooy, Jr. (to Hoffman-La Roche). Dutch Patent 6,601,926 (Aug. 17, 1966).

32b. R. M. Sacher, R. L. Metcalf, and T. R. Fukuto. *J. Agr. Food Chem.* **16**: 779 (1968).

33. R. L. Metcalf and T. R. Fukuto. *J. Econ. Entomol.* **55**: 340 (1962).

34. A. M. Mahfouz, R. L. Metcalf, and T. R. Fukuto. *J. Agr. Food. Chem.* **17**: 917 (1969).

35. J. Fraser, I. R. Harrison, and S. B. Wakerly. *J. Sci. Food Agr.* Suppl. p. 8 (1968).

36. D. K. Lewis. *Nature* **213**: 205 (1967).

37. R. P. Miskus, T. L. Andrews, and M. Look. *J. Agr. Food Chem.* **17**: 842 (1969).

38. M. A. H. Fahmy and T. R. Fukuto. *J. Econ. Entomol.* **63**: 1783 (1970).

39. M. A. H. Fahmy, T. R. Fukuto, R. O. Myers, and R. B. March. *J. Agr. Food Chem.* **18**: 793 (1970).

40. F. Barlow and A. B. Hadaway. *Pestic. Sci.* **1**: 117 (1970).

41. A. B. Hadaway, F. Barlow, J. E. M. Grose, C. R. Turner, and L. S. Flower. *Bull. WHO* **42**: 369 and **42**: 377 (1970).

42. R. B. Roberts, R. P. Miskus, C. R. Duckles, and T. T. Sakai. *J. Agr. Food Chem.* **17**: 107 (1969).

43. C. H. Schaeffer and H. H. Wilder, *J. Econ. Entomol.* **63**: 480 (1970).

44. J. C. Felton. *J. Sci. Food Agr.* Suppl. p. 32 (1968).

45. R. L. Metcalf, T. R. Fukuto, C. Collins, K. Borck, J. Burk, H. T. Reynolds, and M. F. Osman, *J. Agr. Food Chem.* **14**: 579 (1966).

46. M. H. J. Weiden and H. H. Moorefield. *J. Agr. Food Chem.* **13**: 200 (1965).

47. D. L. Bull, D. L. Lundquist, and J. R. Coppedge. *J. Agr. Food Chem.* **15**: 610 (1967).

48. K. Hellenbrand and R. M. Krupka. *Biochemistry* **9**: 4665 (1970).

49. L. E. Gaudette and B. B. Brodie. *Biochem. Pharm.* **2**: 89 (1959).

50. T. R. Fukuto and R. L. Metcalf. *J. Agr. Food Chem.* **4**: 930 (1956).

51. C. Hansch and E. W. Deutsch. *Biochim. Biophys. Acta* **126**: 117 (1966).

52. T. Fujita, J. Iwasa, and C. Hansch. *J. Amer. Chem. Soc.* **86**: 5175 (1964).

53. L. P. Hammett. *Physical Organic Chemistry.* McGraw-Hill, New York, p. 188.

54. G. G. Swain and E. C. Lupton. *J. Amer. Chem. Soc.* **90**: 4328 (1968).

55. R. J. Kitz, S. Ginsberg, and I. B. Wilson. *Biochem. Pharm.* **16**: 2201 (1967).
56. F. L. Hastings, A. R. Main, and F. Iverson. *J. Agr. Food Chem.* **18**: 497 (1970).
57. H. H. Moorefield. *Misc. Publ. Entomol. Soc. Amer.* **2**: 145 (1960).
58. K. Wilhelm and M. Vandekar. In XVth International Conference on Occupational Health, Verlag der Wiener Medizinischen Akademie, H. Egermann, Vienna, Vol. 2. p. 517, 1966.
59. M. S. Quraishi. *Can. Entomol.* **104**: 1191 (1972).
60. H. W. Dorough. *J. Agr. Food Chem.* **18**: 1015 (1970).
61. R. M. Hollingworth. *Bull. WHO* **44**: 155 (1971).
62. J. B. Knaak. *op. cit.* **44**: 121 (1971).
63. R. J. Kuhr. *J. Agr. Food Chem.* **18**: 1023 (1970).
64. A. J. Ryan. *Crit. Rev. Toxicol.* **1**: 33 (1971).
65. E. S. Oonnithan and J. E. Casida. *J. Agr. Food Chem.* **16**: 28 (1968).
66. G. D. Paulson, R. G. Zaylskie, M. V. Zehr, C. E. Portnoy, and V. J. Feil. *J. Agr. Food Chem.* **18**: 110 (1970).
67. A. Strother. *Toxicol. Appl. Pharm.* **21**: 112 (1972).
68. H. W. Dorough, R. B. Davis, and G. W. Ivie. *J. Agr. Food Chem.* **18**: 135 (1970).
69. H. W. Dorough and G. W. Ivie. *op. cit.* **16**: 460 (1968).
70. A. M. Abdel-Wahab, R. J. Kuhr, and J. E. Casida. *J. Agr. Food Chem.* **14**: 290 (1966).
71. C. P. Carpenter, C. S. Weil, P. E. Palm, M. W. Woodside, J. H. Nair, and H. F. Smyth, Jr. *J. Agr. Food Chem.* **9**: 30 (1961).
72. H. W. Dorough and J. E. Casida. *op. cit.* **12**: 294 (1964).
73. M. E. Eldefrawi and M. W. Hoskins. *J. Econ. Entomol.* **54**: 401 (1961).
74. J. B. Knaak, J. Marilyn, M. J. Tallant, W. J. Bartley, and L. J. Sullivan *J. Agr. Food Chem.* **8**: 198 (1965).
75. T. Y. Ku and J. L. Bishop. *J. Econ. Entomol.* **60**: 1328 (1967).
76. N. C. Leeling and J. E. Casida. *J. Agr. Food Chem.* **14**: 281 (1966).
77. H. W. Dorough and J. E. Casida. *op. cit.* **12**: 294 (1964).
78. E. Hodgson and J. E. Casida. *Biochim. Biophys. Acta* **42**: 184 (1961).
79. E. Hodgson and J. E. Casida. *Biochem. Pharm.* **8**: (1961).
80. H. W. Dorough, N. C. Leeling, and J. E. Casida. *Science* **140**: 170 (1963).
81. G. D. Paulson, R. G. Zaylskie, M. V. Zehr, C. E. Portnoy, and V. J. Feil. *J. Agr. Food Chem.* **18**: 110 (1970).
82. I. Y. Mostafa, A. Hassan, and S. M. A. D. Zayed. *Zeitschr. Naturforsch.* **216**: 1060 (1966).
83. R. J. Kuhr and J. E. Casida. *J. Agr. Food Chem.* **15**: 814 (1967).
84. N. R. Andrawes, W. H. Dorough, and D. A. Lindquist. *J. Econ. Entomol.* **60**: 979 (1967).
85. B. W. Hicks, H. W. Dorough, and H. M. Mehendale. *J. Agr. Food Chem.* **20**: 151 (1972).
86. J. B. Knaak, M. J. Tallant, and L. J. Sullivan. *op. cit.* **14**: 573 (1966).
87. N. R. Andrawes, W. P. Bagley, and R. A. Harrett. *J. Agr. Food Chem.* **19**: 727 (1971).
88. N. R. Andrawes, W. P. Bagley, and R. A. Harrett. *op. cit.* **19**: 73 (1971).
89. W. J. Bartley, N. R. Andrawes, E. L. Chancey, W. P. Bagley, and H. W. Spurr. *J. Agr. Food Chem.* **18**: 466 (1970).
90. H. W. Dorough. *J. Agr. Food Chem.* **16**: 319 (1968).
91. G. W. Ivie and H. E. Dorough. *op. cit.* **16**: 849 (1968).
92. J. B. Knaak, D. M. Munger, J. F. McCarthy, and L. D. Satter. *op. cit.* **18**: 832 (1970).
93. R. L. Metcalf, T. R. Fukuto, C. Collins, K. Borck, S. A. El-Aziz, R. Munoz, and C. C. Cassil. *op. cit.* **16**: 300 (1968).
94. R. J. Ashworth and T. J. Sheets. *op. cit.* **20**: 407 (1972).
95. H. W. Dorough. *Bull. Environ. Contam. Toxicol.* **3**: 164 (1968).
96. J. B. Knaak, D. M. Munger, and J. F. McCarthy. *J. Agr. Food Chem.* **18**: 827 (1970).
97. M. A. Abdellatif, H. P. Hermanson, and T. H. Reynolds. *J. Econ. Entomol.* **60**: 14 (1967).
98. R. F. Cook, R. P. Stanovick, and C. C. Cassil. *J. Agr. Food Chem.* **17**: 277 (1969).

# SYNTHETIC ORGANIC INSECTICIDES

# CHAPTER 5 | DDT and Related Compounds

Of all the synthetic organic insecticides, DDT, 2,2-bis-(p-chlorophenyl)-1,1,1-trichloroethane, is the best known and the most widely discussed. It was synthesized for the first time by Othmar Zeidler in 1873 (1) and its insecticidal properties were first discovered by Dr. Paul Müller of J. R. Geigy Company in Switzerland in 1939. Because of its molecular configuration the DDT molecule was earlier called dichlorodiphenyltrichloroethane and hence its name.

Pure DDT is a white crystalline substance with needle-shaped crystals and a melting point of 108.5–109°C. Its insecticidal properties combined with its physical and chemical properties—low vapor pressure, $1.5 \times 10^{-7}$ mm at 20°C; low water solubility, 1.2 ppb at 25°C (2), high fat solubility, and stability to photooxidation—make it a highly persistent insecticide. DDT in thin films has remained effective as an insecticide for up to 18 months (3). The technical grade product, however, is a mixture of several compounds in which the $p,p'$ isomer amounts to 75% or more. It has a lower melting point than the pure compound.

DDT occupied a significant position in the fields of public health and agriculture from the 1940s until the 1960s. It is estimated that about 4 billion pounds of the chemical was used during this period and that there is now (1972) about 1 billion pounds of DDT in the biosphere (4, 5).

The impact of DDT on a war-ravaged Europe in the 1940s, especially the conquest of typhus in Naples, and its success in public health programs has been well documented (6). Only those who have taken part in malaria eradication campaigns and have followed the improvements not only in the health but also in the productivity of the protected population can appreciate the benefits that DDT brings. In Bangladesh in 1950, without any change in agricultural practices, agricultural production in the malaria control areas increased by about 10% (7).

In agriculture, for the first time a potent weapon had been discovered for the effective control of insects as diverse in habits as the codling moth of apples, *Laspeyresia pomonella* (L.); the gypsy moth, *Porthetria dispar* (L.); and the wire worms. In fact, until the 1960s DDT was registered for use in the United States on 334 agricultural commodities, which made DDT a household word.

The initial success of DDT ushered in an era of complacence, but this complacence soon gave rise to caution when the discovery was made in Italy in 1946 that house flies could not be effectively controlled by DDT (3); soon afterwards came the reports that house flies in various parts of United States and mosquitoes in California, Florida, Italy, and Greece were no longer as susceptible to DDT as they were initially. Further reports of DDT-resistant lice in Korea left no doubt that resistance to DDT was an important phenomenon to be reckoned with (8–11). The mode of action and the metabolism of DDT therefore acquired added significance and are discussed in this chapter.

## MODE OF ACTION OF DDT

Despite the well-known status of DDT and decades of intensive investigations, the perplexing fact remains that our knowledge of the mechanism of the action of this chemical or the critical biochemical lesion produced as a result of DDT poisoning is far from satisfactory. Chemically speaking, DDT is a comparatively inert molecule and hence the residue problem and the magnification in food chains; yet it is a potent broad-spectrum insecticide that is also toxic to mammals, fish, and birds. The symptoms of DDT poisoning indicate that, broadly speaking, the site of action is the nervous system. The symptoms start with hyperexcitability followed by tremors that increase in severity and lead to convulsions followed by incoordinated movements, ataxia, prostration, and ultimately death.

Yeager and Munson (12) first showed the multiplication effects of a single impulse on a DDT-treated nerve, that is, the production of prolonged volleys as a result of a single impulse; the finding has since been confirmed by several workers. As a result of DDT poisoning in insects the frequency of spontaneous discharge in the central nerve is increased and synaptic transmission is facilitated. The action of DDT and other insecticides on excitable tissues has been ably summarized by Narahashi (13).

Sternburg and Kearns (14) reported that the blood of DDT-poisoned cockroaches contained a toxic factor other than DDT. This factor was toxic to flies and produced DDT-like effects in cockroach nerve cords. Since then it has been shown by other workers (15–17). that a variety of stresses, including electrical stimulation, produce similar toxic factors in cockroach blood. This factor, which has been called "autotoxin," is possibly released from the nerve as a result of hyperactivity of the nervous tissue (13). The "negative temperature coefficient of intoxication" as mentioned in the case of pyrethrum is also found in the case of DDT-treated insects. Insects that show marked signs of poisoning at 15°C become apparently normal when brought to 30°C. This condition can be repeatedly reproduced. Narahashi (13) has proposed that there is an increase in nerve sensitivity at low temperature. It is possible that DDT forms a complex of the charge transfer type and that when the temperature is raised this complex is dissociated and hence DDT, though present at the site of action, is less effective. Holan's modification of Mullin's theory (26), discussed later, explains this hypothesis.

## THEORIES OF TOXIC ACTION

The earlier theories of the toxic action of DDT were speculative and only took into account the twodimensional formula of DDT. Thus Martin and Wain (18) suggested that HCl was released at the site of action by the dehydrochlorination of the $CHCCl_3$ group and that the nature of ring substituents affected insecticidal properties either because of electronic effect or because of solubility factors. However, insecticidal properties of DDT analogs that do not contain chlorine cannot be explained on the basis of this theory.

Lauger et al. (19) based their theory on their previous experience with moth-proofing agents represented by the "Mitin FF" type in which toxicity was

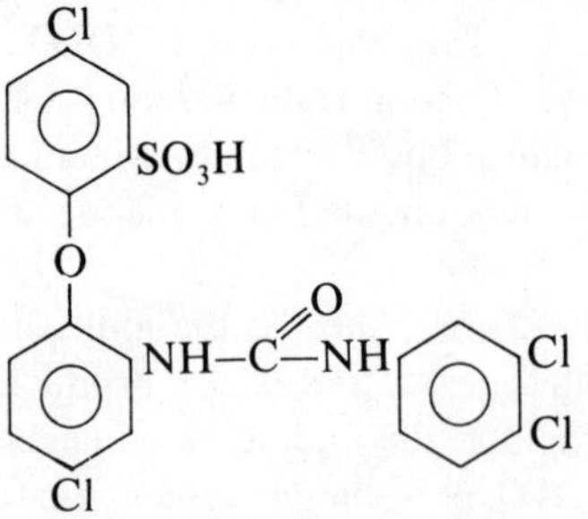

"Mitin FF"

presumed to be due to the $p,p'$-dichlorophenyl ester moiety of the molecule and the fat-solubilizing properties of chloroform. They surmised that the combination in one molecule of both these moieties (conductophore and toxophore) renders DDT an effective insecticide. Like the previous theory, this theory also fails to explain the insecticidal properties of DDT analogs that do not contain chlorine, that is, Prolan, Bulan, or dianysil neopentane.

Rogers et al. (20), on the basis of work with DDT analogs, proposed a theory that explained, among other things, the high activity of Prolan. They considered the optimal stereochemical configuration of DDT and suggested that the bulk $CCl_3$ group was important because it interacted with $p$-chlorophenyl groups to impart to the aryl rings the position of maximum clearance, the "butterfly" configuration of DDT. In this nonrotatable configuration the phenyls assume the position of maximum coplanarity that the authors consider a necessary condition for insecticidal activity. They supported their argument by citing the strong insecticidal activity of 1,1-bis-($p$-methoxyphenyl)-2,2-dimethylpropane and the decreased activity of 1,1-bis-($p$-methoxyphenyl)-2-methyl propane.

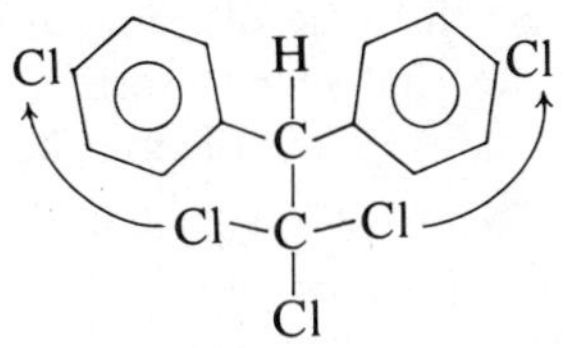

Reimschneider and Otto (21), on the basis of relationships between toxicity and structure of various DDT analogs and isosteres, proposed a theory for the toxic action of DDT. Their theory relates the toxicity of such chemicals to the rotatability of the benzene rings about the alkyl carbon ($C_2$) or $\alpha$-carbon to which they are attached. From observations made on Stuart models, it was argued that when the auxophoric groups (groups present on the benzene rings)

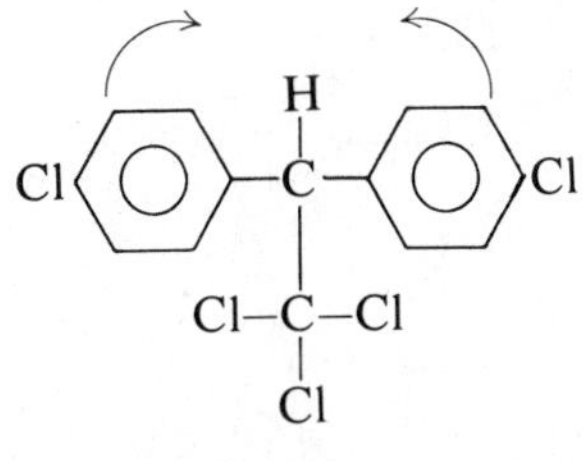

are attached in the $p,p'$ positions the rotatability of the benzene rings is the greatest, and hence the maximum, insecticidal activity of $p,p'$-DDT. The movement of these chlorines to *ortho* or *meta* positions creates a steric hindrance, thereby interfering with the free rotation of the rings, and hence the comparative inactivity of such DDT analogs as $o,o$-DDT, 2,2-bis-($o$-chlorophenyl)-1,1,1-trichloroethane, in which free rotation is inhibited. Similarly, DDE, 2,2-bis-($p$-chlorophenyl)-1,1-dichloroethylene, in which the ethylene double bond prevents the free rotation of the benzene rings, is insecticidally inactive. However, in the case of o-Cl-DDT the rings are also nonrotatable, but it is nearly as potent an insecticide as DDT.

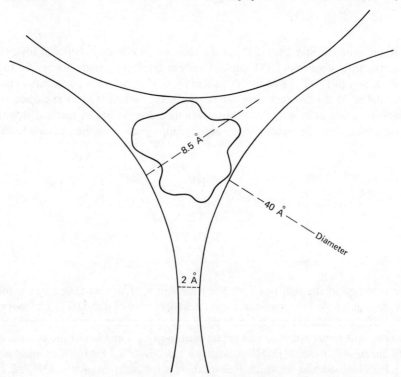

In the case of DDT-poisoning it appears that the impairment of the nerve is brought about by the intact DDT molecule itself, because in the nervous system of the poisoned insect the molecule remains unchanged. Possibly DDT is bound to some biologically vital sites to form a tight complex, thereby impairing the normal cell functions. Some theories have speculated on the dovetailing of the DDT molecule at a critical lipoprotein interface. Mullins (22)

**Fig. 5.1.** Lindane molecule in the membrane interspace.

theorized that the axonic membrane consisted of cylindrical lipoprotein strands about 40 Å in diameter and packed in a hexagonal array such that they are held 2 Å apart forming spaces 8.5 Å in diameter. These spaces are regarded as hypothetical pores in the membrane lattice. Compounds that fit tightly in end-on positions are capable of distorting the membrane structure and causing excitation (the reasons for this excitability were not explained by Mullins, however refer to Holan's concept of "Sodium Gate" which follows this theory).

Mullins' theory, initially propounded to account for the variations in the insecticidal activity of the isomers of hexachlorocyclohexane Fig. 5.1, successfully explains the activity of many of the isosteres of DDT Fig. 5.2, like methoxychlor and dianisyl neopentane, or the inactivity of iodo-DDT, because $p,p'$ substituents are too large; or the insecticidal inactivity of DDE where the tetrahedral bond angle is changed. But it cannot fully explain the moderate toxicity of compounds with fused ring systems, like 2,7-dichloro-9-trichloromethyl-9,10-dihydroanthracene (23–25).

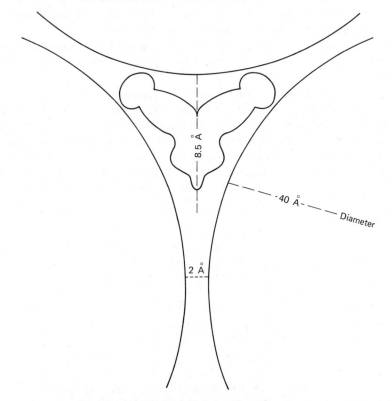

**Fig. 5.2.** DDT molecule in the membrane interspace.

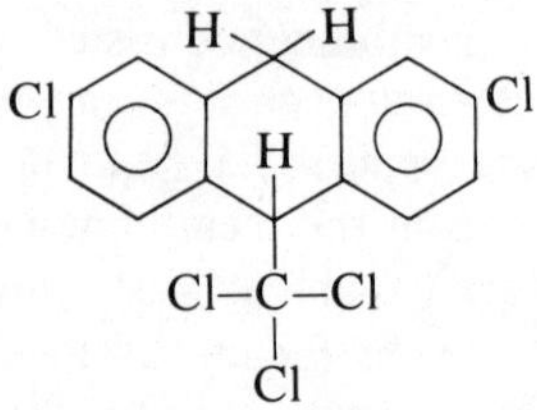

A modification of Mullin's original theory was proposed by Holan (26–28). He studied the structure–activity relationships of 1,1-di-(p-chlorophenyl)-2,2-dichlorocyclopropane, a DDT analog, and its derivatives. Holan envisions the phenyl ring moiety of the molecule as the dovetailing portion of the chemical that locks into the overlaying protein layer in the nerve membrane by forming a molecular complex, whereas the remaining portion of the molecule fits into the pore channel. As a result of this locking of the molecule in the membrane, Holan has attempted to explain the prolongation of the sodium current (13, 29). According to Holan the DDT molecule and molecules similar in shape and dimensions act as wedges in $Na^+$ cavities in the nerve membrane thereby expanding the sodium "springs" and keeping them open.

The alteration of ion transport in nerve membrane also has been suggested by Matsumura and Patil (30), though their hypothesis is that DDT inhibits adenosine triphosphatases.

Gunther et al. (30a) considered the role of bonding energies and similar van der Waals radii of Cl (1.85 Å) and $CH_3$ (2 Å) and suggested a close fit into a receptive protein cavity, such as an epoenzyme, of isosteres possessing comparable dimensions. They surmised that closely fitting molecules would provide optimal insecticidal activity. The mosquito, *Culex pipiens quinquefasciatus* Say, was used as a test insect and 30 DDT analogs were tested. In these the X's were replaced by Cl, $CH_3$, or H in the aromatic (head) and ethane (tail)

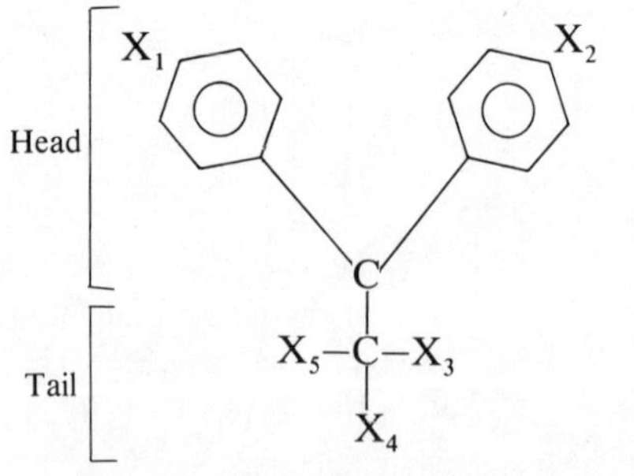

portions of the molecule. Thus for each given tail one could have six different heads. The $LC_{50}$ values were correlated with $\sum \log K_0$ for Cl, $CH_3$, and H van der Waals radii. Lines of similar slope were obtained for those analogs that were insecticidally active. From these data the authors concluded that both moieties of the molecule, namely, head and tail, were important and essentially equivalent in the interaction with the receptor site.

Fahmy et al. (31) used a modification of Holan's (26) model and started with the premise that the receptor site, possibly a protein or a lipoprotein, in a nerve membrane may be visualized as a cavity or a pouch with a limited amount of flexibility. They determined this flexibility with respect to four key substituents L, X, Y, and Z in DDT-type molecules with respect to van der Waals forces (30a). Multiple regression analysis using Taft's steric substituent

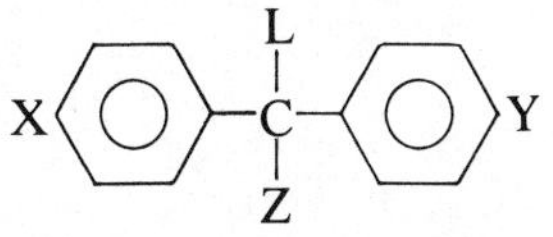

parameter $E_s$ (31a), was done "owing to the belief that fit at the receptor site should be governed primarily by steric effects." It was concluded that when substituents were varied in the L, X, Y, and Z positions, the steric substituent constant $E_s$ was the single most important parameter for the correlation of insecticidal activity when mosquito larvae and house flies were used as test insects. The receptor site was flexible and was capable of accomodating DDT analogs of varying dimensions in positions L, X, Y, and Z, provided the overall size of the molecule did not vary much from that of the DDT molecule. They also examined organosilicon analogs of DDT and found them ineffective, possibly because of the larger size of silicon as compared to carbon.

Though it is hard to envison the exact mechanism of action of DDT, it is not difficult to visualize that the mechanism is highly stereospecific, as is evidenced by the inactivity of such close analogs of DDT as DDE and α-chloro-DDT and by the DDT-like toxic effects produced by the DDT isostere 3,3-bis-(*p*-methylphenyl)-2,2-dimethyl propane which contains no chlorine and is completely methylated.

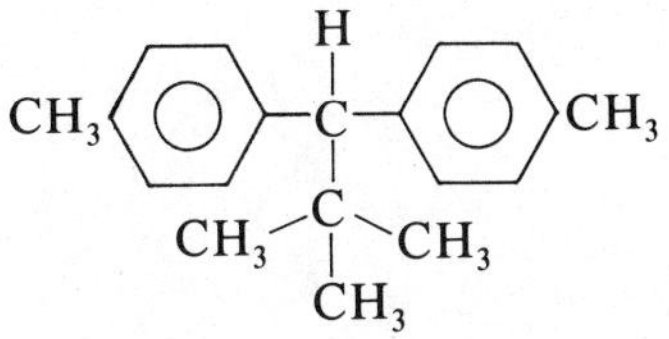

## METABOLISM OF DDT

Knowledge concerning the metabolism of DDT (Fig. 5.3) in living and nonliving environment has been slow in acquisition. Earlier investigations revealed the conversion of DDT by oxidation to DDA bis-[*p*-chlorophenylacetic] acid

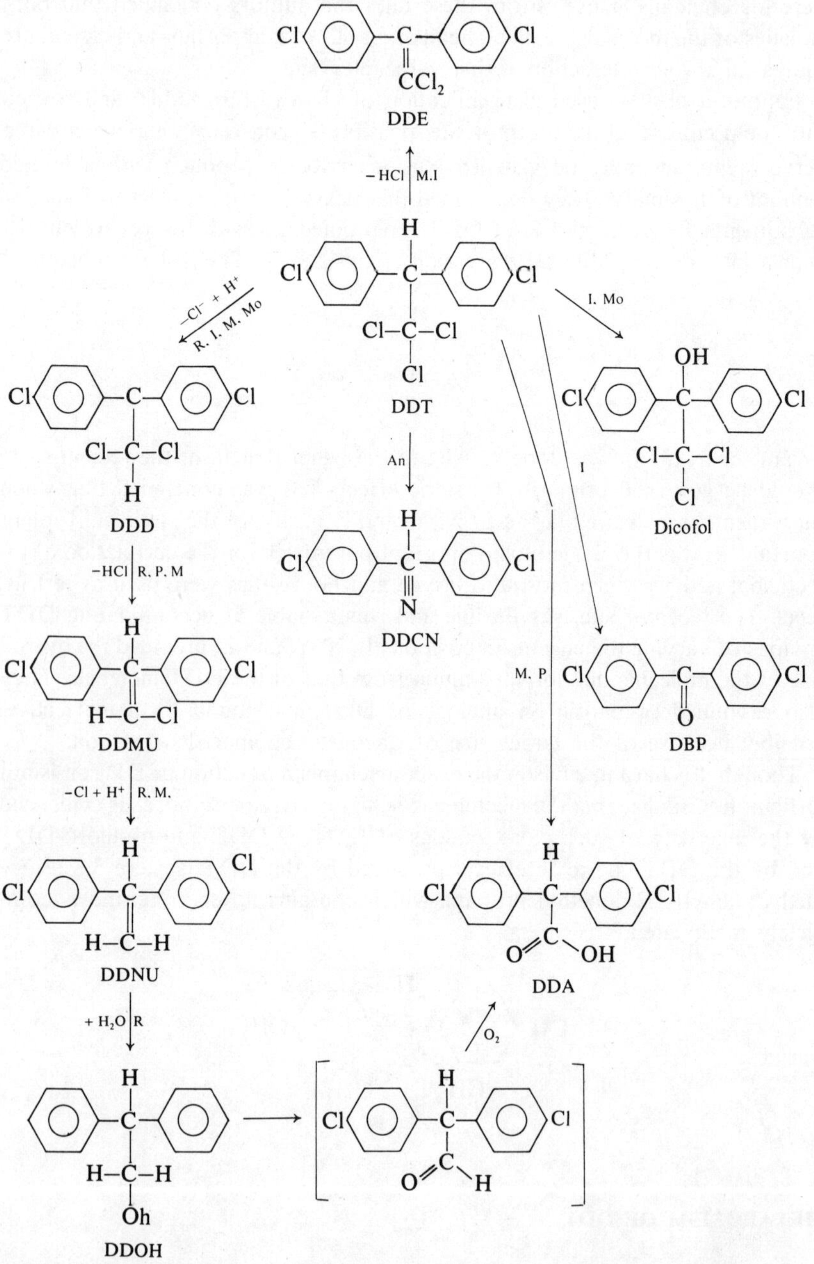

**Fig. 5.3** Metabolic pathways of DDT. An anaerobic; I, insects; M, mammals; Mo, micro-organisms; P. plants; R, rat.

in mammals and the excretion of DDA in feces and urine (32, 33). Conversion of DDT to DDA also has been reported in the body louse (11), which also converts DDT to DBP (dichlorobenzophenone) and DDE. The conversion of DDT to DDE was first demonstrated by Sternburg and Kearns (34) as the biochemical step in the resistance of house flies to DDT. The enzyme responsible for this conversion, DDT-dehydrochlorinase (DDT-ase), was greatly purified and concentrated (46,000-fold purification) by Lipke and Kearns (35), who found it to be a simple protein with a molecular weight of 36,000 and an isoelectric point of 6.5. Man also has the capability of converting DDT to DDE (36), which is the principal form of ingested DDT. However Morgan and Roan (37) opine that this conversion does not take place in man and that DDE in tissues originates mainly from dietary $p,p'$-DDE. Rats convert only a small amount of DDT to DDE and monkeys lack this capability. The reductive dechlorination of DDT to DDD was first reported in the yeast by Kallman and Andrews (38). Later Datta et al. (39) and Peterson and Robinson (40) demonstrated it in the rat. In the gastrointestinal tracts of monogastric animals this conversion possibly results from bacterial action (41). The bovine rumen fluid also converts DDT to DDD (42, 43). Sink et al. (43a) have tentatively identified DDMU 1-chloro-2,2-bis-[$p$-chlorophenyl]ethylene, in addition to DDD and DDE, as a metabolite produced by ovine (*Ovis aries*) rumen fluid. The conversion of DDT to DDE by lake water has also been reported by Miskus et al. (43).

Tsukamoto (44) first described the microsomal oxidation of DDT at the $\alpha$-carbon atom to dicofol (1,1-bis-[$p$-chlorophenyl]-2,2,2-trichloroethanol) in the fruit fly, *Drosophila melanogaster*. In the larvae of Drosophila, dichlorobenzophenone is the other major metabolite of DDT.

DDCN, bis-[$p$-chlorophenyl]acetonitrile, a major new anaerobic degradation product from sewage sludge and lake sediment has been identified by Albone et al. (45) and Jensen et al. (46). Under these conditions DDT was also converted to DDMU.

The quantitative metabolic pattern of DDT-$C^{14}$ was worked out for the rat by Kapoor et al. (47). Over a period of 48 hours they found that 73 and 27% of the recovered activity was in feces and in urine, respectively. Their work indicates that DDT is to a small extent degraded biologically in the mammalian system, but some of the metabolites, especially DDE which is stored in fat together with DDT, cause concern.

## ANALOGS OF DDT

The phenomenal success of DDT led several scientists to synthesize DDT analogs and to test their effectiveness for insecticidal activity. Though none of these chemicals surpassed or even equaled DDT in its overall performance and cost of synthesis, some useful chemicals were made and have proved to be valuable insecticides; a few have found commercial use. The history of DDT analogs and insecticides can be divided into three periods: the first decade after the discovery of the properties of DDT when numerous analogs and chlorinated hydrocarbons were synthesized and tested; the second decade when interest in other chemical structures was greatly aroused; and the third decade when DDT became a dirty word and chlorinated hydrocarbons and DDT-type chemicals came under increasing criticism. However, in the late sixties and early seventies, Metcalf and his colleagues revived interest in the DDT-type structure by their efforts to synthesize biodegradable analogs of DDT. These are discussed later.

### DDT Analogs of the First Decade

An extensive treatment of the analogs developed during the first decade after the discovery of the insecticidal properties of DDT has been done by Brown (3) and Metcalf (48); a few analogs that have found commercial use are discussed here.

*DDD or TDE*, 2,2-bis-(*p*-chlorophenyl)-1,1-dichloroethane is white crystalline solid with mp 109°C. The commercial product has a setting point of 86°C and contains 85% DDD. This chemical has lower mammalian toxicity as compared to DDT. Its oral $LD_{50}$ for rat is 3400 mg/kg; however it causes the degeneration of adrenal cortex in the dog when fed at 50 mg/kg. Such an effect has not been discovered among other mammals. It was formerly used on fruits and vegetables, but its use has been discontinued and it is no longer in production.

In general, DDD is less effective against most insects than DDT, but it has given better results than the latter against mosquito larvae, leaf rollers, and hornworms. The buildup of DDD in food chains in Clear Lake, California, has been well documented and is discussed later.

*Dilan*®, is a commercial insecticide consisting of 52% by weight of *Bulan*®, 1,1-bis-(*p*-chlorophenyl)-2-nitrobutane, 26% of *Prolan*®, 1,1-bis-(*p*-chlorophenyl)-2-nitropropane, and 19% of related compounds. Dilan is more effective than DDT against some insects like Mexican bean beetle, *Epilachna varivestis* Mulsant, and some thrips and aphids. Its registration in the United States was cancelled in 1968. Its oral $LD_{50}$ for rat is 475–600 mg/kg.

*Methoxychlor*, 2,2-bis-(*p*-methoxyphenyl)-1,1,1-trichloroethane or dianisyl trichloroethane is, like DDT, insoluble in water but soluble in common organic solvents. Its oral $LD_{50}$ for rat is 6000 mg/kg. Methoxychlor is especially effective against the Mexican bean beetle and some other insects. One of the most important properties that make this chemical more attractive than other insecticides of the DDT group is its rapid elimination by mammals. Unlike DDT it is not accumulated in fatty tissues. It is, however, more expensive to manufacture than DDT. It is used for the control of black flies and animal pests like flies, fleas, and lice on dairy and beef cattle.

Recently, Harris et al. (49) studied the effect of several dietary levels of technical methoxychlor on reproduction in rats and found that 1000 ppm of

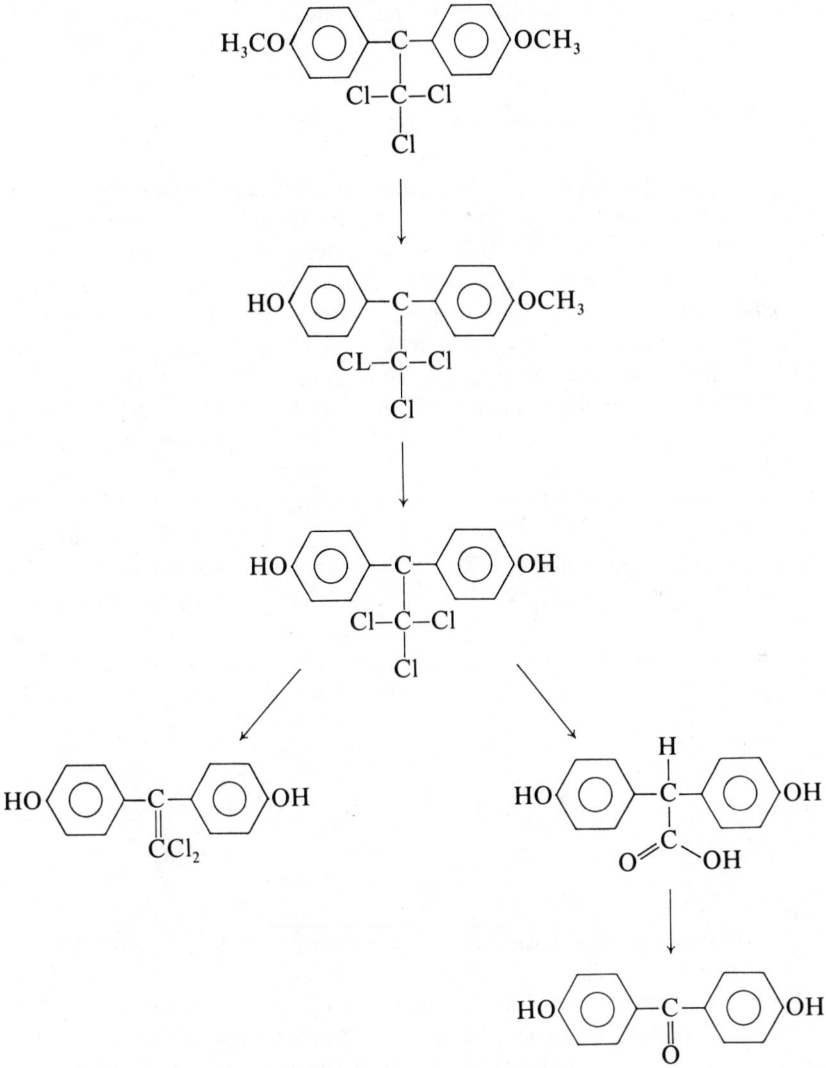

**Fig. 5.4** Metabolism of methoxychlor.

the chemical had no effect on reproduction of female rats, but the reproductive capabilities of the progeny were adversely affected. At high doses (2500 and 5000 ppm) there were reduced matings and only one of the experimental animals littered.

Metabolism of methoxychlor (Fig. 5.4) has been reported by Kapoor et al. (50). The p-CH₃O groups are degraded *in vivo* by O-dealkylation yielding first the monophenol and then the bisphenol. The water-soluble groups help in the excretion of the metabolites from the body.

## BIODEGRADABLE ANALOGS OF DDT

The development of biodegradable analogs of DDT is principally the work of Metcalf and co-workers who started with the studies of the metabolism of DDT-type molecules (47, 50, 51) and speculated that chemical stability of an insecticide is not so undesirable; in fact it may even be desirable if it can be easily degraded in the living system. Chemicals like DDT and DDE have highly stable aryl–chlorine bonds that are not cleaved to any extent by the multifunction oxidases of living organisms. In order to be biodegradable the analogs like methoxychlor should have sites for attack by multifunction oxidases. Metcalf et al. further surmised that (*a*) since the mechanism of action is highly stereospecific but also flexible, the molecule has to conform in general shape and size to the DDT molecule; (*b*) since DDT has no sites of attack by microsomal oxidases, provision of such "handles" to a DDT-type molecule would impart biodegradability to it; (*c*) to be good insecticides the compounds should be readily absorbed by the insect cuticle and transported to the site of action in an active form. Also, these workers have shown that the diphenyl moiety is not uniquely essential for DDT-like insecticidal activity and that a heteroatom N, O, or S can be interposed between the two aryl moieties (53).

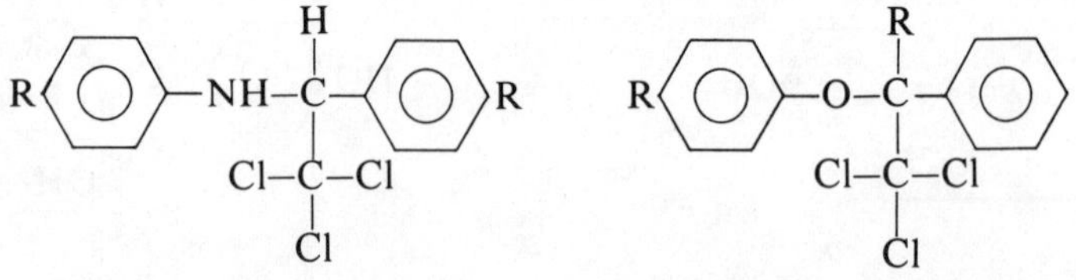

α-trichloromethylbenzylanilines          α-trichloromethylbenzylphenyl ethers

With this premise in view, these workers have synthesized several symmetrical and asymmetrical analogs of DDT and tested their insecticidal properties alone and in combination with synergists. Some of these analogs include methylchlor, ethoxychlor, and methiochlor (21, 53, 54).

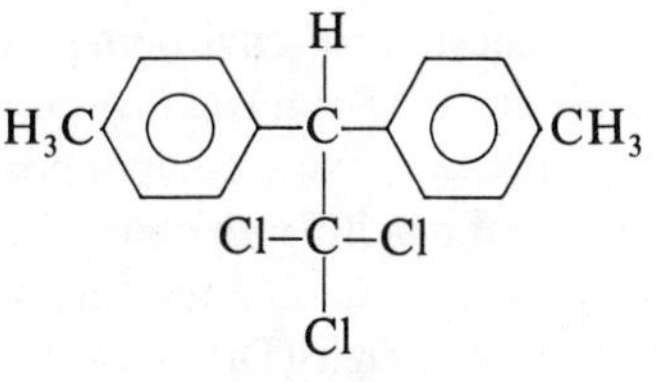

Methylchlor

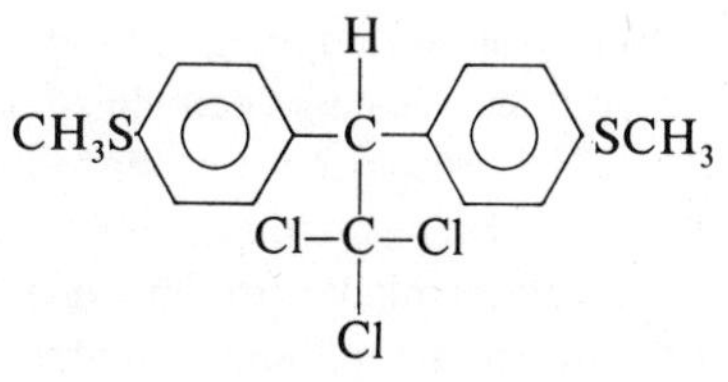

Methiochlor

Some asymmetrical DDT analogs synthesized by Metcalf and his group are given below.

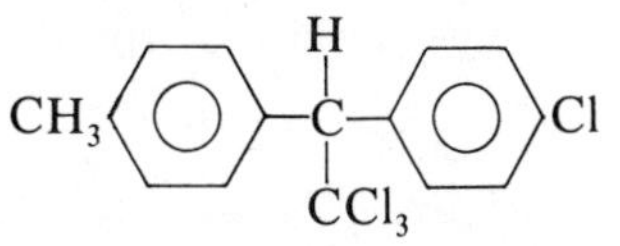

Chloro-methylchlor

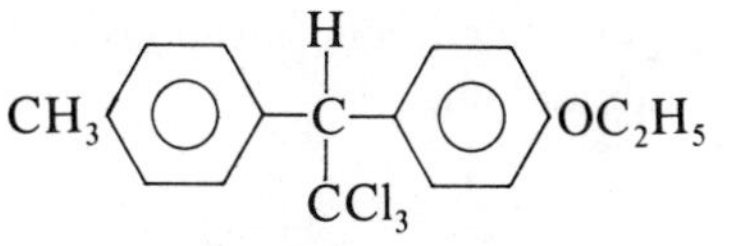

Methyl-ethoxychlor

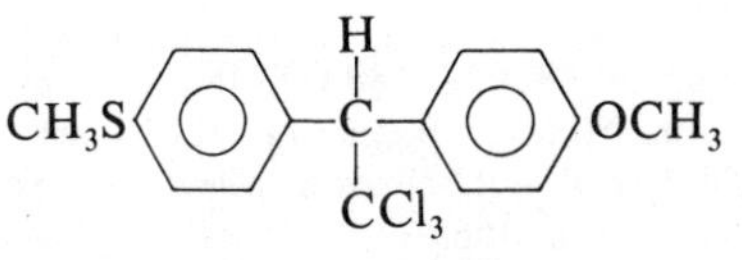

Methoxy-methiochlor

Metcalf and co-workers found that a number of these compounds possessed insecticidal action comparable to DDT and were more effective against a DDT-resistant strain ($R_{sp}$) of house flies. It was surmised that this was due to a lower rate of dehydrochlorination of these compounds by DDT-ase because aryl substituents like $CH_3-$, $CH_3O-$, and $CH_3S$ do not have as much affinity for electrons as Cl or Br and therefore the relative availability of electrons at the α-carbon atom is different, leading to a difference in dehydrochlorination. These compounds also are less toxic to mammals than DDT, possibly because of the high titer of mixed-function oxidases in mammals. The $CH_3O$, $CH_3$, and $CH_3S$ compounds were highly synergized by piperonyl butoxide, and $CH_3O$ and $C_2H_5O$ analogs possessed the highest synergized toxicity to house flies, possibly because of their better fit at the site of action as compared to DDT.

With these investigations and with the research on pyrethroids mentioned earlier, we are entering a new era of insecticide biochemistry. On the one hand, efforts are being made to create more persistent pesticides out of easily degradable but highly potent insecticides and, on the other hand, attempts are being made to provide "handles" for biodegradability to chemicals that are otherwise persistent. It is too early to speculate which of these two endeavors would yield the best results, but we have already entered a stimulating phase in insect control and the next decade promises to be an exciting one in insect toxicology.

With the development of new analogs arises the question of environmental acceptibility of new chemicals, and Metcalf and his group (54) have devised a model laboratory ecosystem to assess the biodegradibility of new compounds. It has a terrestrial aquatic interface and a seven-element trophic web consisting of Sorghum plants grown on a terrestrial shelf of white sand. These plants provide food for the salt marsh caterpillars, *Estigmene acrea* (Drury); other elements in the food chain are plankton, Daphnia, *Culex* larvae, *Gambusia*, fish and algae, and *Physa* snails. The entire system is housed in a small aquarium $10 \times 12 \times 20$ in. with 180 in.$^2$ of surface area. This is housed in an environmental chamber. Plants are treated with radioactive chemicals under investigation and concentrations in the food chain are determined at specified periods of time.

## REFERENCES

1. O. Zeidler. *Ber. Dtsch. Chem. Ges.* **7**: 1180 (1874).
2. M. C. Bowman, J. Acree, and M. K. Corbett. *J. Agr. Food Chem.* **8**: 406 (1960).
3. A. W. A. Brown. *Insect control by Chemicals.* Wiley, New York, 1951, 817 pp.
4. J. Bitman. *Agr. Sci. Rev.* **7**(4): 6 (1969).
5. R. L. Metcalf. *J. Agr. Food Chem.* **21**: 511 (1973).
6. S. W. Simmons. *DDT—The Insecticide Dichlorodiphenyl Trichloroethane and Its Significance* (P. Müller, ed.), Vol II, Birkhauser-Verlag, Basel, 1959.
7. Final Comprehensive Report, Pakistan, East Bengal Malaria Control Demonstration Team. UN WHO Regional Office. *Pak. J. Health* **2**: 61 (1952).
8. G. Sacca. *Riv. Parasitol.* **8**: 141 (1947).
9. W. V. King and J. B. Gahan. *J. Econ. Entomol.* **42**: 405 (1949).
10. R. B. March and R. L. Metcalf. *Univ. Calif. Citrus Exp. Stn. News.* **38**: 7 pp (1949).
11. A. S. Perry, S. Miller, and A. J. Buchner. *J. Agr. Food Chem.* **11**: 457 (1963).
12. J. Yeager and S. Munson. *Science* **102**: 305 (1945).
13. M. Narahashi. *Adv. Insect Physiol.* **8**: 1 (1971).
14. J. Sternburg and C. W. Kearns. *Science* **116**: 144 (1952).
15. J. W. L. Beament. *J. Insect Physiol.* **2**: 199 (1958).
16. S. E. Lewis, J. B. Waller, and K. S. Fowler. *J. Insect Physiol.* **4**: 128 (1960).
17. J. Sternburg. *Annu. Rev. Entomol.* **8**: 19 (1963).
18. H. Martin and R. L. Wain. *Nature* **154**: 512 (1944).
19. P. Lauger, H. Martin, and P. Müller. *Helv. Chem. Acta* **27**: 892 (1944).
20. E. F. Rogers, H. D. Brown, I. M. Rasmussen, and R. E. Neal. *J. Amer. Chem. Soc.* **75**: 2991 (1953).
21. R. Reimschneider and H. D. Otto. *Z. Naturforsch.* **B9**: 95 (1954).
22. L. Mullins. *J. Amer. Inst. Biol. Sci. Publ.* **1**: 123 (1956).
23. F. A. Vingiello and P. E. Newallis. *J. Org. Chem.* **25**: 905 (1960).
24. R. L. Metcalf. *J. Agr. Food Chem.* **21**: 511 (1973).
25. A. S. Hirwe and R. L. Metcalf. Unpublished data, quoted in ref. 24 above.
26. G. Holan. *Nature* **221**: 1025 (1969).
27. G. Holan. *Bull WHO* **44**: 355 (1971).
28. G. Holan. *Nature* **232**: 644 (1971).

29. T. Narahashi and H. G. Hass. *J. Gen. Physiol.* **51**: 177 (1968).
30. F. C. Matsumura and K. C. Patil. *Science* **166**: 121 (1969).
30a. F. A. Gunther, R. C. Blinn, G. E. Carman, and R. L. Metcalf. *Arch. Biochem. Biophys.* **50**: 504 (1954).
31. M. A. H. Fahmy, T. R. Fukuto, and R. L. Metcalf. *J. Agr. Food Chem.* **21**: 585 (1973).
31a. R. W. Taft. Jr. In *Steric Effects in Organic Chemistry.* John Wiley and Sons, New York. 1956.
32. W. C. White and T. R. Sweeny. *Public Health Rep.* (US) **60**: 66 (1945).
33. J. D. Judah. *Brit. J. Pharmacol.* **4**: 120 (1949).
34. J. Sternburg and C. W. Kearns. *J. Econ. Entomol.* **45**: 597 (1952).
35. H. Lipke and C. W. Kearns. *J. Biol. Chem.* **234**: 2123 (1959).
36. W. J. Hayes, G. E. Quinby, K. C. Walker, J. W. Elliott, and W. M. Upholt. *A. M. A. Arch. Ind. Health* **18**: 398 (1958).
37. D. P. Morgan and C. C. Roan. *Arch. Environ. Health* **20**: 452 (1970).
38. B. J. Kallman and A. K. Andrews. *Science* **141**: 1050 (1963).
39. P. R. Datta, E. P. Laug, and A. K. Klein. *Science* **145**: 1052 (1964).
40. J. E. Peterson and W. H. Robinson. *Toxicol. Appl. Pharmacol.* **6**: 321 (1964).
41. J. L. Mendel and M. S. Walton. *Science* **151**: 1527 (1966).
42. G. R. Fries, G. S. Marrow, and C. H. Gordon. *J. Agr. Food Chem.* **17**: 860 (1969).
43. R. P. Miskus, D. P. Blair, and J. E. Casida. *J. Agr. Food Chem.* **13**: 481 (1965).
43a. J. Sink, H. Varela-Alvarez, and C. Hess. *op. cit,* **20**: 7 (1972).
44. M. Tsukamoto. *Botyu-Kagaku* **24**: 141 (1959).
45. E. S. Albone, G. Elington, N. C. Evans and M. M. Rhead. *Nature* **240**: 420 (1972).
46. S. Jensen, R. Göthe, and M. O. Kindstedt. *Nature* **240**: 422 (1972).
47. I. P. Kapoor, R. L. Metcalf, A. S. Hirwe, P.-Y. Lu, J. R. Coats, and R. F. Nystrom. *J. Agr. Food Chem.* **20**: 1 (1972).
48. R. L. Metcalf. *Organic Insecticides.* Interscience, New York, 1955, 392 pp.
49. S. J. Harris, H. C. Cecil, and J. Bitman. *J. Agr. Food Chem.* **22**: 969 (1970).
50. I. P. Kapoor, R. L. Metcalf, R. F. Nystrom, and G. K. Sangha. *J. Agr. Food Chem.* **18**: 1145 (1970).
51. I. P. Kapoor, R. L. Metcalf, A. S. Hirwe, P.-Y. Lu, J. R. Coats, and R. F. Nystrom. *J. Agr. Food Chem.* **21**: 310 (1973.
52. A. S. Hirwe, R. L. Metcalf, and I. P. Kapoor. *op. cit.* **20**: 818 (1972).
53. M. A. H. Fahmy, T. R. Fukuto, and R. L. Metcalf. *op. cit.* **21**: 585 (1973).
54. R. L. Metcalf, G. K. Sangha, and I. P. Kapoor. *Environ. Sci. Technol.* **5**: 709 (1971).

# SYNTHETIC ORGANIC INSECTICIDES

| | Cyclodiene (Diene-Organochlorine) Insecticides |
|---|---|
| **CHAPTER 6** | |

This is a group of highly chlorinated insecticides prepared by the diene-synthesis or Diels-Alder reaction. In this reaction a compound containing a double bond is added to the 1,4-positions of a conjugated system.

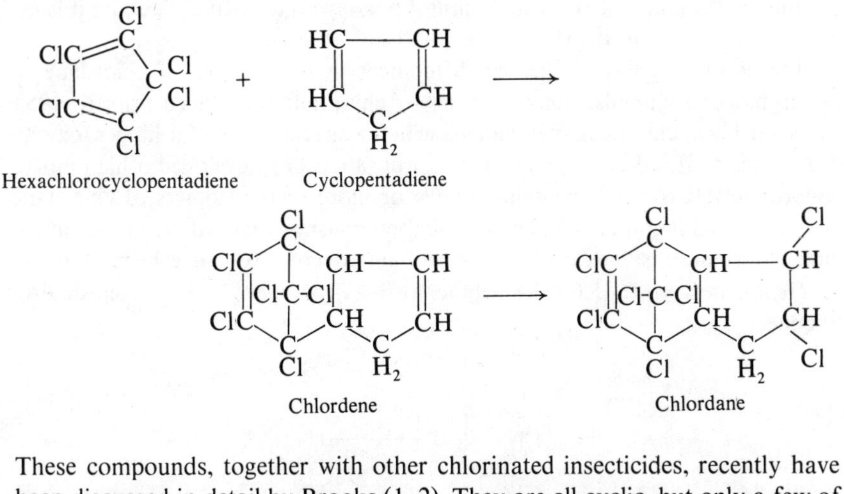

Hexachlorocyclopentadiene     Cyclopentadiene

Chlordene        Chlordane

These compounds, together with other chlorinated insecticides, recently have been discussed in detail by Brooks (1, 2). They are all cyclic, but only a few of them are dienes. Their syntheses and development were based mostly on the work of Julius Hyman and associates dating from the mid 1940s. The first successful insecticidal chemical of the series, chlordane ($C_{10}H_6Cl_8$) or 1068 was discovered almost simultaneously in the United States by Julius Hyman and by Kearns, Ingle and Metcalf, and in Germany by Reimschneider. In Germany it was known as M 410. As shown above, chlordane is prepared by reacting hexachlorocyclopentadiene with cyclopentadiene to yield chlordene which is further chlorinated to give the insecticide.

*Chlordane*,     2,3,4,5,6,7,8,8-octachloro-2,3,3a,4,7,7a-hexahydro-4,7-methanoindene, can exist in *exo* and *endo* configurations. Reactions and dipole measurements indicate that it is in the form of the *endo* isomer. Depending on the disposition of the chlorines on the five-membered ring, there can be two isomers, *cis*, or *trans*. Both isomers have been isolated. The *cis* isomer α-chlordane is a white crystalline substance, mp 106.5–108°C; the *trans* isomer has a slightly lower melting point. The technical insecticide, however, is a viscous amber-colored liquid that is a mixture of polychlorinated bicyclohydrocarbons of varying degrees of insecticidal and physiological properties. Of these the *cis* and *trans* isomers comprise between 60 and 70% of the total. Chlordane is a contact and stomach poison; it is insoluble in water but highly soluble in organic solvents. Chlordane is toxic to warm-blooded animals, its acute oral $LD_{50}$ for the rat being between 457 and 590 mg/kg. Since it is easily absorbed through mammalian skin, skin contact should be avoided. In the presence of alkali, chlordane loses HCl to form chemicals that are not insecticidal; traces of iron catalyze this reaction. Steel containers therefore have a baked protective coating on the inside.

Chlordane is used for household pests, pests of man and domestic animals, termite control, and also for agricultural and lawn pests. In agriculture it is especially useful against grasshoppers and soil insects.

Earlier reports have indicated differences in the toxicity of chlordane to warm-blooded animals. Some samples of chlordane poroduced prior to 1951 contained hexachlorocyclopentadiene which was responsible for higher toxicity (3). Recently, the Velsicol Chemical Corporation has developed a high-purity chlordane (HCS 3260) containing 95% or more of the isomers of chlordane (4). With this high-purity chemical the metabolism of chlordane in the rat (5) and chlordane residues in the milk and fat of cows (6) have been studied.

*Heptachlor*,     1,4,5,6,7,8,8-heptachloro-4,7-methano-3a,4,7,7a-tetrahydroindene,

is a white crystalline substance, mp 95–96°C; it is nearly insoluble in water but highly soluble in kerosene and aromatic hydrocarbons. The technical grade material is a waxy mass, mp 46–74°C, that contains about 70% heptachlor and 30% related compounds. Heptachlor is highly toxic, acute oral $LD_{50}$ for male rat being 100 mg/kg; it is easily absorbed through mammalian skin.

Heptachlor is used for the control of soil insects like wireworms, cutworms, and grubs. The insecticidal properties of this chemical appear to be due to its conversion to its epoxide which is more toxic than heptachlor to all living organisms. The epoxidation easily takes place in treated animals and in soil. In UV and sunlight heptachlor is converted to photoheptachlor.

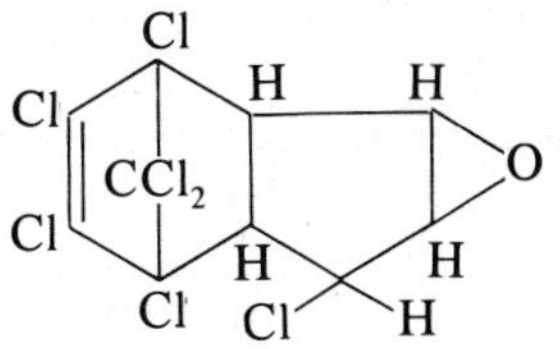

Heptachlor epoxide 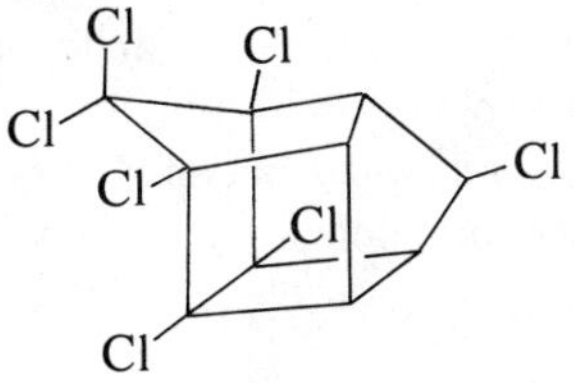                        Photoheptachlor

*Aldrin*, 1,2,3,4,10,10-hexachloro-1,4,4a,5,8,8a-hexahydro-1,4-*endo-exo*-5,8-dimethanonaphthalene,

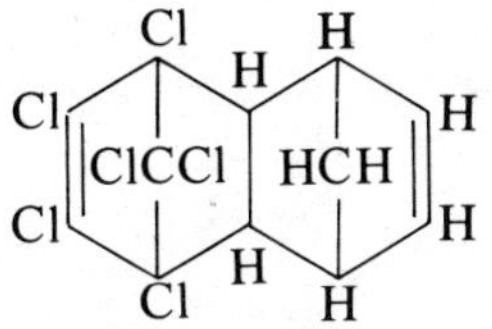

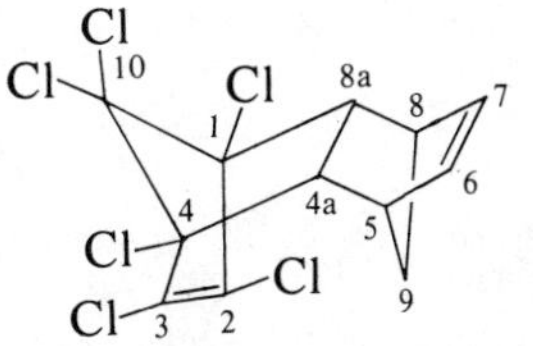

is a white crystalline solid, mp 104–105°C; its vapor pressure at 25°C is 6 × $10^{-6}$ mm Hg. The technical material contains not less than 95% aldrin and melts at 45–60°C. Aldrin is almost insoluble in water but soluble in organic solvents. It is highly toxic, acute oral $LD_{50}$ for rat being 55 mg/kg; it is easily absorbed through skin. Aldrin is a contact and stomach poison that also possesses fumigant action. It is used primarily to control soil insects cotton rootworms, wireworms, white grubs, cotton insects, and turf pests.

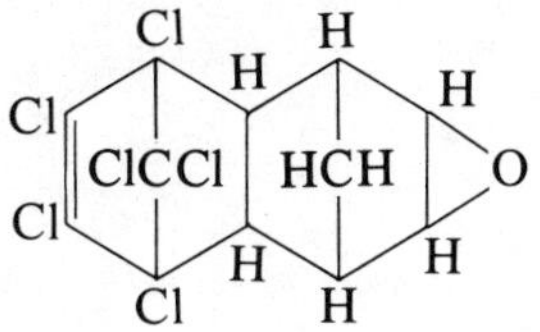

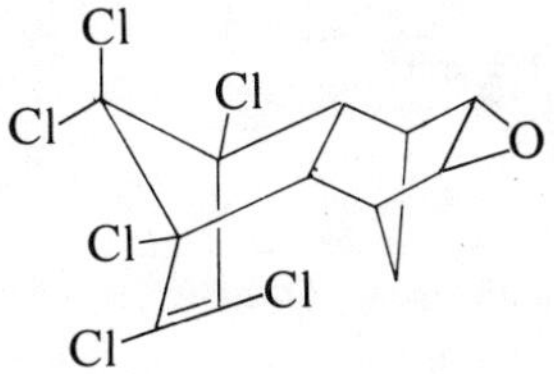

*Dieldrin*,   1,2,3,4,10,10-hexachloro-6,7-epoxy-1,4,4a,5,6,7,8,8a-octahydro-1,4-*endo-exo*-5,8-dimethanonaphthalene, is an epoxide of aldrin. It is a white crystalline substance, mp 175–176°C, with vapor pressure of $1.8 \times 10^{-7}$ mm Hg at 25°C. Dieldrin is insoluble in water but highly soluble in most organic solvents. The technical grade material contains not less than 85% dieldrin and is a light brown material. It is highly toxic, acute oral $LD_{50}$ for rat being 46 mg/kg. Dieldrin is a contact and stomach poison that gives slow knockdown of treated insects. However its residual properties make it a good insecticide for the control of soil insects. Dieldrin is also used for the control of cotton pests, grasshoppers, cutworms, armyworms, and insects of public health importance, especially in the form of residual sprays.

Aldrin and dieldrin can be converted to photoaldrin and photodieldrin in the presence of sunlight and also by microorganisms (Fig. 6.1) (7–9). Living

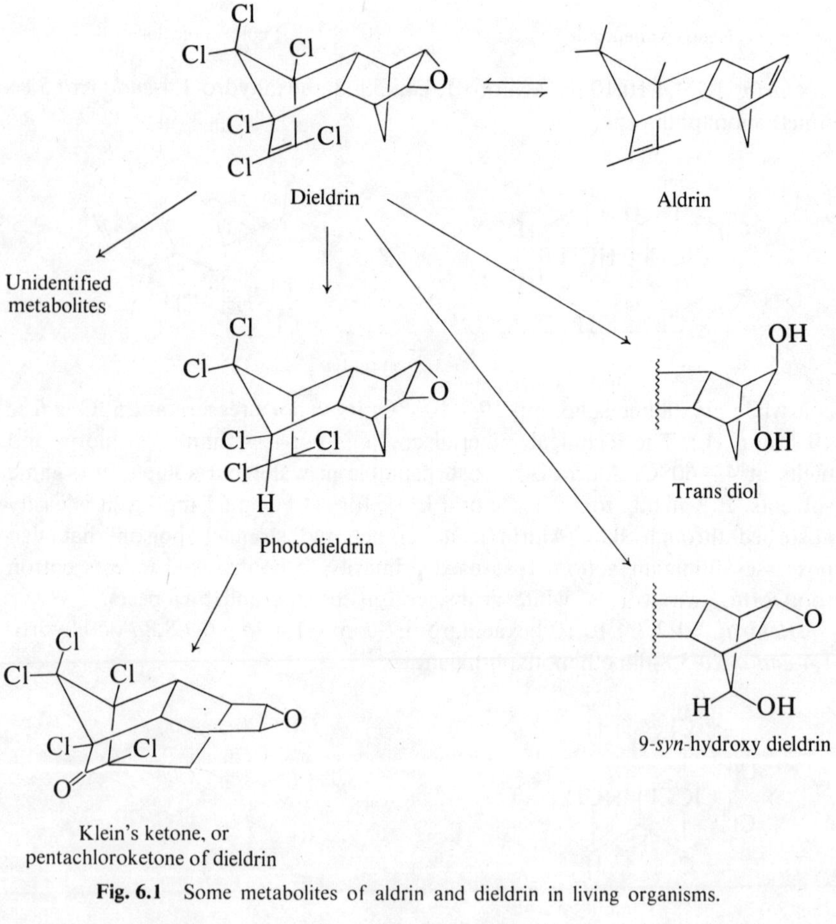

Fig. 6.1  Some metabolites of aldrin and dieldrin in living organisms.

organisms can also convert aldrin and photoaldrin to photodieldrin (10), which can thus be the terminal residue of aldrin and dieldrin; photodieldrin is more toxic to aquatic and terrestrial organisms than aldrin or dieldrin (11–13). Reddy and Khan (14) report that the microsomal mixed-function oxidase of rat, mice, and house fly metabolize photodieldrin at very low levels. Photodieldrin is converted in insects to Klein's ketone, which is even more toxic than its precursor.

*Isodrin*, 1,2,3,4,10,10-hexachloro-1,4,4a,5,8,8a-hexahydro-1,4,-*endo-endo*-5,8-dimethanonaphthalene,

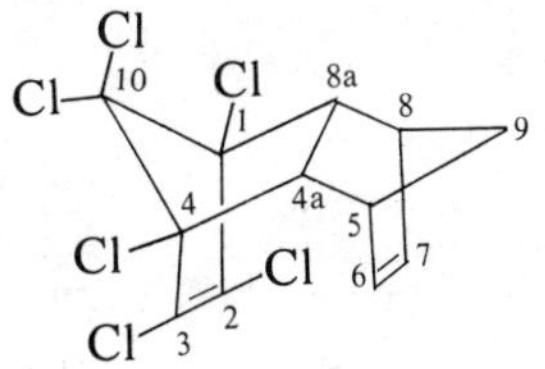

is the stereoisomer (*endo-endo*) of aldrin. It is a white crystalline compound, mp 240–242°C. Isodrin is highly toxic to mammals, $LD_{50}$ for rat 7–15 mg/kg. It is no longer manufactured for use as an insecticide.

*Endrin*, 1,2,3,4,10,10-hexachloro-6,7-epoxy-1,4,4a,5,6,7,8,8a-octahydro-1,4-*endo-endo*-5,6-dimethanonaphthalene,

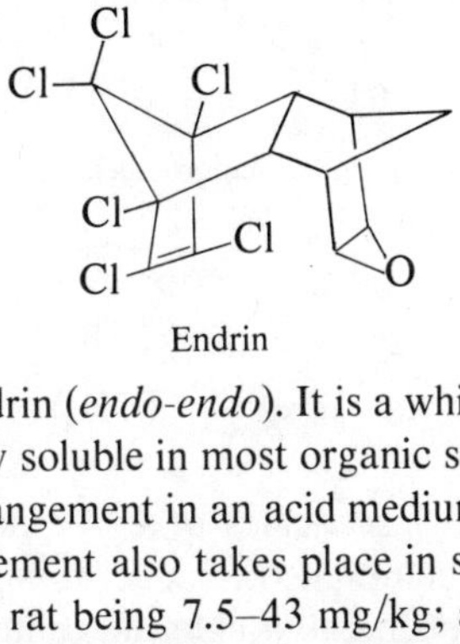

Endrin

is the stereoisomer of dieldrin (*endo-endo*). It is a white crystalline chemical, insoluble in water and highly soluble in most organic solvents. Endrin is stable to alkali but undergoes rearrangement in an acid medium to an insecticidally inactive ketone. This rearrangement also takes place in strong sunlight. It is highly toxic, acute oral $LD_{50}$ for rat being 7.5–43 mg/kg; acute dermal $LD_{50}$ for the same animal is 5.6–94 mg/kg. Contact of endrin with skin should be avoided. The technical insecticide is brownish in color and contains not less than 85% of the active material.

Endrin is used for the control of cotton insects, cutworms, armyworms, plant bugs, lygus bugs, and other pests. It is especially useful for the control of the blackcurrant mite, which is naturally resistant to most insecticides and miticides. Endrin is also used for the control of pine mice in orchards. Because of its rearrangement to ketone in the presence of sunlight, it is not highly persistent.

Endrin is highly toxic to fish; 24 hour $LC_{50}$ for rainbow trout exposed to endrin at 1.6°C, 7.2°C, and 12.7°C was found to be 15, 5.3, and 2.8 ppb, respectively (14a).

The rearrangement products in the isodrin–endrin series, unlike photoaldrin and photodieldrin, tend to be less toxic than the parent compounds. In the case of endrin there are two rearrangement products, namely, endrin hexachloroaldehyde and Δ-keto-endrin (the endrin hexachloro half-cage ketone); isodrin forms photoisodrin or photodrin.

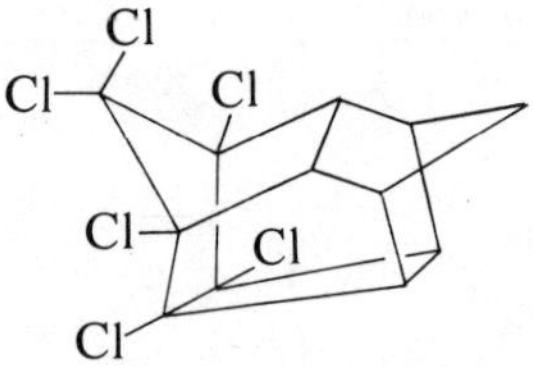

Photodrin 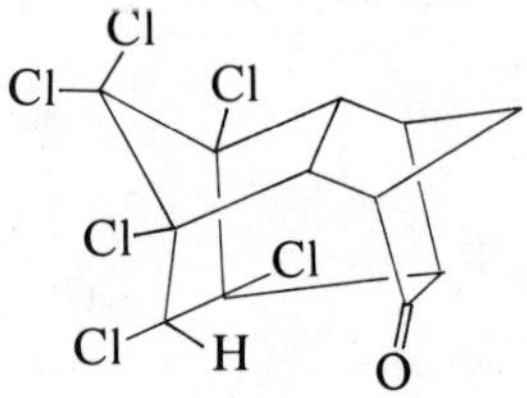                                            Δ-keto-endrin

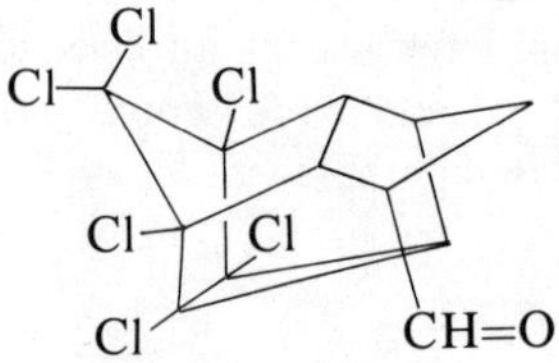

Endrin hexachloroaldehyde

*Toxaphene,*

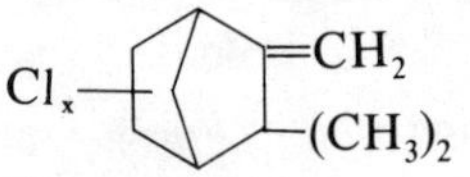

is one of the best known insecticides of this group, though it is not truly a cyclodiene but, rather, a chlorinated terpene. It is prepared by chlorination of the bicyclic terpene, camphene, to a chlorine content of 67–69%. It is a complex mixture of polychlorocamphenes and camphenes of different structures and has an empirical formula $C_{10}H_{10}Cl_8$. Toxaphene is a waxy white material that melts between 65 and 90°C. Its acute oral $LD_{50}$ for rat is 69 mg/kg. It is effective against all major pests of cotton and has low toxicity for bees. It acts as a stomach and contact poison and dehydrochlorinates in the presence of alkalis and on prolonged exposure to sunlight. Toxaphene is very effective against grasshoppers, armyworms, and cutworms. It is also used for livestock pests, mites, mange, horn flies, lice, ticks, and sheep ked. Toxaphene is highly toxic to fish.

*Alodan*, 1,2,3,4,7,7-hexachloro-5,6-bis(chloromethyl)-2-norborene,

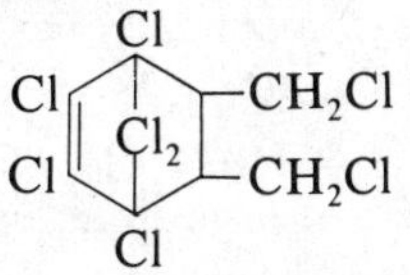

is a white crystalline compound, mp 104–106°C. It is insoluble in water but soluble in most organic solvents. It has very low toxicity to vertebrates; its acute oral $LD_{50}$ for rat is 1500 mg/kg. Alodan is phytotoxic and is not used on plants, but in Germany it is used in granaries.

*Endosulfan* or *Thiodan*®, 6,7,8,9,10,10-hexachloro-1,5,5a,6,9,9a-hexahydro-6,9-methano-2,4,3-benzodioxathiepin 3-oxide, can also be regarded as a cyclic ester of sulfurous acid.

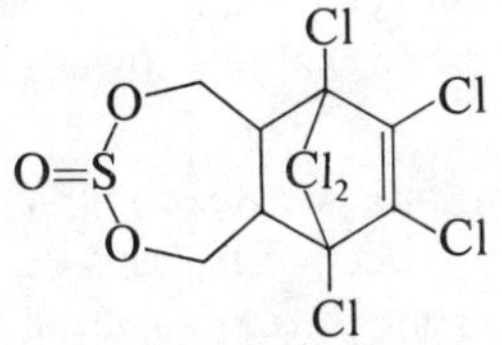

It is a white crystalline substance that exists in two isomeric forms. Technical endosulfan is a mixture of these forms in the ratio of 4:1. It is insoluble in water but soluble in organic solvents. Endosulfan is effective for the control of beetles, cutworms, and aphids. It is highly toxic to fish.

*Mirex* has been used since 1962 as 0.3% bait in soyabean oil sprayed onto

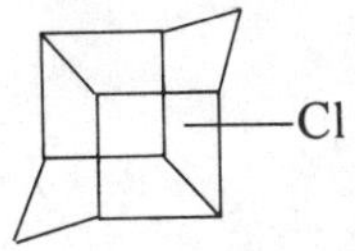

corncob grits for the control of the imported fire ant, *Solenopsis saevissimia richteri* Forel. The bait is applied at the rate of 1.2 lb acre containing 1.7 grams of mirex.

## METABOLISM OF CYCLODIENES

Earlier investigations on the metabolism of cyclodienes demonstrated the epoxidations of heptachlor, aldrin, and isodrin (15–20). This is an activation process since the epoxides are more toxic than the parent compounds (for a different view see refs. 1 and 2). Later, production of more hydrophilic and less toxic metabolites in insects and mammals was reported (2, 21–27). These products were shown to be produced by hydrolytic processes or microsomal oxidation. The hydroxylated derivatives are either conjugated or excreted without conjugation. In the case of dieldrin at least six metabolites have been isolated (27–32) in the urine and feces of rabbits, rats, and sheep. Two of these, namely, *trans*-6,7-dihydroxydihydroaldrin (*trans*-diol), and 9-(*syn*-epoxy) hydroxy-1,2,3,4,10,10-hexachloro-6,7-epoxy-1,4, 4a5,6,7,8,8a-octahydro-1,4-*endo*-5,8-*exo*-dimethanonaphthalene (9-hydroxy-HEOD), have been identified. In addition, Matthews and Matsumura (31) isolated and identified as a ketone an oxidized, dechlorinated metabolite that did not have an intact dieldrin ring system.

The fate of aldrin-$C^{14}$ in maize, wheat, and soils recently has been studied by Korte and his colleagues (32). The main radioactive products were dieldrin and dihydrochlordene dicarboxylic acid. Trace amounts of aldrin, photodieldrin, unknown acidic compounds, and an unknown polar metabolite were also found.

The metabolism of endosulfan is also complicated; it forms endosulfan alcohol by hydrolysis and oxidizes to endosulfan sulfate. Hydrophilic metabolites are also formed. Though α-endosulfan decomposes fairly rapidly (50% in ∼ 60 days) with the formation of endosulfan sulfate (33) which was found to be relatively more stable, β-Endosulfan was found to be much more persistent (∼50% in 800 days) as compared to the α-compound.

Mirex is relatively very stable and is not easily metabolized by animals (34, 35). Its photodecomposition is slow and has been studied by Ivie et al. (36).

In mammals isodrin is epoxidized to endrin, which is oxidized without skeletal rearrangement and excreted.

## REFERENCES

1. G. T. Brooks. Chlorinated Insecticides Vol. 1, *Technology and Application* CRC Press, Cleveland, 1974, 249 pp.
2. G. T. Brooks. Chlorinated Insecticides Vol. II, *Biological and Environmental Aspects*, CRC Press, Cleveland, 1974, 197 pp.
3. L. Ingle. A monograph on chlordane, University of Illinois, Urbana, 1975.
4. Velsicol Corp., Chicago, Illinois, Bull. No. E-1-1-3, 1970.
5. J. R. Barnett and H. W. Dorough. *J. Agr. Food Chem.* **22**: 612 (1974).
6. H. W. Dorough and R. W. Hanken. *Bull. Environ. Contam. Toxicol.* **10**: 208 (1973).
7. J. Robinson, A. Richardson, and B. Bush. *op. cit.* **1**: 127 (1966).
8. E. P. Lichstenstein, K. R. Schulz, T. W. Fuhremann, and T. T. Liang. *J. Agr. Food Chem.* **18**: 100 (1970).
9. J. Kohli, S. Zarif, I. Weisgerber, W. Klein, and F. Korte. *op. cit.* **21**: 855 (1973).

10. H. J. Eagan. *Assoc. Off. Anal. Chem.* **52**: 299 (1969).
11. J. D. Rosen and D. J. Sutherland. *Bull. Environ. Contam. Toxicol.* **1**: 133 (1966).
12. Food and Agricultural Organization, and World Health Organization. "Evaluation of Some Pesticide Residues in Food." FAO/PI. 1967; M/11/1, Rome, 1968, 104 pp.
13. M. A. Q. Khan, R. H. Stanton, D. J. Sutherland, J. D. Rosen, and N. Maitra. *Arch. Environ. Contam. Toxicol.* **1**: 159 (1973).
14. G. Reddy and M. A. Q. Khan. *J. Agr. Food Chem.* **22**: 910 (1974).
14a. K. J. Macek, C. Hutchinson, and O. B. Cope. *Environ. Contam. Toxicol.* **3**: 174 (1969).
15. N. Gannon and J. H. Bigger. *J. Econ. Entomol.* **51**: 1 (1958).
16. N. Gannon and G. C. Decker. *op. cit.* **51**: 8 (1958).
17. O. Giannotti, R. L. Metcalf, and R. B. March. *Ann. Entomol. Soc. Amer.* **49**: 588 (1956).
18. G. T. Brooks. *Nature* **186**: 96 (1960).
19. B. Davidow and J. L. Radomski. *J. Pharm. Exp. Ther.* **107**: 259 (1953).
20. A. S. Perry, A. M. Mattson, and A. J.    Buckner. *J. Econ. Entomol.* **51**: 346 (1958).
21. P. Gerolt. *op. cit.* **58**: 849 (1965).
22. E. S. Oonnithan and R. Miskus. *op. cit.* **57**: 425 (1961).
23. C. Cueto jr. and W. J. Hayes, Jr. *J. Agr. Food Chem.* **10**: 366 (1962).
24. P. R. Datta, E. P. Luag, J. O. Watts, A. K. Klein, and M. J. Nelson. *Nature* **208**: 289 (1965).
25. D. E. Heath and M. Vandekar. *Br. J. Ind. Med.* **21**: 269 (1964).
26. G. T. Brooks. *World Rev. Pest control* **5**: 62 (1966).
27. F. Korte and H. Arent. *Life Sci.* **4**: 2017 (1965).
28. R. D. Hedde, K. L. Davison, and R. D. Robbins. *J. Agr. Food Chem.* **18**: 116 (1970).
29. V. J. Feil, R. D. Hedde, R. G. Zaylskie, and C. H. Zachrison. *op. cit.* **18**: 120 (1970).
30. M. S. Walton, V. Beck, and R. L. Baron. *Toxicol. Appl. Pharm.* **17**: 278 (1970).
31. H. B. Matthews and F. Matsumura. *J. Agr. Food Chem.* **17**: 845 (1969).
32. I. Weisgerber, J. Kohli, R. Kaul, W. Klien, and F. Korte. *op. cit.* **22**: 609 (1974).
33. D. K. R. Stewart and K. G. Cairns. *op. cit.* **22**: 984 (1974).
34. J. R. Gibson, G. W. Ivie, and H. W. Dorough. *op. cit.* **20**: 1246 (1972).
35. H. W. Dorough and G. W. Ivie. *J. Environ. Qual.* **3**: 65 (1974).
36. G. W. Ivie, H. W. Dorough, and E. G. Alley. *J. Agr. Food Chem.* **22**: 933 (1974).

# SYNTHETIC ORGANIC INSECTICIDES

| Lindane

*Lindane,* 1,2,3,4,5,6-hexachlorocyclohexane ($\gamma$-isomer), in earlier literature was erroneously called benzenehexachloride (BHC). The insecticidal properties of hexachlorocyclohexane (HCH) were discovered simultaneously in England and France in 1942. This chemical was first prepared by Michael Faraday in 1825 and the existence of the $\gamma$-isomer was first demonstrated by Van der Linden after whom lindane is named. Chemically speaking the name benzene-hexachloride should be given to the compound in which the six hydrogen atoms of benzene are replaced by six chlorine atoms, leaving the ring unsaturated. However, if a chlorine atom is added to each double bond of benzene so that the resultant compound is saturated, we obtain hexachlorocyclohexane, which is a derivative of the saturated compound cyclohexane in which six hydrogen atoms have been replaced by six chlorine atoms.

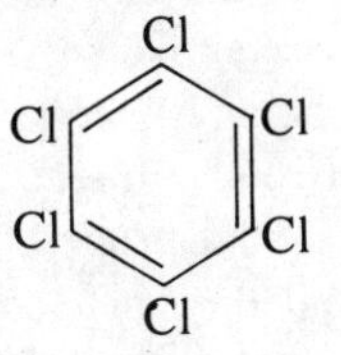

| The chemical<br>Benzenehexachloride | Cyclohexane | Hexachloracyclohexane<br>erroneously called BHC |
|---|---|---|

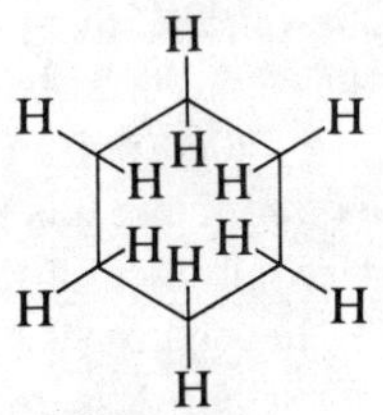

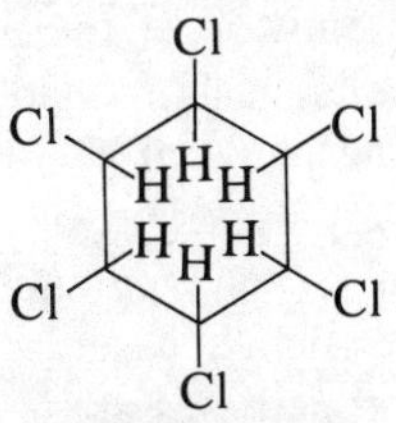

HCH is prepared by the chlorination of benzene in ultraviolet light. The crude product is a grayish amorphous solid with a characteristic musty odor that is attributed to the presence of heptachlorocyclohexanes and other cyclohexanes; it is a mixture of several isomers. It is now common practice to use the abbreviation HCH for the crude mixture and $\gamma$-HCH for lindane or the $\gamma$-isomer, though BHC and $\gamma$-BHC are still used in the literature. Use of HCH on some edible crops imparts an "off-flavor" or "taint," this is particularly true of potatoes. Pure $\gamma$-isomer or lindane does not impart this taint. However the flavor of cereals does not seem to be affected by the crude product, nor is the taint noticeable in sugar refined from treated beets. Both DDT and lindane are phytotoxic to cucurbits. For the control of codling moth neither DDT nor lindane is recommended, because they are ineffective against red spider mites and kill their predators, thereby greatly increasing the population of mites in treated orchards. Unlike DDT, lindane is effective in controlling various mites infesting man and livestock. HCH in many ways proved to be a very useful insecticide soon after its introduction. It was remarkably toxic to locusts and grasshoppers that were tolerant of DDT. Because of its high vapor pressure it also possesses fumigant properties and in warmer climates it is very effective against insects that hide in corners and crevices. For instance, DDT is relatively ineffective against boll weevil larvae because they remain in the cotton boll; lindane is effective against them possibly because of fumigant action.

The cyclohexane ring can exist in two forms, the boat form or the chair or Z form. The chair form has the minimum potential energy at room temperature and X-ray diffraction studies have shown that $\alpha$, $\beta$, $\gamma$, $\delta$, and $\epsilon$ HCH isomers exist in this form. In the chair form there are two types of C—Cl bonds; six lie

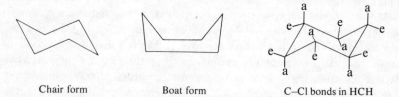

Chair form                      Boat form                      C—Cl bonds in HCH

radially outward at angles of 109.28° parallel to the plane of the ring and are called equatorial (e) bonds, and of the remaining six, three are above and three below the plane of C atoms and are called axial (a) or polar (p) bonds.

In many ways lindane is a unique insecticide; it is the only highly insecticidal chemical of its type and the steric requirement for insecticidal properties is very specific. No other isomer of HCH even comes close to lindane in its insecticidal properties. Pure lindane is almost odorless and is stable to the action of heat. It has a comparatively high vapor pressure, $9.4 \times 10^{-6}$ mm Hg at 20°C. Quraishi (1) found that it volatilizes easily from treated surfaces and the rate of volatilization follows the theoretical equation

$$y = \frac{2.2x + 200}{2x/41 + 2}; \text{ where } x < 250 \text{ minutes}$$

It is very slightly soluble in water but soluble in organic solvents. Lindane is toxic, oral $LD_{50}$ for male rat being 88 mg/kg.

Principal uses of lindane were for cotton, beef cattle, and hogs.

## MODE OF ACTION

Symptoms of typical lindane poisoning in insects include tremors, ataxia, convulsions, and prostration. Compared to DDT, lindane is quicker in action, the symptoms of poisoning occurring much more rapidly after the administration of a toxic dose. These would indicate that, like DDT, it is neurotoxic agent, but it is more rapid in action. In the case of mammals lindane stimulates the central nervous system (CNS) and as a consequence the blood pressure rises and the rate of heart beat decreases. This is followed by increased rate of respiration, tremors, salivation, and finally convulsions. Later, respiration slows down and death may result from either respiratory failure or cardiac arrest.

Slade (2a) originally suggested that the toxic action of lindane was due to its similarity in structure to myoinositol, one of the isomers of hexahydrocyclohexane (one of the B vitamins). He hypothesized that lindane antagonized the action of this compound at some vital site. But careful investigations later revealed that myoinositol is the isostere of $\delta$-HCH and hence this hypothesis does not seem plausible. Also, there is hardly any evidence to

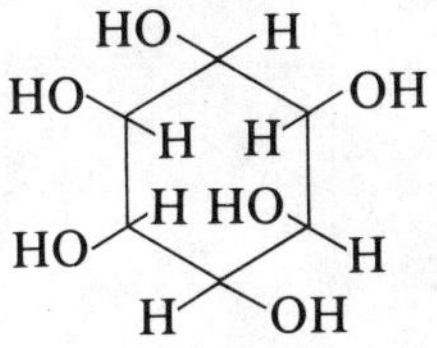

Myoinositol

indicate that lindane acts by interacting with a specific enzyme, though as compared to DDT it is a better inhibitor of the $Na^+$, $K^+$, and $MG^{2+}$ ATP-ase from American cockroach nerve cord (2b). For theories of toxic action of lindane, see Chapter 5.

## METABOLISM OF LINDANE

Earlier studies on the metabolism of lindane indicated a fairly rapid breakdown to unknown metabolites by mice (2). In the case of insects, Bradbury and co-workers and other investigators (3–13) have reported the mechanism of aromatization to yield $\gamma$-1,3,4,5,6-pentachlorocyclohexene (PCCH), some water-soluble metabolites, all six isomers of dichlorothiophenol, 1,2,3- and 1,2,4-trichlorobenzene (TCB), 1,2,3,4- and 1,2,4,5-tetrachlorobenzene (TECB) (14), pentachlorobenzene, and an isomer of PCCH, "iso PCCH" (12, 15–19). The interesting aspects of lindane metabolism (Fig. 7.1) are the mechanisms of aromatization to yield various chlorobenzenes and their derivatives and the ultimate fate of these products. It has not been established whether $\gamma$-PCCH is a labile intermediate or a secondary product; in house flies it appears to be a secondary product. $\gamma$-PCCH is readily metabolized by this insect to give 1,2,4-TCB and 1,2,4,5-TECB. Iso PCCH is metabolized comparatively slowly by the house fly and produces 1,2,4-TCB, 1,2,3-TCB, 1,2,4,5-TECB, and pentachlorobenzene.

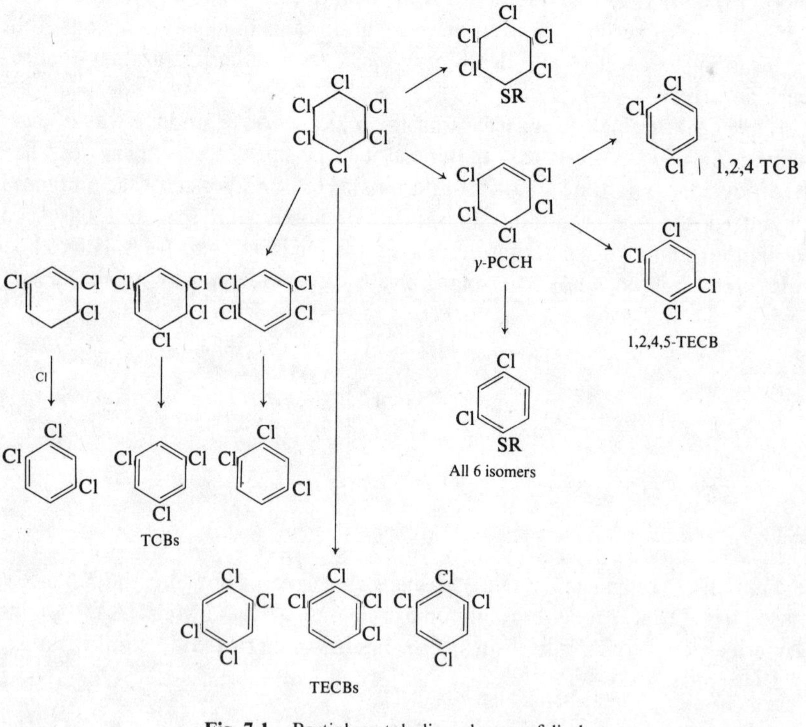

**Fig. 7.1** Partial metabolic scheme of lindane.

In mammals, as in insects, several metabolites are formed, but only the following have been identified: 1,2,4-trichlorobenzene, 2,3 and 2,4-dichlorophenols, 2,4-dichlorophenyl mercapturic acid, 2,3,5-, 2,4,5-, and 2,4,6-trichlorophenol. Others tentatively identified include pentachlorobenzene;

2,3,4,5- and 2,3,4,6-tetrachlorophenols; 2,5-, 2,6-, and 3,4-dichlorophenols; 2,3,4-, 2,3,6-, and 3,4,5-trichlorophenols; pentachlorophenol; 1,2-dichloroben-zene; 1,2,4-trichlorobenzene; 1,2,3,4- and 1,2,4,5-tetrachlorobenzene (20, 21).

## LINDANE AND HEMATOLOGIC REACTIONS

There has been some indication that lindane vaporizers may be suspect in caus-ing serious blood dyscrasias, primarily aplastic anemia, though a definite rela-tionship has not been established (22).

## REFERENCES

1. M. S. Quraishi. *Can. Entomol.* **102**: 1190 (1970).
2. K. van Asperen. *Bull. Entomol. Res.* **46**: 837 (1956).
2a. R. E. Slade. *Chem. Ind. (Lond.)* **64**: 314 (1945).
2b. R. B. Koch, L. K. Cutkomp, and F. M. Do. *Life Sci.* **8**: 289 (1969).
3. F. R. Bradbury and P. Nield. *Nature* **172**: 1052 (1955).
4. F. R. Bradbury and H. Standen. *J. Sci. Food Agr.* **6**: 90 (1955).
5. F. R. Bradbury and H. Standen. *J. Sci. Food Agr.* **7**: 389 (1956).
6. F. R. Bradbury and H. Standen. *Nature* **178**: 1053 (1956).
7. F. H. Bradbury and H. Standen. *J. Sci. Food Agr.* **9**: 203 (1958).
8. F. H. Bradbury and H. Standen. *Nature* **183**: 983 (1959).
9. R. G. Bridges. *Nature* **184**: 1337 (1959).
10. M. Ishida and P. A. Dahm. *J. Econ. Entomol.* **58**: 383 (1965).
11. Ibid. **58**: 602 (1965).
12. W. T. Reed and A. J. Forgash. *Science* **160**: 1232 (1968).
13. J. Sternburg and C. W. Kearns. *J. Econ. Entomol.* **49**: 548 (1956).
14. W. T. Reed and A. J. Forgash. *J. Agr. Food Chem.* **17**: 896 (1969).
15. *Ibid.* **18**: 475 (1970).
16. K. van Asperen and F. J. Oppenoorth. *Nature* **173**: 1000 (1954).
17. P. L. Grover and P. Sims. Biochem. J. **96**: 521 (1965).
18. W. R. Jondorf, D. V. Parke, R. T. Williams. *Biochem. J.* **61**: 512 (1955).
19. W. Koranski, J. Portig, H. W. Wohland, and L. Klempan. *Exp. Pathol. Pharmacol.* **247**: 49 (1964).
20. R. W. Chadwick and J. J. Freal. *Bull. Environ. Contam. Toxicol.* **7**: 137 (1972).
22. K. Kay. *Industrial Medicine.* **38**: 52 1969.

# SYNTHETIC ORGANIC INSECTICIDES

<table>
<tr><td rowspan="3">CHAPTER 8</td><td>Pharmacodynamics of</td></tr>
<tr><td>Chlorinated Hydrocarbons</td></tr>
<tr><td>and Their Present Status</td></tr>
</table>

The sum total of complex changes that an administered chemical undergoes in a living organism is known as pharmacodynamics or biodynamics. These terms may be used to include the biological transformations, distribution patterns of metabolites within the organism, amounts retained in the body and their dynamics, excretion and the chemical and biochemical nature of the excreted products, and the time required for these changes. A number of recent articles and reviews on the pharmacodynamics of organochlorine insecticides is available (1–4).

One important point to remember is that biodynamics is a continuous process and the metabolites isolated at any given time represent the picture at that particular moment; the detoxication process and excretion are continuously taking place and metabolic detoxication may have occurred long before the identification of metabolites at a particular time. One of the important questions that concern us is whether ingestion of small amounts of persistent chlorinated hydrocarbons have a cumulative effect or will an equilibrium or steady state be reached, the final concentration in the organism being a function of daily intake. That the latter is the case has been demonstrated in several investigations (5–7).

## PRESENT STATUS OF CHLORINATED HYDROCARBONS

Once a most trusted and respected chemical, DDT is no longer enjoying the reputation it once had. In 1964 the production of DDT was over 400,000 metric tons. Since then its production has been declining and now it is only between 200,000 and 250,000 metric tons (a little over 60% of its peak in 1959). About 25% of the present production is used for the control of vectors of human diseases, mostly mosquitoes (malaria control). Cyclodienes have made their greatest contribution as soil insecticides and are useful in the protection of grain crops that constitute more than 50% of the total global food supply. About 1.5 billion people in the Far East depend on rice as the main dietary source. Grains in most countries provide more than 50% of the total protein source in the diet. In the Far East this figure is 80%, in Africa and the Near East more than 70%, in Europe more than 60%, and in North America about 40%. The United States is among the largest grain producers in the world, producing about half of the global corn. Aldrin is mainly used as soil treatment for corn.

The ability to detect minute quantities of chlorinated hydrocarbons (in parts per billion) and the mortality in birds due to chlorinated hydrocarbons in the United States and England started the controversy surrounding these insecticides. Up to 1969 the primary concern was environmental, and a legal tolerance limit for chlorinated hydrocarbons in fish had not been established. In 1969 the U.S. Food and Drug Administration stopped the release of some canned salmon after finding samples that contained 0.3 ppm of dieldrin. These shipments were released later, but for the first time this established an administrative action level for dieldrin in fish (8). In the same year the FDA declared an interim tolerance limit of 5 ppm for DDT and analogs in fish. This set a precedent and was accompanied by the confiscation of 39,000 lb of frozen coho salmon from interstate commerce. The residues in these fish were in the range of 13–19 ppm of DDT and its analogs.

At about the same time restrictions on the use of chlorinated hydrocarbons were under consideration in several states; Arizona had already put a ban for 1 year on the use of DDT. In Illinois the State Department of Agriculture banned use of this chemical and Michigan put a ban on DDT the same year (1969).

A federal commission headed by Emil M. Mrak was appointed in April 1969 and "charged with the responsibility of gathering all available evidence on both the benefits nad (sic) risks of using pesticides, evaluating it thoroughly, and reporting their findings and recommendations to Secretary Finch" (7). The commission submitted its report in December 1969, and the commission's 14 recommendations included the following:

"Eliminate within two years all uses of DDT and DDD in the United States . . .". Exception was made in the cases of uses "essential to the preservation of human health or welfare . . . ." Unanimous approval by the Secretaries of the Departments of Health, Education, and Welfare, Agriculture, and Interior was required for such uses.

Development of suitable standards for pesticide content in food, water and air and other aspects of the environment.

In July 1969 the Federal Pest Control Program of the United States Department of Agriculture (USDA) temporarily suspended the use of organochlorine pesticides. The USDA Agriculture Research Service (ARS) also banned the use of DDT on tobacco, for the prevention of Dutch elm disease, around the house, and over water. The use of DDT on cotton (the principal use of DDT in United States) was not banned. These bans were prompted by a court action brought by the Environmental Defense Fund (EDF) and other environmentalist groups. However the exemption of cotton and some other uses from this ban was appealed against by the EDF. The ARS during this time had published a notice (35 F.R. 12293) on July 31, 1970 requesting submission of views with respect to the uses of certain pesticide chemicals, including chlordane and heptachlor. In December 1970 the Environmental Protection Agency (EPA) was established and the authority of administering the Federal Insecticide, Fungicide, the Rodenticide Act (FIFRA) of 1947 was transferred from the USDA to the EPA. At about this time the EDF won its appeal referred to earlier, and in January 1971 the EPA issued a notice that it was considering a total ban on DDT and related pesticides. This further extended the restrictions on chlorinated hydrocarbons that had been introduced during the previous 2 years. Chemical manufacturers and agricultural interests immediately started a series of legal objections because FIFRA provides an opportunity for manufacturers to appeal against a decision by calling for independent review by a scientific advisory committee or a public hearing. A consolidated hearing on DDT took place in Washington, D.C. (August 1971–March 1972). The EPA appointed Mr. Edmund M. Sweeny as hearing examiner. In these hearings the petitioners against the ban included the USDA and 27 corporate registrants. The EPA was the respondent and was aided by the EDF and other environmental agencies as intervenors. The consolidated DDT hearings, including Hearing Examiner's recommended findings, conclusions, and orders (40 CFR 164.32) (9), were issued by the EPA on April 25, 1972. Mr Sweeney's conclusions were that it could not be proved that the hazards of continuing to use DDT outweighed the benefits. Mr. Ruckelshaus also appointed a Scientific Advisory Committee for the purpose. However the Advisory Committee with the same facts took the opposite view. Mr. Ruckelshaus, Administrator put a total ban on the use of DDT in the United States after December 31, 1972, except in Public Health. Exports of DDT were not affected by the ban (I.F. and R. Docket Nos. 63 et al. June 1972) (10).

Some interesting aspects of these hearings were the cross examinations. It was suggested by one scientist that DDT might persist in seawater for 40 years. However this scientist admitted he was unaware of findings indicating that 90% of DDT and its metabolites disappear from seawater in 38 days. The calculations of the National Academy of Sciences that indicated that one-quarter of all the DDT manufactured had already found its way into the oceans were questioned and discredited. These calculations were arrived at by extrapolating the finding of DDT in rain water by two British scientists, Tarrant and Tatton (11). In the light of Sondgren's research (12) this figure was reduced to $\frac{1}{32}$ by extrapolation. The DDT hearings of 1972 have been summarized in *Bioscience* and *Nature* (13, 14).

The Environmental Protection Agency also appointed an Aldrin/Dieldrin Advisory Committee under the Chairmanship of R. D. O'Brien; the report of this committee was made available on March 28, 1972 (15). In 1974 the use of aldrin and dieldrin was banned by the EPA except for termite control, dips for roots and tops of nonfood plants, and use against clothes moths under certain circumstances (16).

On October 6, 1971, the EPA requested views with respect to uses of pesticides containing chlordane and heptachlor (FR Doc. 71-14620). On July 29, 1975, the EPA announced a Notice of Intent to Suspend production and sale by manufacturers of chlordane and heptachlor. The use of these chemicals for subterranean termite control was exempted. The reasons for this ban are, "imminent hazard to the environment," because of cancer threat. The use of existing stocks according to label instructions was allowed.

The hearings concerning the registered uses of chlordane and heptachlor began on August 12, 1975, under Presiding Judge Herbert Perlman. Forty days (17 days for supporters of suspension, and 23 days for defenders of the pesticides) are planned for presentation of the testimony and cross examination (Government Affairs Bulletin No. 6, dated August 4, 1975) (17).

Despite the bans on the use of pesticides, exceptions have been made. The EPA allowed the use of DDT for the control of the pea-weevil, *Bruchus pisorium* L., in Washington and Idaho in early 1973 because $5 million worth of crop was threatened.

## REFERENCES

1. J. Robinson. *Annu. Rev. Pharmacol.* **10**: 353 (1970).
2. G. T. Brooks. *Residue Rev.* **27**: 81 (1969).
3. G. T. Brooks. In Proceedings of the 2nd International IUPAC Congress (A. S Tahori, ed.), Vol. 6, Gordon and Breach, New York, p. 11.
4. W. Klein. In *Environmental Quality and Safety*, (F. Coulston and F. Korte, eds.). Vol. 1, Academic Press, 1972, p. 164.
5. J. Robinson. *Can. Med. Assoc. J.* **100**: 180 (1969).
6. M. L. Quaife, J. S. Winbush, and O. G. Fitzhugh. *Food Cosmet. Toxicol.* **5**: 39 (1967).
7. E. M. Mrak, Chairman. Report of the Secretary's Commission on Pesticides and Their Relationship to Environmental Health. U.S. Department of Health, Education, and Welfare, Washington, D. C., 1969, 677 pp.

8. C. C. Johnson, Statement before the Subcommittee on Energy, Natural Resources and the Environment of the Committee on Commerce, "Effects of Pesticides on Sports and Commercial Fisheries," Part 1, Ser. No. 91–15, 1969.

9. Consolidated DDT hearings, hearing examiner's recommended findings, conclusions, and orders, 40 CFR 164.32, Environmental Protection Agency, Washington, D.C., April 25, 1972.

10. Consolidated DDT hearings before the Environmental Protection Agency In re: Stevens Industries, Inc., et al. I. F. & R. Docket Nos. 63 et al. Opinion and Order of the Administrator. June 2, 1972.

11. K. B. Tarrant and J. O'G Tatton. *Nature* **219** : 725 (1968).

12. P. Sondgren. *Nature* **236**: 397 (1972).

13. T. H. Jukes. *Bioscience* **22**: 670. (1972).

14. Washington Correspondent. *Nature* **237**: 422 (1972).

15. Report of the aldrin/dieldrin advisory committee to William D. Ruckelshaus, Administrator, Environmental Protection Agency (R. D. O'Brien, Chairman), March 28, 1972.

16. U.S. Environmental Protection Agency, Information Section, News Release R-558. Washington, D.C., 1974.

17. Government Affairs Bulletin No. 6, August 4, 1975 Washington, D.C.

# OTHER CHEMICALS

# CHAPTER 9

Organic and Inorganic Insecticides
of Minor Importance

## DINITROPHENOLS

Among other organic chemicals the dinitrophenols have been used as insecticides. Potassium-dinitro-$o$-cresylate was marketed in Germany in 1892 and has the distinction of being the first commercial synthetic organic insecticide. Among the compounds used on a small scale at present are DNOC, 4,6-dinitrocresol; DNOSB, dinoseb, 4,6-dinitro-$o$-$sec$ butylphenol; and DNOCHP, 4,6-dinitro-$o$-cyclohexylphenol.

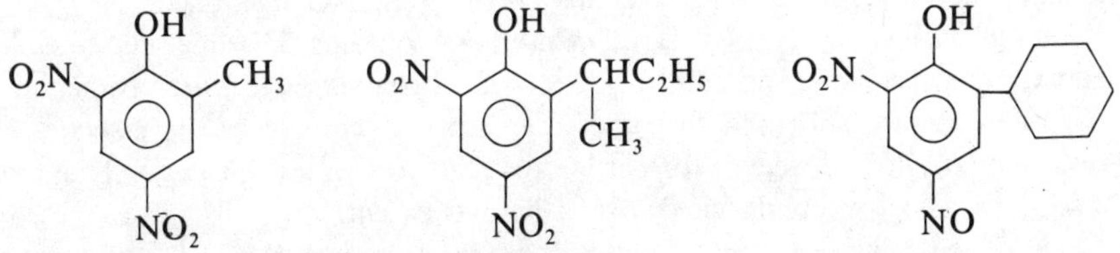

DNOC  DNOSB  DNOCHP

These compounds uncouple oxidative phosphorylation, thus greatly increasing the rate of respiration in the poisoned insect; and death is caused as a result of metabolic exhaustion. These chemicals also are highly toxic to mammals; in the case of DNOC, $LD_{50}$ oral for rat is 26–65 mg/kg. A number of agricultural workers have died as a result of poisoning by this chemical.

137

## ORGANOTHIOCYANATES

The first widely used synthetic organic insecticide was Lethane 384, $\beta$-butoxy-$\beta'$-thiocyanodiethyl ether. It was marketed in 1932 and gave a rapid knock-down and control of household insects. Several compounds containing the thiocyano (SCN) or isothiocyano (CNS) groups possess insecticidal properties. *Thanite*, isobornyl thiocyano acetate, has also been marketed as an insecticide.

## INORGANIC INSECTICIDES

### Arsenicals

Compounds of arsenic have been used as stomach poisons for insects, and to a limited extent, are being used even now. Before the advent of synthetic organic insecticides, they were of considerable importance. The first arsenical *paris green*, $Cu(C_2H_3O_2)_2 \cdot 3Cu(AsO_2)_2$, was used against the Colorado potato beetle, *Leptinotarsa decemlineata* (Say), around 1865.

The biological activity of an arsenical is due to the arsenic content and is proportional to the metallic arsenic it contains. Two acids of arsenic, the tri-valent arsenious acid $[(As(OH)_3]$ forming the arsenites and the pentavelent arsenic acid $[O{=}As(OH_3]$ forming the arsenates, are known. The arsenites are comparatively more toxic to insects, but they are also more phytotoxic. Their use therefore is confined to baits. The arsenates are used against agricultural pests.

*Lead arsenate*, $PbHAsO_4$, has 21.6% arsenic and was developed in the early 1890s for control of the gypsy moth. It has been used against the codling moth of apples and other chewing insects. It, however, produces plant burns. For foliage, therefore, basic lead arsenate $Pb_4(PbOH)(AsO_4)_3$ is used.

*Calcium arsenate*, $(Ca_3(AsO_4)_2)_3 \cdot Ca(OH)_2$, contains 25% metallic arsenic and was used against the cotton boll weevil, *Anthonomus grandis* Boheman.

Arsenicals are still used to some extent for the control of the insecticide-resistant cotton boll weevil, for cattle dips, and for some fruit pests. They are also used as mosquito larvicides to a limited extent.

The arsenicals are generally highly toxic to mammals, their $LD_{50}$ for mammals varies between 18 and 25 mg/kg.

### Fluoride Insecticides

Salts of hydrofluoric acid (HF), fluosilicic acid $(H_2SiF_6)$, and fluoaluminic acid $(H_3AlF_6)$ have been used as insecticides. As with arsenic compounds, the insecticidal action of these compounds is approximately proportional to their fluorine content and the extent to which the digestive enzymes of the insect solubilize them. Water-soluble fluorides are phytotoxic.

*Cryolite*, $Na_3AlF_6$, a naturally occurring mineral, has been used as a stomach poison. It is also produced synthetically, in which case it has a lower density and is fluffy.

Sodium fluoride, NaF, and sodium fluosilicate, $Na_2SiF_6$, have been used as insecticides against nonagricultural pests. They are phytotoxic.

Fluorine compounds vary in their mammalian toxicity; for instance, *fluenetil*, 2-fluoroethyl-biphenyl-4-ylacetate, a commercial insecticide and acaricide marketed in Europe, has an $LD_{50}$ for rat of 8.7 mg/kg, for mice 57 mg/kg, and for monkeys 75 mg/kg.

## SORPTIVE DUSTS AND OTHER CHEMICALLY INERT SUBSTANCES

Fine dusts and ashes have long been used by natives in several countries for the protection of grain. For detailed discussion of sorptive dusts for pest control see the fine review by Ebeling (1). Commercial materials like diatomaceous earth and pyrophyllite (a hydrous aluminium silicate) have been used for insect control, though their principal use now is as diluents for pesticides. Some of the inert dusts include *Dri-Die*®, a silica aerogel with ammonium fluosilicate to 3% fluorine content; *Cal-O-Sil*® (Silica, fumed), 99+% pure, ultrafine particle dry powder; *Micro-Cel-C*® (Silicates, synthetic dry); precipitates of hydrated calcium silicate, and silicon dioxide.

The gels are amorphous powders consisting mostly of $SiO_2$. They have low density and high porosity and therefore are called aerogels. When used against insects, they cause abrasion and disruption of the wax layer resulting in loss of water from the insect. Ebeling (1) believes that disruption by physical abrasion is not necessary, but powder particles resting on an insect cuticle or a synthetic film can cause disruption of the wax layer. Only those powders that possess this property of disrupting wax films are insecticidal. These powders usually possess a large surface area per unit weight and their pore sizes are greater than 20 Å. There is a possibility that these pores may also disrupt the wax layer by capillary action.

Ebeling and Wagner (2) also surmise that silica aerogels containing fluorides possibly add fluoride toxicity to dust desiccation. Ebeling and Rierson (3), Ebeling and Wagner (2), and Ebeling et al. (4) have shown the efficacy of dusts in repelling and controlling cockroaches and termites.

Boric acid, $H_3BO_3$, has also been used for the control of cockroaches (1, 5, 6).

Ebeling et al. (7) report that when American and German cockroaches were treated with powders containing silica gel, aerogel, and boric acid, they ingested large quantities of the powder while preening their appendages. The powder was mainly concentrated in the crop because the dentricles of the armarium of the proventriculus prevented the posterior movement of the powder. This resulted in the distention of the crop.

## REFERENCES

1. W. Ebeling. *Annu. Rev. Entomol.* **16**: 123 (1971).
2. W. Ebeling and R. E. Wagner. *J. Econ. Entomol.* **52**: 190 (1959).
3. W. Ebeling and D. A. Rierson. *Calif. Agr.* **23**: 4 (1969).
4. W. Ebeling, R. E. Wagner, and D. A. Rierson. *Pest Control Operators News* **25**: 16 (1965).
5. R. C. Moore. *J. Econ. Entomol.* **65**: 458 (1972).
6. C. C. Wright and R. C. Hillman. *op. cit.* **66**: 1075 (1973).
7. W. Ebeling, D. A. Rierson, R. J. Pence, and M. S. Viray. *Pestic. Biochem. Physiol.* **5**: 81 (1975).

# SYNTHETIC INSECT CONTROL AGENTS

# CHAPTER 10 | Insect Chemosterilants

The possibility of using sterile males for the control of insects having a low population density was first considered by Knipling (1). This was experimentally established when the screwworm was eradicated from Florida and the southeastern United States (2). However the sterilization in this case was achieved by the use of gamma rays from a $Co^{60}$ source; this eradication provided an impetus for research workers, and the U.S. Department of Agriculture initiated an extensive program for the investigation of chemosterilants. Prior to this work, during war time research on mustard gas, fruit flies had been accidentally sterilized by chemical mutagens in England (3). In the early 1950s investigations on chemosterilants were done with antimetabolites that caused sterility in females by inhibiting egg production. However in the late 1950s several radiomimetic chemicals were known and became candidates for insect chemosterilization and this was the period when the USDA initiated an extensive screening program during which alkylating agents were also tested. Earlier studies on chemosterilants have been ably summarized (4–9).

The alkylating agents act by replacing the hydrogen atom of a molecule by an alkyl group. They bear no similarity in structure but share the property of combining with electronrich centers, that is, they are electrophiles and therefore accept an electron pair from carbon in an organic reaction. Alkylating agents readily react with nucleophiles which are polar compounds that donate an electron pair to the carbon in an organic molecule. The agents used in biological experiments include azaridines, methane sulfones, and mustards. They possibly react with some target nucleophile(s), especially in the insect gametes, essential in insect reproduction. Thus insect chemosterilants have a mode of action different from human birth control pills.

Some sterilants, which are not necessarily alkylating agents, attack rapidly dividing cellular systems by preference; since in adult insects these are found almost exclusively in the reproductive organs, chemosterilization is brought about. Affinity for actively proliferating tissue is characteristic for most antineoplastic agents; therefore several insect chemosterilants have been found among them.

## ALKYLATING AGENTS, THE AZARIDINES

Some derivatives of azaridines in which the ring nitrogen is attached to a PO, PS, or PN moiety have proved to be effective insect sterilants. Generally di- and polyaziridines are more effective than the monoazaridines. Three aziridinyl compounds, namely, apholate, tepa, and aphomide were the first successful compounds of this series. Of these the first two were very active. Later metepa and thiotepa were also added to the group.

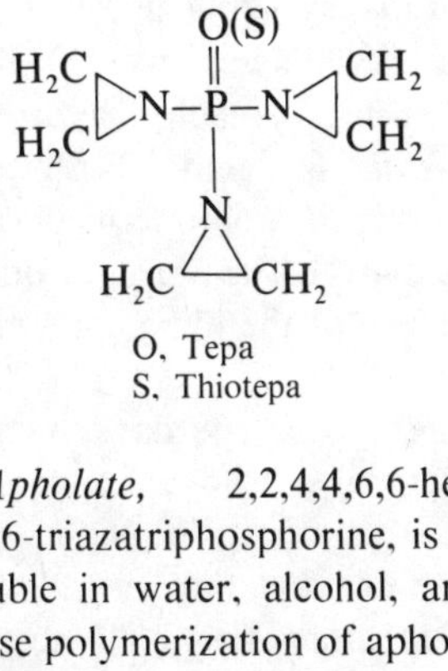

O. Tepa
S. Thiotepa

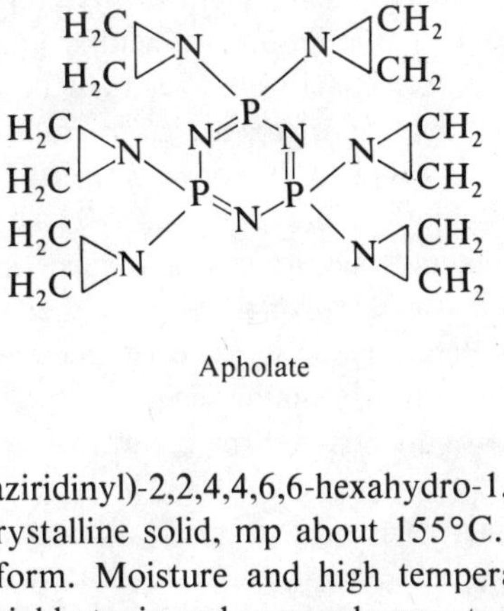

Apholate

*Apholate,*   2,2,4,4,6,6-hexakis-(1-aziridinyl)-2,2,4,4,6,6-hexahydro-1,3,5,-2,4,6-triazatriphosphorine, is a white crystalline solid, mp about 155°C. It is soluble in water, alcohol, and chloroform. Moisture and high temperature cause polymerization of apholate. It is highly toxic and causes damage to bone marrow and germinal tissue. Apholate is easily absorbed through skin and extreme care is necessary in handling it. When fed at 0.5% in food apholate completely eliminated hatch in the house fly. It sterilizes both sexes.

*Tepa,* tris-(1-aziridinyl) phosphine oxide, or aphoxide, is a colorless, odorless, crystalline solid with a melting point of 41°C. It is hygroscopic and therefore highly soluble in water. It is soluble in organic solvents. In aqueous solutions it is unstable. Tepa is highly toxic and when used for cancer treatment at 1.6 mg/kg it possibly contributed to the deaths of patients (10). It causes sterility in the house fly in doses as low as 0.1% in food. Both sexes are sterilized.

*Metepa*, tris-(2-methyl-1-aziridinyl)phosphine oxide, is a straw-colored liquid soluble in water and all organic solvents. It is comparatively stable in storage and is effective against both sexes.

*Thiotepa*, the sulfur analog of tepa, is also an insect chemosterilant.

## NONALKYLATING CHEMOSTERILANTS

Borkovec (11) has divided the nonalkylating chemosterilants into three major groups of structurally related compounds and a fourth group of miscellaneous compounds: phosphoramides; melamines; dithiobiurets and dithiazolium salts; and miscellaneous compounds. We follow the same classification in our discussion.

### Phosphoramides

When the aziridinyl groups in alkylating sterilants are replaced by alkylamino groups, a new class of chemosterilants is obtained (12). The two important members of this group are hempa and thiohempa.

*Hempa* effectively sterilizes male house flies and its physiological effects on the reproductive organs are comparable to those of alkylating agents (13). This is probably due to its conversion in the insect to a highly reactive hydroxymethyl intermediate that possibly acts as an alkylating agent;. Terry and Borkovec (14) synthesized and successfully tested one of such intermediates. Another possible explanation of the sterilizing activity of hempa and other nonalkylating analogs could be the steric resemblance of the dimethyl amino group to the aziridinyl ring.

When administered to rats in low dosages (2 mg/kg) for 169 days, hempa produced no apparent toxic effects (15). However at higher dosages its toxicity resembled that of tepa in qualitative and quantitative aspects, though bone marrow depression, which is a characteristic effect of alkylating agents, was not observed with hempa (15, 16). Its acute oral $LD_{50}$ for rats is 2600 mg/kg. Tepa is about 50 times as effective in sterilizing house flies as hempa, but the toxicities of both compounds to the house fly are about equal (17).

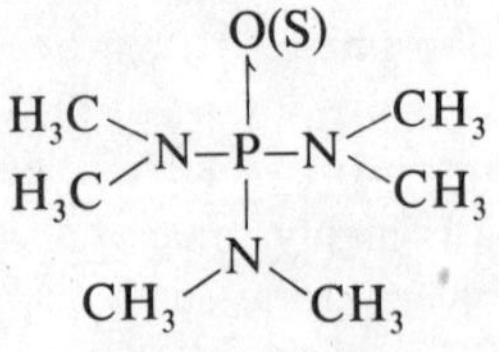

$$O(S)$$
$$H_3C \diagdown N-P-N \diagup CH_3$$
$$H_3C \diagup \quad | \quad \diagdown CH_3$$
$$N$$
$$CH_3 \diagup \diagdown CH_3$$

O. Hempa
S. Thiohempa

Hempa has also found increasing use as a solvent in industry in the last few years, and recent investigations carried out on its chronic toxicity (17a) indicate that hempa is a possible carcinogen. In Charles River CD strain rats, two groups receiving 400 and 4000 ppm (by volume) dose of the chemical by inhalation, 6 hours a day, 5 days a week, developed squamous cell carcinoma in 8 months. However an excess of animals in these groups had died between the sixth and eighth month as a result of degenerative changes in the convoluted tubules of the kidney.

A fatal pneumonia in rats is also produced by inhalation of hempa. Possibly this chemical activates the usually rather benign organisms responsible for the disease which is occasionally found in laboratory reared animals.

*Thiohempa* is the sulfur analog of hempa and is an effective insect chemosterilant.

## Melamines, Substituted Triamino-1,3,5-triazines

The active compounds in this group are "... in some respect analogs of well-known alkylating agents in which the aziridinyl groups were replaced by dimethylamino group ..." (11). They are most active against the house fly. Possibly activation, as indicated in the case of methyl phosphoramides, may take place (18–20).

## Dithiobiurets and Dithiazolium salts

The Hercules Powder Company first developed a compound belonging to this group. It is a dimethylamino compound, 3,5-bis(dimethylamino)1,2,4-dithiazolium chloride. A series of homologs has since been synthesized and some active compounds have been discovered.

## Miscellaneous Compounds

Other compounds like folic acid antagonists, such as aminopterin; mitotic poisons, for example, the nitrogen mustard; and compounds that interfere with the division of the cell nucleus have been found to possess sterilizing properties. The sterilizing action of triphenyltin hydroxide is in doubt. Other compounds reported to possess this property include 2,6-dimethylhydroquinone and 5-fluoroorotic acid, a folic acid inhibitor.

The antibiotic anthramycin sterilizes both drosophila and the house fly (21–23); it inhibits the biosynthesis of nucleic acids and therefore is a carcinostatic agent.

Chang et al. (24) have tested the chemosterilant activity of several antineoplastic agents and found L-asparaginase to be a good chemosterilant. This, according to these authors, is the first reported sterilant activity of an enzyme or a protein.

Borkovec and co-workers have tested a number of chemicals for their sterilizing effects against the boll weevil and other insects; these include the azaridines, s-triazines, phosphorus amides, antineoplastic agents, and other miscellaneous chemicals (25–28).

In general, the nonalkylating chemosterilants are more species specific and have a narrower range of activity as compared to their alkylating analogs.

## METHODS FOR TESTING CHEMOSTERILANTS

Introducing the chemosterilant into the diet of the candidate is one of the commonest ways of testing a chemosterilant. In the case of mosquitoes immature stages have been treated to produce sterile adults. Topical application, injection, and exposing insects to treated surfaces also have been successfully tried. Dipping of pupae or puparia in chemosterilant solution is another method that has been employed. LaBrecque has dealt with the subject in detail (29).

In the case of chemosterilants with high biological activity and volatility that can be absorbed through the insect, fumigation can be effectively utilized for sterilization (30–32). Chang et al. (33) reported on the chemosterilization of male house flies by fumigation and found that fumigation in a circulating atmosphere could treat a large number of insects in a uniform and rapid manner. Chang et al. (34) further examined the effectiveness of 41 bis-(1-aziridinyl) phosphine oxides and sulfides by circulatory fumigation and found that in the homologous series, the methyl and ethyl compounds were more effective than the higher alkyl or aryl compounds. They reported that in analogous pairs, the phosphine sulfides were more active than the oxides.

## POSSIBLE FIELD USES OF CHEMOSTERILANTS

Whereas chemosterilants can be used to sterilize mass-produced insects, they can be more efficiently employed in conjunction with attractants and food in nature. Thus a large number of individuals can be successfully sterilized with mimimum effort to control insect population. Investigations of this sort have been conducted, with encouraging results, in the case of the house fly on refuge dumps on an isolated island (35), in small towns (36), and in pit privies (37). However, because of the highly toxic nature of chemosterilants, extreme care is necessary even in the planning of pilot tests. In fact, the author feels that their carcinogenic property and the fact that they alkylate DNA makes them highly undesirable even for pilot tests.

# REFERENCES

1. E. F. Knipling. *J. Econ. Entomol.* **48**: 459 (1955).
2. E. F. Knipling. *Sci. Amer.* **203**: 54 (1960).
3. C. Auerbach and J. M. Robson. *Nature* **147**: 302 (1946).
4. K. R. S. Ascher. *World Rev. Pest Control* **3**: 7 (1964).
5. G. C. LaBrecque and A. B. Borkovec. *Annu. Rev. Entomol.* **9**: 269 (1964).
6. G. C. LaBrecque, and C. N. Smith, eds. *Principles of Insect Chemosterilization*, Appleton Century-Crofts, New York, 1968.
6a. M. D. Proverbs. *Annu. Rev. Entomol.* **14**: 81 (1969).
7. A. B. Borkovec. *Insect Chemosterilants*, Interscience, New York, 1966.
8. A. B. Borkovec. *Ann. N.Y. Acad. Sci.* **163** (Art. 2): 860 (1969).
9. K. R. S. Ascher. *World Rev. Pest Control* **9**: 140 (1970).
10. G. C. LaBrecque. *J. Econ. Entomol.* **54**: 684 (1961).
11. A. B. Borkovec. In *Insecticides* (A. S. Tahori, ed.), Vol. 1 Gordon and Breach, New York, 1972, 496 pp.
12. S. C. Chang, P. H. Terry, and A. B. Borkovec. *Science* **144**: 57 (1967).
13. A. B. Morgan. *Ann. Entomol. Soc. Amer.* **60**: 812 (1967).
14. P. H. Terry and A. B. Borkovec. *J. Med. Chem.* **13**: 782 (1970).
15. L. D. Shott, A. B. Borkovec, and W. A. Knapp, Jr. *Toxicol. Appl. Pharm.* **18**: 499 (1971).
16. R. Kimbrough and T. B. Gaines. *Nature* **211**: 146 (1969).
17. S. C. Chang and A. B. Borkovec. *J. Econ. Entomol.* **62**: 1417 (1969).
17a. J. Zapp, Jr. *HMPA: a Possible Carcinogen*, Letters to the Editor, *Science* **190**: 422 (1975).
18. A. B. Borkovec. In *Residue Rev.* **53**: 67 (1974).
19. S. C. Chang, A. B. DeMilo, C. W. Woods, and A. B. Borkovec. *J. Econ. Entomol.* **61**: 1357 (1968).
20. S. C. Chang, C. W. Woods, and A. B. Borkovec. *J. Econ. Entomol.* **63**: 1510 (1970).
21. S. B. Horowitz. "Anthramycin", In *Progress in Molecular and Subcellular Biology* (F. Hahn, ed.), Vol. 2. Springer, New York, 1972.
22. J. R. Barnes, J. Felig, and M. Mitrovic. *J. Econ. Entomol.* **62**: 902 (1969).
23. A. B. Borkovec, S. C. Chang, and S. B. Horowitz. *J. Econ. Entomol.* **64**: 983 (1971).
24. S. C. Chang, A. B. Borkovec, and B. H. Braun. *Trans. N.Y. Acad. Sci. Ser. II* **36**: 101 (1974).
25. A. B. Borkovec, C. W. Woods, and D. G. McHaffey. *J. Econ. Entomol.* **65**: 1543 (1972).
26. A. B. DeMilo, A. B. Borkovec, and D. G. McHaffey. *op. cit.* **65**: 1548 (1972).
27. P. H. Terry and A. B. Borkovec. *op. cit.* **65**: 1550 (1972).
28. A. B. DeMilo, R. L. Fye, and A. B. Borkovec. *op. cit.* **66**: 1007 (1973).
29. G. C. LaBrecque. In ref. 6 above.
30. H. Kido and E. M. Stafford. *J. Econ. Entomol.* **59**: 1064 (1966).
31. A. B. Borkovec, S. Nagasawa, and H. Shinohara. *J. Econ. Entomol.* **61**: 695 (1968).
32. A. J. Gadallah and E. M. Stafford. *op. cit.* **64**: 1521 (1971).
33. C. S. Chang, A. B. Borkovec, C. W. Woods, and P. H. Terry. *op. cit.* **66**: 23 (1973).
34. S. C. Chang, A. B. Borkovec, C. W. Woods, and B. H. Braun. *op. cit.* **67**: 1 (1974).
35. G. C. LaBreque, C. N. Smith, and D. W. Meifert. *J. Econ. Entomol.* **55**: 449 (1962).
36. G. Saccà, A. Scirocchi, G. M. DeMeo, and M. I. Mastrilli. *Atti Soc. Peloritana Sci. Fis. Mat. Nat.* **12**: 457 (1966).
37. D. W. Meifert, G. C. LaBreque, C. N. Smith, and P. B. Morgan. *J. Econ. Entomol.* **60**: 480 (1967).

# SYNTHETIC INSECT CONTROL AGENTS

# CHAPTER 11

Attractants, Repellents,
and Antifeedants

The attraction of insects to food, egg-laying sites, and mates is well known, and recently great interest has been generated in using attractants for insect control. The two main approaches in the search for insect attractants have been empirical screening, and extraction and identification of natural pheromones. In the former approach, a large number of chemicals are tested and those found attractive are further investigated. Related compounds are synthesized and tested again to find a potent lure. Some of the synthetic attractants are given in Table 11.1 (1).

## PRACTICAL APPLICATION OF ATTRACTANTS

At present the first three of these synthetic attractants are being used for the detection of the three worst insect pests in the world, and their detection and subsequent eradication has been greatly facilitated. In 1929, before the discovery of the attractant for the Mediterranean fruit fly, the introduced pest had spread over a large part of central Florida before it was discovered. To achieve eradication, fruit crops and other hosts of the fly were destroyed, and molasses with arsenicals was used as bait. Synthetic attractants have greatly facilitated the detection of pests, though their direct use in control has yet to be established. However Steiner et al. (2) eradicated the oriental fruit fly, *Dacus dorsalis*, from a Pacific island, Mariana Islands, (33 miles² in area) by using a

## Table 11.1

| Common and/or scientific name | Formula | Insect attracted |
|---|---|---|
| Trimedlure (t-Butyl-2-methyl-4-chlorocyclo-hexanecarboxylate) | 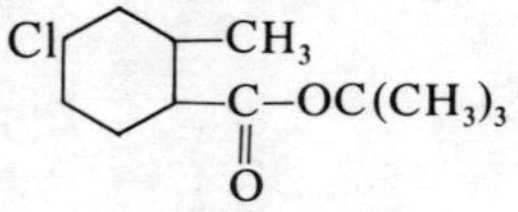 | Mediterranean fruit fly *Ceratitis capitata* (Wiedemann) |
| Cue-lure (p-Acetoxyphenethyl-methyl ketone) | 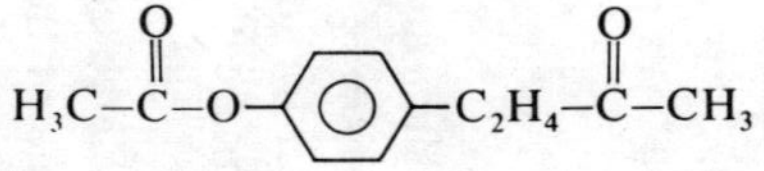 $H_3C-C-O-\underset{\phantom{O}}{\underset{O}{\parallel}}\!\!-C_6H_4-C_2H_4-C-CH_3$ | Melon fly *Dacus cucurbitae* Coq. |
| Methyl eugenol | 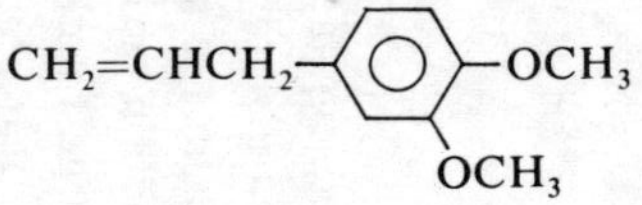 $CH_2{=}CHCH_2{-}C_6H_3(OCH_3)(OCH_3)$ | Oriental fruit fly *Dacus dorsalis* Hendel |
| Heptyl butyrate | $CH_3(CH_2)_5CH_2O(O)CCH_2CH_2CH_3$ | Yellow jacket *Vespula* spp. |
| Ethyl 3-isobutyl-2,2-dimethyl cyclopropane carboxylate | 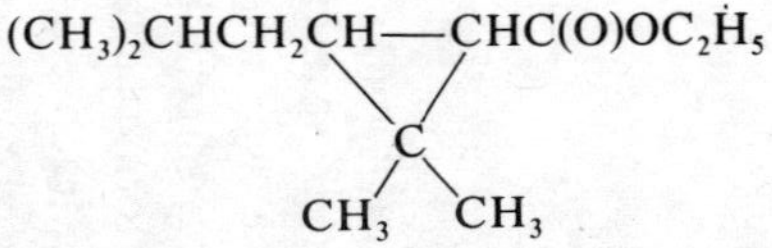 $(CH_3)_2CHCH_2CH{-}CHC(O)OC_2H_5$ with cyclopropane C(CH_3)(CH_3) | Coconut rhinoceros beetle *Oryctes rhinoceros* (L.) |

combination of naled and methyl eugenol, a powerful attractant for the male fly.

In southern California, traps baited with methyl eugenol are now permanently maintained in a grid of about one per 1.5 km. Whenever male oriental fruit flies are found the number of traps is increased to more effectively assess the density of the introduced pest. Appropriate elimination measures are taken by using a combination of methyl eugenol and a quick-acting insecticide. Such eliminations were achieved in 1966–1967, 1969, and 1970 (3).

Attractants are highly specific and thus work for the pest species only; they are therefore ecologically most acceptable. But their specificity also makes them difficult to discover and, other than the fact that most attractants are esters, there appears to be no similarity in their structures. In the absence of any clues, patience, perseverance, and empirical testing of a large number of chemicals are the only known criteria. Even then some attractants can be missed because vapor concentration may not be right, release rate may affect the response of the insect, or the environmental or physiological conditions may not be ideal for the attraction of the insect.

Screening programs have, however, produced successful attractants. McGovern et al. (4) found that 1,4-benzodioxan-2-carboxylate was a potent attractant for the European chafer, *Amphimallon majalis* (Razoumowsky), and was more persistent and effective than butyl sorbate which had been used previously. Other attractants discovered in this manner include diethyl dioxaspiro undecane for the yellow fever mosquito, *Aedes aegypti*, and methyl cyclohexanepropionate for the Japanese beetle, *Popillia japonica* Newman.

Sometimes mixtures of chemicals work better as attractants than each chemical used alone and considerable research is underway to discover active combinations. Thus a mixture of geraniol, phenethyl butyrate, and eugenol in the ratio of 9:9:2 works better than the standard mixture of phenyl butyrate and eugenol (9:1) for the Japanese beetle. In the case of pink bollworm it was reported that propylure, 10-*n*-propyl-*trans*-5,9-tridecadienyl acetate, was the sex pheromone, that deet, *N,N*-diethyl-*m*-toluamide, was the sex pheromone activator, and that they worked in combination. It was also reported that the synethetic material hexalure, *cis*-7-hexadecenyl acetate, proved better in attracting the males than a combination of the two chemicals mentioned earlier. However Hummel et al. (4a) investigated the matter further and identified the sex pheromone of the insect as a mixture of *cis,cis* and *cis,trans* isomers of 7,11-hexadecadienyl acetate.

## ANTIFEEDANTS AND REPELLENTS

Antifeeding agents do not kill insects by their action, but either prevent an insect from feeding or inhibit the continuation of feeding. In the presence of antifeeding compounds, an insect would not feed even if ample food were present right next to it. Locusts would starve to death in the presence of food if the food were treated with azadirachtin, a product extracted from the Indian lilac, *Melia azadirachta* L., syn. *Melia indica* A. Juss. (5). Haskell and Mordue (6) tested azadirachtin at a concentration of 1 ppm on filter papers and found that a 100% antifeedant effect was obtained in the case of fifth instar hoppers of the desert locust, *Schistocerca gregaria* Forsk. In the United States a closely related tree, *Melia azedarch* (locally called chinaberry tree), was tested for its antifeedant activity by McMillian and Starks (7) and McMillian et al. (8). They found that chloroform extract of the leaves possessed pronounced antifeedant and retardant effect against the larvae of the corn earworm, *Heliothis zea* (Boddie). and the fall armyworm, *Spodoptera frugiperda* (J. E. Smith), at a concentration of 0.03% in diet. The feeding of the leafworm *Spodoptera littoralis* is inhibited by an alkaloid, insoboldine, obtained from the climbing shrub *Cocculus trilobus* (9).

Organotin compounds have been tested as antifeedants with limited success and some pronounced failures (10). However Marfurt and Toscani (11) in Argentina reported successful protection of apples with fentin acetate against

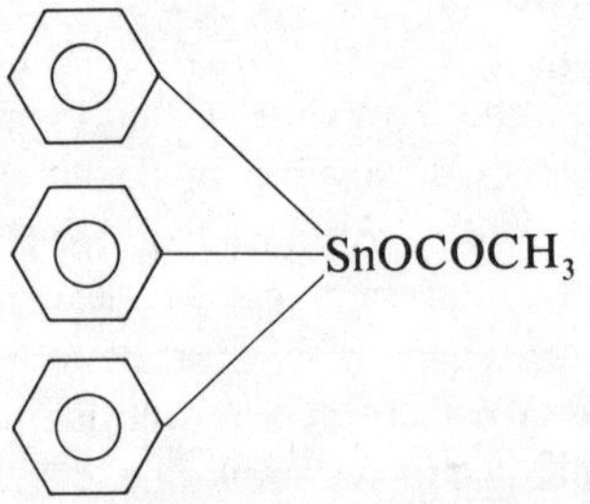

the codling moth and the oriental fruit moth *Grapholitha molesta* (Busck). Ascher has reviewed the antifeeding properties of organotin compounds (12). The compound 4'-(3,3-dimethyl-1-triazeno) acetanilide or AC 24055 was introduced by American Cynamid as an antifeedant, but it has enjoyed mixed fortunes. Loschiavo (13) found it a somewhat effective antifeedant for *Tribolium confusum* Jacquelin du Val, but it was also toxic. However an interesting aspect was that *Oryzaephilus surinamensis* (L.) was inhibited from ovipositing on treated surfaces. David and Loschiavo also report the toxicity of this chemical for *Drosophila melanogaster* Meigen (14).

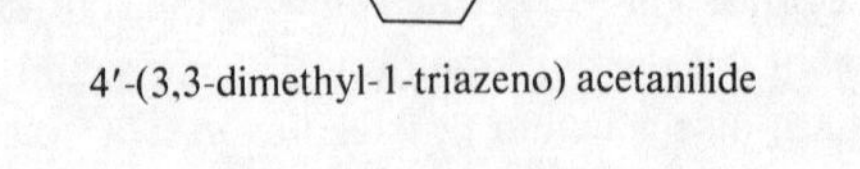

4'-(3,3-dimethyl-1-triazeno) acetanilide

Beck (15) and Klun and Brindley (16) have extensively studied the corn borer-resistance factor in corn plants and initially found two factors, resistance factor A, RFA (ether soluble), and resistance factor B, RFB (ether insoluble). From RFA two chemicals, namely, 6-methoxybenzoxazolinone (6-MBOA) and 2,4-dihydroxy-7-methoxy-1,4-benzoxazine-3-one (DIMBOA), were isolated and found to be effective in the protection of the young plant and some lines of older plants that retained their juvenile resistance against the borer.

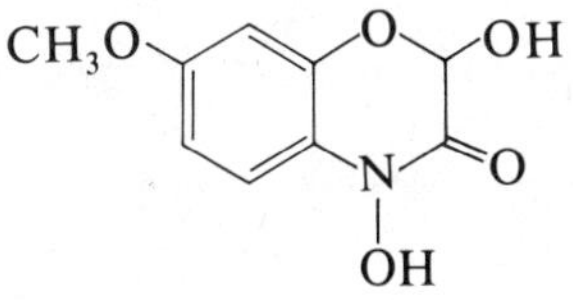

DIMBOA 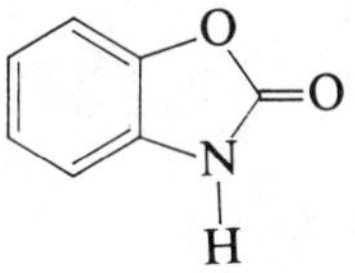                                              6-MBOA

Though insect repellents have not been used for insect control per se their advantages in the protection of man and materials are obvious. These chemicals have played a very important role in the protection of man against blood-sucking insects and vectors of diseases. During World War II dimethyl phthalate and 2-ethyl-1,3-hexanediol (612) were widely used by the United States Armed Forces in the South Pacific as repellents against mosquitoes for the prevention of malaria and other mosquito-borne diseases. Dibutyl phthalate and benzyl benzoate, both mite repellents, were used for the impregnation of clothes as a prevention against scrub typhus carried by *Trombicula* spp. (17, 18).

*Deet*, *N,N*-diethyl *m*-toluamide is another useful repellent that is the active ingredient in commercial repellents used for the protection of man against mosquitoes, biting flies, and fleas.

## REFERENCES

1. M. Beroza. In *Pest Control Strategies for the Future*, National Academy of Sciences. Washington, D.C. 1972, p. 226.
2. L. F. Steiner, W. C. Mitchell, E. J. Harris, T. T. Kozuma, and F. S. Fujitomo. *J. Econ. Entomol.* **58**: 961 (1965).
3. D. L. Chambers, R. T. Cunningham, R. W. Lichty, and R. B. Thrailkill. *Bioscience* **24**: 150 (1974).
4. T. P. McGovern, B. Friori, M. Beroza, and J. C. Ingangi. *J. Econ. Entomol.* **63**: 168: (1970).
4a. H. E. Hummel, L. K. Gaston, H. H. Shorey, R. S. Kaae, K. J. Byrne, and R. M. Silverstein. *Science* **181**: 873 (1973).
5. J. H. Butterworth and E. D. Morgan. *Chem. Commun.* **23**: (1968).
6. P. T. Haskell and J. Mordue. *Entomol. Exp. Appl.* **12**: 591 (1969).
7. W. W. McMillian and K. J. Starks. *Ann. Entomol. Soc. Amer.* **59**: 516 (1966).

8.  W. W. McMillian, M. C. Bowman, R. L. Burton, K. J. Starks, and B. R. Wiseman. *J. Econ. Entomol.* **62**: 708 (1969).

9.  K. Munakata. "Insect Antifeedants in Plants". In *Control of Insect Behavior by Natural Products*. Academic Press, New York, 1970.

10.  J. B. R. Findlay. "The Use of Antifeeding Compounds as Protectants against Damage to Plants". M.Sc. Thesis, University of Pretoria, 1968.

11.  T. A. Marfurt and H. A. Toscani. *Delta Paraná* **6**: 45 (1966–1967).

12.  K. R. S. Ascher. *World Rev. Pest Control* **9**: 140 (1970).

13.  S. R. Loschiavo. *J. Econ. Entomol.* **62**: 102 (1969).

14.  J. David and S. R. Loschiavo. *Can. Entomol.* **106**: 1009 (1974).

15.  S. D. Beck. *Annu. Rev. Entomol.* **10**: 207 (1965).

16.  J. A. Klun and T. A. Brindley. *J. Econ. Entomol.* **59**: 711 (1966).

17.  C. N. Smith. *Misc. Pub. Entomol. Soc. Amer.* **7**: 199 (1970).

18.  R. L. Metcalf and R. A. Metcalf. In *Introduction to Insect Pest Management* (R. L. Metcalf and W. Luckman, eds.). Wiley, New York, 1975.

# INSECT CONTROL AGENTS OF INSECT, PLANT AND MICROBIAL ORIGIN

# CHAPTER 12

# Pheromones and Related Chemicals

From the early days in the history of natural science the powerful attraction an unmated female provides for the male insect has fascinated scientists. Coppel et al. (1), in field investigations, found that one introduced female pine sawfly, *Diprion similis,* attracted over 11,000 males. Besides sex attractants there are other chemicals, released externally by insects that influence the behaviour of individuals belonging to the same species. Such chemicals, including sex attractants, are called pheromones. In this category also belong alarm substances of various species of ants and the "queen substance" (9-oxodec-*trans*-2-enoic acid) secreted by the queen bee, which is a sex pheromone and, when passed on to workers, prevents the development of their ovaries. Karlson and Butenandt (2) have adequately covered the literature on pheromones through 1957. Later reviews (3–9) have amply summarized the research on the subject. Since important literature has been covered in these reviews only research pertinent to the present discussion is cited in this chapter.

Among orders of insects in which pheromonal communication has been established are included Orthoptera, Isoptera, Hemiptera, Homoptera, Coleoptera, Lepidoptera, Diptera, and Hymenoptera. Some recent pheromones in insects belonging to these orders have been described by MacConnell and Silverstein (10). Study in controlling insects by the use of attractants did not receive much impetus until the shortcomings of the use of the synthetic organic insecticides became apparent in the late 1950s.

## SEX ATTRACTANTS

Among pheromones, sex attractants present the greatest potential for pest control. They are emitted for the purpose of mating. Hundreds of insects have been shown to possess sex pheromones and indeed many more will be added to the list as research continues. The chemical identification of pheromones is extremely difficult because a large number of insects are needed to obtain even a minute quantity of the chemical, and when working with minute quantities, isolation and characterization are not easy. Recent improvements in techniques and instrumentation like gas and thin layer chromatography, infrared, ultraviolet, nuclear magnetic resonance spectroscopy (NMR), and mass spectrometry have made possible the identification of minute quantities (fractions of milligrams) of chemicals. Naturally, when one is dealing with such small quantities, exceptional skills are needed. It is therefore not surprising that the structures of most of the pheromones have been determined only during the last 10 years.

Despite these techniques the tasks of isolation and characterization of pheromones are further complicated because of other biological factors. For instance, the sex attractant of boll weevil has four exotic constituents (11). Bark beetles present an interesting case of the difficulties involved (12–14). The females release pheromones after landing on a tree; these, in combination with the exudates from the tree as a result of the attack, attract other beetles, with males predominating. The males release another pheromone which discourages the immigration of other males. Finally a concentration of chemicals is released that inhibits further response from both sexes. This device finally limits the population on individual trees.

An interesting situation in the case of the gypsy moth has been reported by Cardé et al. (15). The sex attractant of this insect is *cis*-(Z)-7,8-epoxy-2-methyloctadecane (disparlure), however the sex pheromone-producing glands of the moth also contain 2-methyl-(Z)-7-octadecane, possibly a precursor of disparlure. The latter chemical is a potent inhibitor of the pheromone. A similar inhibitor was found in the spruce budworm, *Choristoneura fumiferana* (16).

Besides being used for the evaluation of the density of a pest, sex attractants can be used for the control of pests. Male insects can be trapped, thereby reducing the probability of matings. This approach has given some results in the case of orchard pests. The U.S. Department of Agriculture is attempting to stop the spread of the gypsy moth from New England to parts of Virginia by using an extensive system of baited traps dropped by an airplane to trap the males (the female moth does not fly). Also, permeating the environment with pheromones may either confuse the males or so fatigue the receptors that they may fail to find mates. Another approach could be the use of chemicals that may inhibit the perception of the pheromone by the male.

Some of the earlier reported identifications of insect pheromones were those of the silkworm moth (17) and the female gypsy moth. The latter was reported to be 10-acetoxy-7-hexadecen-1-ol (gyptol). A homolog, 12-acetoxy-9-octadecen-1-ol (gyplure), was also reported to attract males (18, 19). However both compounds were cases of mistaken identity. In 1970 Bierl et al. (20) correctly identified and synthesized the sex pheromone of gypsy moth as *cis*-7,8-epoxy-2-methyloctadecane (disparlure). This chemical is active in concentrations as low as $2 \times 10^{-12}$ gram. In a 1970 survey, traps baited with synthetic disparlure were compared with those baited with natural pheromone. Captures with disparlure were 9- to 37-fold greater than those with the natural lure (21).

Carlson et al. (22) in 1971 isolated and characterized the houe fly sex attractant as *cis*-9-tricosene (muscalure). Because it was a weak attractant, its development was not actively pursued. However in 1972 a private industry (Thuron Industries, a subsidiary of Zoecon Corp.) developed a fly bait containing small amounts of muscalure and obtained the first registration of a pheromone for insect control from the Environmental Protection Agency in 1973 (23).

## REFERENCES

1. H. C. Coppel, J. E. Casida, and W. C. Dauterman. *Ann. Ent. Soc. Amer.* **53**: 510 (1960).
2. P. Karlson and A. Butenandt. *Annu. Rev. Entomol.* **4**: 39 (1959).
3. M. Jacobson. *Adv. Chem. Ser.* **41**: 1 (1963).
4. M. Beroza and M. Green. *op. cit.* **41**: 11 (1963).
5. M. Beroza, ed. *Chemicals Controlling Insect Behavior* Academic Press, New York, 1970.
6. M. Jacobson. *Insect Sex Attractants*, Wiley—Interscience, New York, 1965.
7. M. Jacobson. *Annu. Rev. Entomol.* **11**: 403 (1966).
8. D. L. Wood, R. M. Silverstein, and M. Nakajima, eds. *Control of Insect Behavior by Natural Products*, Academic Press, New York, 1970.
9. N. C. Birch, ed. *Pheromones*, American Elsevier, New York, p. 495 (1974).
10. J. G. MacConnell and R. M. Silverstein. *Angew Cheme.* **12**: 644 (1973).
11. J. H. Tumlinson, D. D. Hardee, R. C. Gueldner, A. C. Thompson, P. A. Hedin, and J. P. Minyard. *Science* **166**: 1010 (1969).
12. J. A. Renwick and J. P. Vité. *Nature* **224**: 1222 (1969).
13. J. P. Vité. *Contrib. Boyce Thompson Inst. Plant Res.* **24**: 343 (1970).
14. R. M. Silverstein, R. G. Brownlee, and J. O. Rodin. *Science* **164**: 1284 (1969).
15. R. T. Cardé, W. L. Roelofs, and C. C. Doane. *Nature* **241**: 474 (1973).
16. J. Weatherston and W. Maclean. *Can. Entomol.* **106**: 281 (1974).
17. A. Butenandt, R. Beckman, D. Stamm, and E. Hecker. *Z. Naturforsch. B* **14**: 283 (1959).
18. M. Jacobson, M. Beroza, and W. A. Jones. *Science* **132**: 1011 (1960).
19. M. Jacobson and W. A. Jones. *J. Org. Chem.* **27**: 2523 (1962).
20. B. A. Bierl, M. Beroza, and C. W. Collier. *Science* **170**: 87 (1970).
21. M. Beroza and E. F. Knipling. *Science* **177**: 19 (1972).
22. D. A. Carlson, M. S. Mayer, D. L. Silhacek, J. D. James, and M. Beroza. *Science* **174**: 76 (1971).
23. C. Djerassi, C. Shih-Coleman, and J. Diekman. *Science* **186**: 596 (1974).

# INSECT CONTROL AGENTS OF INSECT, PLANT AND MICROBIAL ORIGIN

| Hormones and analogous chemicals

The complex and diverse changes during the development of insects are controlled with precision by hormones secreted by the neuroendocrine system. The neurosecretory cells of the protocerebrum release a hormone, often referred to as brain hormone, prothoracotropic hormone, or allatotropic hormone. This hormone transmits neural messages to the endocrine glands and other tissues. "Target" endocrine glands, such as corpora allata and/or prothoracic glands are stimulated and secrete hormones. Because of the activation function of brain hormone, Sláma et al. (1) have called it "activation hormone."

The hormone secreted by corpora allata is known as juvenile hormone (JH). There are several natural and synthetic materials that mimic the action of JH; these substances are commonly called JH analogs.

The prothoracic glands secrete the molting hormones, ecdysones or the prothoracic gland hormone. In the case of deficiency or absence of this hormone the molt cycles do not take place. Thus ecdysones are essential for molting, but JH determines the genetic programming of the molt. When JH is present growth and molt without maturation take place, and hence metamorphosis during immature life is prevented. In the absence of JH, metamorphosis to adult reproductive stage takes place.

Insect endocrinology is a field in which there has been great activity in the last two decades; it is therefore not possible to review all important literature in this short chapter. However those interested in the subject are referred to some excellent reviews and books available on the subject (1–8).

## JUVENILE HORMONE

Though as early as 1936 Wigglesworth had discovered the existence of a JH, the idea of using this hormone as an insecticide originated with Williams (9).

He prepared lipid extracts from the abdomens of Cecropia moths, *Platysamia cecropia* (L.), and showed that they possessed JH effect on metamorphosis. Wigglesworth confirmed these results on *Rhodnius* (10). Following these discoveries there was a great interest in Cecropia extracts and finally Röller and co-workers (11) identified an active compound, Cecropia moth JH.

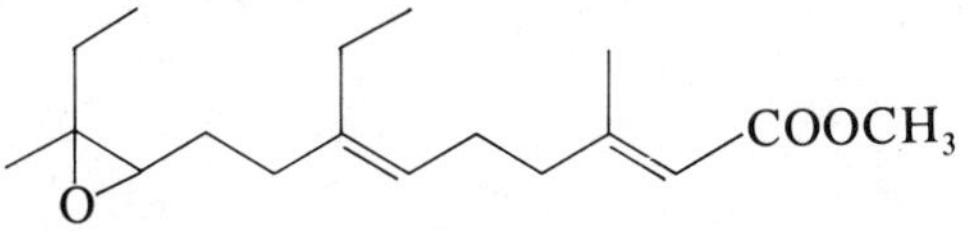

Before this classic work of Röller and his co-workers, Schmialek had discovered the first chemical, farnesol, from the feces of *Tenebrio*, that possessed JH activity (12). Schmialek later (13) found that the methyl ether and the diethyl amine of farnesol possessed even greater biological activity than the parent alcohol. Bowers et al. (14), after considerable investigation, synthesized *trans, trans*-10,11-epoxy methyl farnesenate, which was found to be extremely active. On the basis of high morphogenetic and gonadotropic activity shown by this compound, Bowers et al. (14) predicted a resemblance to JH. The elucidation of the structure of JH by Röller and co-workers therefore created considerable interest among scientists. Later Meyer et al. (15) announced the isolation and the characterization of a second Cecropia moth JH. The work of this group has been summarized by Meyer (16).

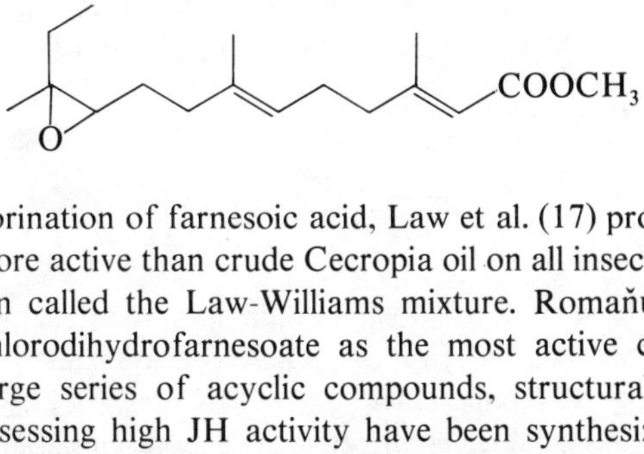

By the hydrochlorination of farnesoic acid, Law et al. (17) produced a mixture that was far more active than crude Cecropia oil on all insects tested. This preparation is often called the Law-Williams mixture. Romaňuk et al. (18) identified 7,11-dichlorodihydrofarnesoate as the most active component in this mixture. A large series of acyclic compounds, structurally related to farnesoic acid, possessing high JH activity have been synthesized.

Bowers and colleagues investigated the possibility of synergising JH analogs (Fig. 13.1) by conventional insecticide synergists and discovered that several synergists, such as piperonyl butoxide and sesamex, in themselves possessed JH activity of natural and synthetic JH analogs (19). These investigators combined chemical features of synergists with terpenoids and found that shortening of the terpenoid portion by one isoprene unit in such compounds greatly increased their JH activity (20).

The fortuituous discovery of "paper factor" by Sláma and Williams (21) has been described by Sláma et al. (1) and Williams (22). In effect, the European

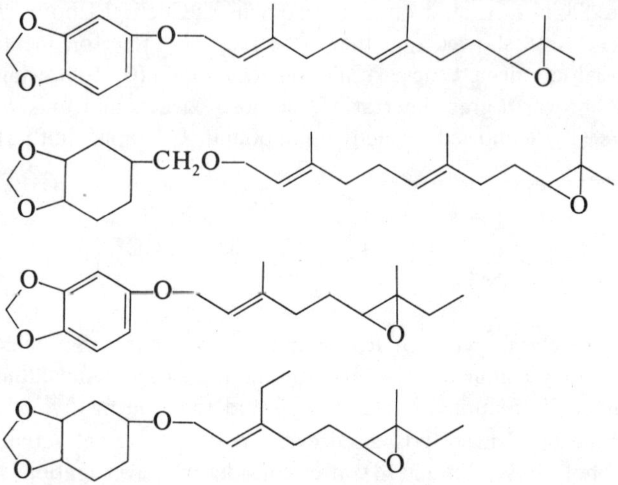

**Fig. 13.1** Some of the compounds synthesized by Bowers (19, 20).

bug, *Pyrrhocoris apterus* (Pyrrhocoridae), failed to develop normally in Harvard Biological Laboratories. It underwent a supernumerary larval molt to form a giant sixth-stage larva rather than metamorphosing into a sexually mature adult. The active component responsible for this physiological effect was tracked down, through the paper towels used in rearing jars, to balsam fir, *Abies balsamea,* and other evergreen trees from which American paper pulp is made. The active compound was identified by Bowers et al. (23) as juvabione.

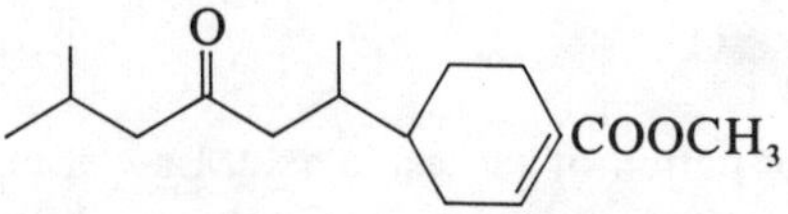

Later the Czech workers isolated a second compound, dehydrojuvabione, with JH activity from Balsam fir (24). These workers also synthesized several aromatic analogs of "paper factor."

Synthesis of JH analogs is an active field of research and several groups in different countries are working in this area. Among some of these are Staal and co-workers at Zoecon Corporation Palo Alto, Calif. J. J. Menn and colleagues at Stauffer Chemical Co., Mountain View, Calif., and K. Mori and his group in Japan (25). Recently Henrick et al. (26) of Zoecon Corporation have synthesized a new class of insect growth regulators, including isopropyl 11-methoxy-3,7,11-trimethyl-*trans*-2-*trans*-4-dodecadienoate and ethyl 3,7,11-trimethyl-*trans*-2-*trans*-4-dodecadienoate. When tested against the yellow fever mosquito, greater wax moth, *Galleria mellonella* (L.), and the yellow mealworm, *Tenebrio molitor* L., these compounds were found to be more potent and more stable than the known natural juvenoids. Henrick et al. also synthesized all four possible stereoisomers of ethyl 3,7,11-trimethyl-2,4-dodecadienoate and tested them against the insects mentioned above and against the house fly, pea aphid *Acyrthosiphon pisum* (Harris), and tobacco budworm, *Heliothis virescens* (Fabricius). The 2*E*,4*E* stereoisomer 1 was found to be biologically most active (27).

## PHYSIOLOGICAL EFFECTS OF JH ANALOGS

Compounds with JH activity comprise a wide spectrum of acyclic or aromatic compounds. With a few exceptions, they have been produced in the laboratory; they have one feature in common—they are hormonomimetic and possibly affect physiological processes at their earlier stages. Whether they start their physiological interactions at the same receptor site that receives its signal from JH has not been established. Also, whether there are some structural features common to the synthetic JH analogs and endogenous JH still remains unclear. These compounds have been known by different names; JH analogs, insect growth regulators, and insect developmental inhibitors. The term juvenoids was first used by Gill et al. (8) in 1972; Sláma et al (1) also have proposed the same terminology, though they have not referred to Gill et al. The term juvenoids corresponds to nictonioids, pyrethroids, and so on and I feel should be adopted for general use.

If JH is applied to an insect during the molting process, cellular differentiation and maturation are prevented. JH also interferes with embryogenesis and is lethal to the embryo, though it also stimulates ovarian development and formation of mature eggs (28–30). Treatment with JH breaks diapause of insects that enter this condition to survive environmental stress (31–32). JH analogs also exert fumigant action on several species of insects (33). It would therefore be evident that JH analogs, properly used, could become useful tools in insect control. The idea was offered by Williams (34) who called them "third generation pesticides."

The control of insects by JH analogs is based on the fact that in the life cycles of insects there are very definite and precisely timed periods during which the presence or absence of JH plays a vital role in the insect's life. Any change in JH level during these precise periods can be fatal to the insect. At present most of our control procedures make use of the phases in the life cycle when the natural hormone is present in very low titer or is absent. Addition of JH, which easily penetrates insect cuticle, results in abnormal development and ultimate death of the insect. The time of application of juvenoids is therefore to be carefully planned in advance. A knowledge of the insect's life cycle under field conditions is very necessary.

Since insects are not killed immediately, procedures used in the evaluation of results are different. Complex analysis of physiological effects on treated population is necessary. For instance, in the case of treatment of mosquito larvae, larval or pupal counts after treatment do not give an estimate of control; adult emergence is an effective measure of control.

Selective toxicity of JH analogs is another factor to be considered. Some like juvabione and dehydrojuvabione, are highly selective, being active exclusively against the family Pyrrhocoridae. Others, like Cecropia JH, show activity against several species of insects. Chemical stability and persistence under field conditions is an additional consideration. Some beetles, higher Diptera, some sawflies, and Dipterous and Hymenopterous parasites are reported to be more resistant to juvenoids; this may prove useful in the overall scheme of insect control.

Some of the juvenile hormone analogs are selective in their action and affect only certain stages of development of a few species. Because of this property and on account of a few reports (35, 36) that some juvenile hormone analogs do not adversely affect parasitoid development or fecundity, it was assumed that they would play an important role in integrated control. However careful investigations by several workers have shown that this may not be true in all cases. Neal et al. (37, 38) reported that when the diapausing alfalfa weevil, *Hypera postica* (Gyllenhal), was treated with an active farnescenic acid derivative its parasitoids *Microtonus aethiops* (Nees) and *M. colesi* Drea emerged prematurely. Recent investigations, (39–40), have confirmed the adverse and even detrimental effects of juvenile hormone analogs on some parasitoids. Vinson treated larvae of tobacco budworm, *Heliothis virescens* (F), with ENT-70221, 1-(6',7'-epoxy-3,7-dimethyl oct-2'enyl) 4-ethylphenyl ether. The larvae were previously parasitized by either *Cardiochiles nigriceps* Viereck (Hymenoptera: Braconidae) or *Campoletis sonorensis* (Cameron) (Hymenoptera: Ichneumonidae). The emergence of the endoparasitoids was delayed and there was a decrease in the number of emerging adults of *C. nigriceps* and a change in sex ratio. McNeil treated parasitized aphids *Macrosiphum euphorbiae*, with methoprene (Altocid), ZR-619, and ZR-777 and reported a high incidence of mortality of the endoparasite *Aphidius nigriceps*. The larval and pupal stages of the endoparasitoid were susceptible.

## SOME COMMERCIAL INSECT REGULATORS

Altocid®, ZR-515, Ent. No. 70460, isopropyl (2$E$,4$E$)-11-methoxy-3,7,11-trimethyl-2,4-dodecadienoate, has been given the common name methoprene. As stated earlier, this compound was synthesized by Henrick et al. (26) of Zoecon Corporation. The technical material is an amber-colored liquid soluble in organic solvents but only very slightly so (1.3 ppm) in water; it has a vapor pressure of $2.37 \times 10^{-5}$ mm Hg at 25°C. Altocid has very low toxicity; acute oral $LD_{50}$ for rat (male and female) is greater than 34,600 mg/kg; for the dog this figure is greater than 5000 mg/kg. No toxicity was demonstrated in rats given 10,500 mg/kg of this chemical (41). It is the first insect growth regulator (IGR) for which the Environmental Protection Agency has granted an experimental permit.

Altocid has shown special promise in the control of mosquitoes when medium to late IV instar larvae were treated (42, 43). It is effective against horn flies and stable flies when administered orally to cattle (44) and also against the imported fire ant (45, 46). Miura and Takahashi (47) reported no adverse effects of methoprene on most of the 35 aquatic organisms tested; however larvae of aquatic Diptera were affected. Schaefer and Dupras (48) found that the half-life of the chemical was only 2 hours in field water and chemical residues of methoprene were not detectable in water 24 hours after treatment even though biological activity persisted for several days.

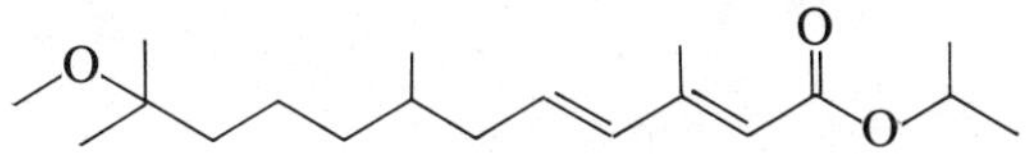

Methoprene

## DEGRADATION AND METABOLISM OF METHOPRENE

The environmental degradation and metabolism of methoprene has been studied by, among others, Quistad et al. (49–51) and Schooley et al. (52, 53). In alfalfa and rice, ester hydrolysis, O-demethylation and oxidative scission of the 4-ene double bond were the major pathways responsible for the degradation of the chemical. 7-Methoxycitronellal was isolated from vapors evolved from the plants. Radiolabeled Altocid was used in these studies, and the possibility of the incorporation of radiolabel from the extensively degraded

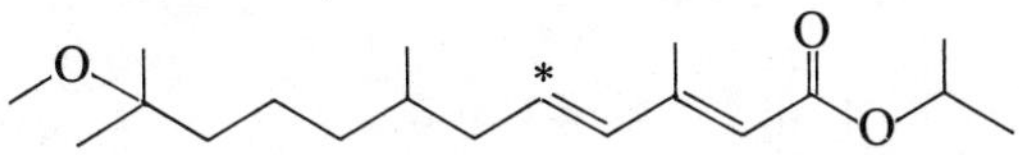

$(2E)$-(5-$^{14}$C) methoprene molecule into carotenoids, chlorophyll, and other higher molecular weight plant constituents was surmised on the basis of chromatographic evidence by Quinstad et al. (49). Schooley et al. (52) studied the biodegradation of methoprene in pond water containing unknown organisms. A half-life of 30 and 40 hours at concentrations of 0.001 and 0.01 ppm, respectively, were reported. Water from different sources gave different metabolites. Photodecomposition of methoprene was studied by Quinstad et al.

(51), who reported that the most abundant photoproduct was 7-methoxy-citronellal (9–14%). Absorption, excretion, and metabolism of radiolabeled Altocid by a guinea pig, a steer, and a cow were examined by Chamberlain et al. (54). These investigators reported that a large percentage of the radiolabel was incorporated in the tissue and was respired by the animals. Methoprene was not found in urine, but approximately 40% of the radiolabel in feces was contributed by unmetabolized methoprene. Some free primary metabolites were found in urine and feces, though more of the chemical was incorporated into simple glucuronides, and polar compounds accounted for a considerable quantity of the radiolabel.

Among other juvenile hormone analogs developed by various companies are the following:

*Zoecon Corporation*: *Altozar*®, Zoecon-512, ethyl (2*E*,4*E*-3,7,11-trimethyl-2,4-dodecadienote; ZR-619, ethyl (2*E*,4*E*)-11-methoxy-3,7,11-trimethyl-2,4-dodecadienethiolate; ZR-777, prop-2-ynyl (2*E*,4*E*)-3,7,11-trimethyl-2,4-dodecadienoate.

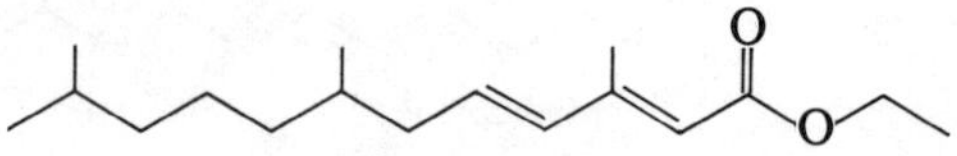

Altozar

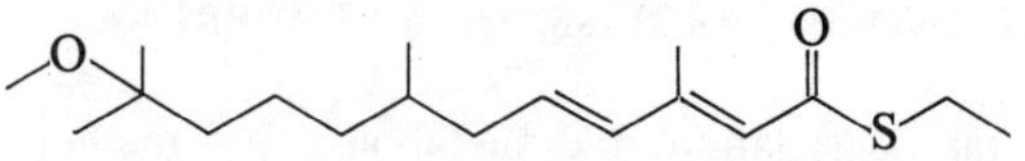

ZR-619

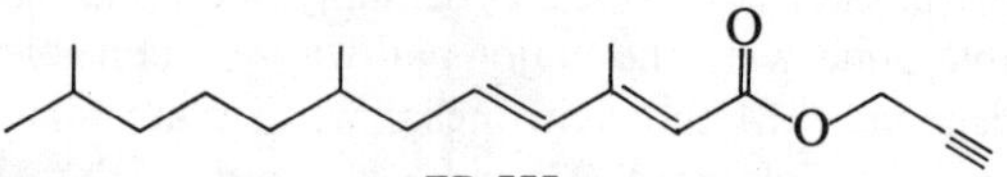

ZR-777

*Hoffman-La Roche*: 5503, 10,11-epoxy-1-ethyoxy-3,7,11-trimethyl-1-2-6-tridecadiene.

*Stauffer*: 20458, 1-(6′,7′-epoxy-3′,7′-dimethyl oct-2′-enyl) 4-ethylphenyl ether.

*Niagara*: 23509, 10,11-epoxy-*N*-ethyl-3,7,11-trimethyl-2,6-dodecadiene-amide.

Sláma et al. have reported a model experiment with a peptidic juvenoid in which 100% control of *Dysdercus cingulatus* F. was obtained in an artificial cotton plantation. The dose applied was 0.1 g/hectare and the juvenoid was effective for several weeks (1).

## RESISTANCE TO JUVENILE HORMONE ANALOGS

Since JH mimics resemble insect hormones, it was hoped that resistance to these chemicals would be less likely than to conventional insecticides, but a measure of cross-tolerance to JH analogs has been reported in the red flour beetle, *Tribolium castaneum* (Herbst); in tobacco budworm, *Heliothis viriscens* (F.); and in house fly (55–58). Brown and Brown (59) induced 13-fold resistance to methoprene in the northern house mosquito, *Culex pipiens pipiens* L., by larval selection. However Schaefer and Wilder (42) failed to induce resistance to methoprene in the southern house mosquito, *Culex pipiens quinquefasciatus* Say, when 20 generations were reared under pressure of the chemical. The cross-tolerance in insects is associated with increased levels of microsomal oxidases (58). The synergism of methoprene by piperonyl butoxide has been reported by Solomon and Metcalf (60).

### TH 6040, DIMILIN®

For want of a better place, the insect growth regulator TH 6040 is being treated here, though chemically, as well as in its mode of action, it is different from JH analogs. TH 6040, ENT 2905, 1-(chlorophenyl)-3-(2,6-difluoro-benzoyl)urea, is a member of 1-(2,6-disubstituted benzoyl)-3-phenylureas. It was synthesized by Philips-Duphar B.V. of the Netherlands; in the United States it is being developed by the Thompson-Hayward Chemical Company.

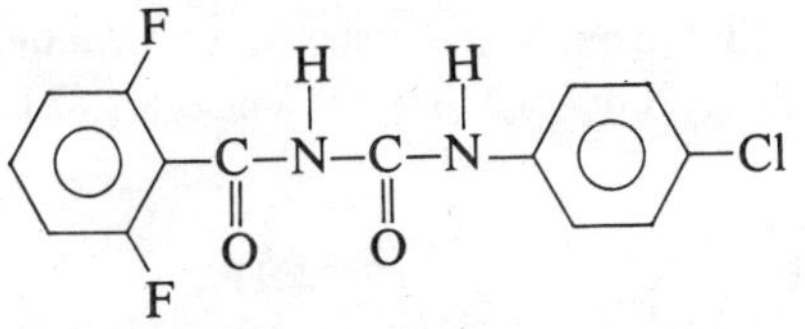

TH 6040 is a white crystalline solid soluble in organic solvents but only barely (0.2 ppm) soluble in water. The oral $LD_{50}$ for rat of the technical material is greater than 10,000 mg/kg. It is active by ingestion and has a wide spectrum of activity against chewing insects, but sucking insects are not affected by this chemical. In small field plots Neal (61) found it effective against alfalfa weevil, *Hypera postica* (Gyllendal). Miller (62) reports that this compound at concentrations of 0.1 and 1 ppm killed 100% of face fly and house fly larvae, respectively, in cow manure. When fed at 0.5 mg/kg to cows, 99% or more of the seeded larvae died in the manure obtained from the treated cows. Foliar application of TH 6040 was found effective in the United States and in Brazil against soybean insects by Turnipseed et al. (63).

**Mode of Action of TH 6040**

TH 6040 interferes with the formation of insect cuticle. The synthesis and
deposition of chitin in the endocuticle is prevented. Studies with radioactive
glucose have shown that no glucose was incorporated in the endocuticle.
Probably acetylation of glucose to form glucosamine, an essential constituent
of chitin, is prevented. There is very little change in the incorporation of
amino acids. The cuticle of a treated larva is therefore weak and can neither
withstand the normal turgor during ecdysis nor provide enough support for the
muscles involved in ecdysis (64–66a).

## INSECT MOLTING HORMONES

Butenandt and Karlson (67) isolated pure crystalline molting hormone $\alpha$-
ecdysone in 1954 from the silkworm, *Bombyx mori* (L.) pupae, and the formula for ecdysone was determined by Huber and Hoppe (68). The stucture
$2\beta,3\beta,14\alpha,22\alpha(R)$, 25-pentahydroxy-$5\beta$-cholest-7-en-6-one was further confirmed by syntheses (69, 70). Ecdysone is readily soluble in polar organic
solvents. In 1959 Stamm isolated another molting hormone, ecdysterone ($\beta$-
ecdysone), from adult locusts, *Dociostaurus maroccanus* (71). Seven years
later (1966) Kaplanis et al. (72) also isolated ecdysterone in addition to ecdysone from the pupae of the tobacco hornworm, *Manduca sexta*. Ecdysterone ($\beta$-ecdysone or crustecdysone) controls the ecdysis of crustacea.
Horn et al. (73) isolated it from a crayfish, *Jasus lalandei*, and ascribed to it
the structure $20\beta$-hydroxyecdysone. Later syntheses (74, 75) confirmed this
formula.

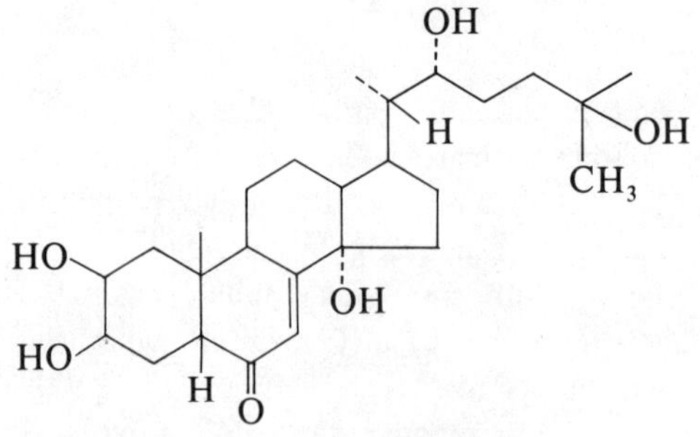

$\alpha$-Ecdysone

While work on the isolation and characterization of insect molting hormones
was being carried out in various laboratories, scientists in others were examin-
ing plants for steroidal compounds, and, surprisingly enough, compounds

related to ecdysones were found among plants. Nakanishi et al. (76) isolated a steroid, ponasterone A, from the plant *Podocarpus nakaii* Hay. It is almost as active as ecdysone and has a structure almost identical with that of the molting hormone. Since this discovery, ecdysone-related chemicals have been isolated from several plants. Takemoto et al. (77) isolated ecdysterone and in-okosterone (an isomer of ecdysterone) from the roots of *Acharynthes* spp. The history and the chemistry of these hormones have been covered by Sláma et al. (1).

Ecdysoids do not readily penetrate insect cuticle and therefore have not found practical application for insect control. However Ittycheriah et al. found that when female *Culex tarsalis* Coquillett were treated topically with $\beta$-ecdysone before a blood meal, there was no appreciable effect on the egg rafts and 90% or more of the eggs were viable. Similar treatment after a blood meal resulted in the production of infertile eggs and raft formation was prevented (78).

Since ecdysoids possess pharmacological action on vertebrate organs, their use in insect control appears to be limited.


## OTHER CHEMICALS

Quraishi and Thorsteinson (79) reported the transient toxicity of synthetic queen substance, 9-oxodec-*trans*-2-enoic acid toward *Aedes aegypti* where the treated larvae died during the pupal stage. Later Quraishi and co-workers (un-published data) found that various batches of the same chemical synthesized by different routes differed in their toxicities. The physiological interactions of chemicals related to queen substance have been discussed by Quraishi and co-workers (80, 81).

These workers also examined the physiological effects of about 3000 ex-tracts from various plant parts on insect development (81, 82). Of particular interest were extracts from the plant *Cynoglossum officinale* L. When *Droso-phila melanogaster* was reared from eggs in a medium containing this extract development proceeded apparently normally, but the imago failed to emerge from the puparium though the operculum was pushed open (83).

## REFERENCES

1. K. Sláma, M. Romaňuk, and F. Šörm. *Insect Hormones and Bioanalogues*. Springer, New York, 1974, 474 pp.
2. F. Engelmann. *Annu. Rev. Entomol.* **13**: 1 (1968).
3. V. J. A. Novak. *Insect Hormones* Methuen, London, 1966, 478 pp.

4. V. B. Wigglesworth. *Insect Hormones* Oliver and Boyd, Edinburgh, 1970. 159 pp.
5. F. Engelmann. *The Physiology of Insect Reproduction* Pergamon Press, New York, 1970 307 pp.
6. K. C. Hingman and L. Hill. *The Contemporary Endocrinology of Invertebrates.* Edward Arnold, London, 1969. 270 pp.
7. J. B. Siddall. *Chemical Ecology*, (E. Sondheimer, and J. B. Simeone eds.), Academic Press, New York, 1970. p. 281.
8. J. J. Menn and M. Beroza, eds. Insect Juvenile Hormones: Chemistry and Action, 341 pp. Academic Press, New York, 1972.
9. C. M. Williams. *Nature (Lond.)* **178**: 212 (1956).
10. V. B. Wigglesworth. *J. Insect Physiol.* **2**: 73 (1958)
11. H. Röller, K. H. Dahm, C. C. Sweeley, and B. M. Trost. *Angew. Chem.* **79**: 190 (1967).
12. P. Schmialek. *Z. Naturforsch.* **16b**: 461 (1961).
13. P. Schmialek. *op. cit.* **18b**: 516 (1963).
14. W. S. Bowers, M. J. Thompson, and E. C. Uebel. *Life Sci.* **4**: 2323 (1965).
15. A. S. Meyer, H. A. Schneiderman, and E. Hanzmann. *Fed. Proc.* **27**: 393 (1968).
16. A. S. Meyer. *Bull. Soc. Entomol. Suisse* **44**: 37 (1971).
17. J. H. Law, C. Yuan, and C. M. Williams. *Proc. Nat. Acad. Sci. U.S.,* **55**: 576 (1966).
18. M. Romanuk, K. Sláma, and F. Šörm. *Proc. Nat. Acad. Sci. U.S.* **57**: 349 (1967).
19. W. S. Bowers. *Science* **161**: 895 (1968).
20. W. B. Bowers. *Science* **164**: 323 (1969).
21. K. Sláma and C. M. Williams. *Proc. Nat. Acad. Sci. U.S.* **54**: 511 (1965).
22. C. M. Williams. In *Chemical Ecology* (E. Sondheimer, and J. B. Simeone, eds.), Academic Press, New York, 1970. p. 122.
23. W. S. Bowers, H. M. Fales, M. J. Thompson, and E. C. Uebel. *Science* **154**: 1020 (1966).
24. V. Černý, L. Dolejš, L. Labler, F. Šörm, and K. Sláma. *Tetrahedron Lett.* **1967**: 1053.
25. K. Mori. *Tetrahedron* **28**: 3747 (1972).
26. C. A. Henrick, G. B. Staal, and J. B. Siddall. *J. Agr. Food Chem.* **21**: 354 (1973).
27. C. A. Henrick, W. E. Willy, B. A. Garcia, and G. B. Staal. *op. cit.* **23**: 396 (1975).
28. L. M. Riddiford and C. M. Williams. *Proc. Nat. Acad. Sci. U.S.* **57**: 595 (1966).
29. K. Sláma and C. M. Williams. *Nature* **210**: 329 (1967).
30. W. F. Walker and W. S. Bowers. *J. Econ. Entomol.* **63**: 1231 (1970).
31. W. S. Bowers and C. C. Blickenstaff. *Science.* **154**: 1673 (1966).
32. R. V. Connin, O. K. Jantz, and W. S. Bowers. *J. Econ. Entomol.* **60**: 1752 (1967).
33. W. S. Bowers. *Science* **164**: 323 (1969).
34. C. M. Williams. *Sci. Amer.* **217**: 13 (1967).
35. J. E. Wright and G. E. Spates. *Science* **178**: 1292 (1972).
36. J. D. Wilkinson and C. M. Ignoffo. *J. Econ. Entomol.* **66**: 643 (1973).
37. J. W. Neal, Jr., W. E. Bickley, and C. C. Blickenstaff. *op. cit.,* **63**: 681 (1970).
38. J. W. Neal, Jr., W. J. Holloway, and W. E. Bickley, *op. cit.* **64**: 338 (1971).
39. S. B. Vinson. *op. cit.* **67**: 335 (1974).
40. J. McNeil. *Science* **189**: 640 (1975).
41. J. B. Siddall and M. Slade. In *Invertebrate Endocrinology and Hormonal Heterophyll* (W. J. Burdette, ed.), Springer, New York, 1974. 345 pp.
42. C. H. Schaefer and W. H. Wilder. *J. Econ. Entomol.* **65**: 1066 (1972).
43. C. H. Schaefer and W. H. Wilder. *Ibid.* **66**: 913 (1973).
44. R. L. Harris, E. D. Frazer, and R. L. Younger. *J. Econ. Entomol.* **66**: 1099 (1973).
45. E. W. Cupp and J. O'Neil. *Environ. Entomol.* **2**: 191 (1973).
46. S. B. Vinson, R. Robeau, and L. Dzuik. *J. Econ. Entomol.* 325 (1974).
47. T. Miura and M. Takahashi. *J. Econ. Entomol.* **66**: 917 (1973).
48. C. H. Schaefer and E. F. Dupras, Jr. *J. Econ. Entomol.* **66**: 923 (1973).

49. G. B. Quinstad, L. E. Steiger, and D. A. Schooley. *J. Agr. Food Chem.* **22**: 584 (1974).
50. G. B. Quinstad, L. E. Steiger, and D. A. Schooley. *Life Sci.* **15**: 1797 (1974).
51. G. B. Quinstad, L. E. Steiger, and D. A. Schooley. *J. Agr. Food Chem.* **23**: 299 (1975).
52. D. A. Schooley, B. J. Bergot, L. J. Dunham, and J. B. Siddall. *op. cit.* **23**: 293 (1975).
53. D. G. Schooley, K. M. Cresswell, L. E. Steiger, and G. M. Quinstad. *op. cit.* **23**: 369 (1975).
54. W. F. Chamberlain, LaW. M. Hunt, D. E. Hopkins, A. R. Gingrich, J. A. Miller, and B. N. Gilbert. *op. cit.* **23**: 736 (1975).
55. C. E. Dyte. *Nature* **238**: 48 (1972).
56. J. Benskin and S. B. Vinson. *J. Econ. Entomol.* **66**: 15 (1973).
57. F. W. Plapp and S. B. Vinson. *Pestic. Biochem. Physiol.* **3**: 131 (1973).
58. D. C. Cerf, G. P. Georghiou, and C. E. Dyte. *Nature* **238**: 48 (1972).
59. T. M. Brown and A. W. A. Brown. *J. Econ. Entomol.* **67**: 799 (1974).
60. K. R. Solomon and R. L. Metcalf. *Pestic. Biochem. Physiol.* **4**: 127 (1974).
61. J. W. Neal, Jr. *J. Econ. Entomol.* **67**: 300 (1974).
62. R. W. Miller. *op. cit.* **67**: 697 (1974).
63. S. G. Turnipseed, E. A. Heinrichs, R. F. P. da Silva, and J. W. Todd *op. cit.* **67**: 760 (1974).
64. K. Wellinga, R. Mulder, and J. J. van Daalen. *J. Agr. Food Chem.* **21**: 348 (1973).
65. K. Wellinga, R. Mulder, and J. J. van Daalen. *J. Agr. Food Chem.* **21**: 993 (1973).
66. R. Mulder and M. J. Gijswijt. *Pestic. Sci.* **4**: 737 (1973).
66a. L. C. Post and R. Mulder. "Insecticidal properties and mode of action of 1-(2,6-dihalogenbenzoyl)-3-phenylureas," In Mechanism of Pesticide Action (G. K. Kohn ed.) American Chemical Society, Washington, 1974, 180 p.
67. A. Butenandt and P. Karlson. *Z. Naturforsch.* **9b**: 389 (1954).
68. R. Huber and W. Hoppe. *Chem. Ber.* **98**: 2403 (1965).
69. J. B. Siddall, A. D. Cross, and J. R. Fried. *J. Amer. Chem. Soc.* **88**: 862 (1966).
70. A. Furlenmeier, A. Furst, A. Langemann, G. Waldvogel, P. Hocks, U. Kerb, and R. Weichert. *Experientia* **22**: 573 (1966).
71. M. D. Stamm. *An. Real Soc. Esp. Fis. Quim.* **B55**: 171 (1959).
72. J. N. Kaplanis, M. J. Thompson, R. T. Yamamoto, W. E. Robbins, and S. J. Louloudes. *Steroids* **8**: 605 (1966).
73. D. H. S. Horn, E. J. Middleton, J. A. Wunderlich, and F. Hampshire. *Chem. Commun.* **1966**: 339.
74. G. Huppi and J. B. Siddall. *J. Amer. Chem. Soc.* **89**: 6790 (1967).
75. U. Kerb, R. Weichert, A. Furlenmeier, and A. Furst. *Tetrahedron Lett.* **1968**, 4277.
76. K. Nakanishi, M. Koreeda, S. Sasaki, M. L. Chang, and H. Y. Hsu. *Chem. Commun.* **1966**: 915.
77. T. Takemoto, S. Ogawa, and S. Nishimoto. *Yakugaku Zasshi* **87**: 325 (1967.
78. P. I. Ittycheriah, E. P. Marks, and M. S. Quraishi. *Ann. Entomol. Soc. Amer.* **67**: 595 (1974).
79. M. S. Quraishi and A. J. Thorsteinson. *J. Econ. Entomol.* **58**: 185 (1965).
80. M. S. Quraishi. *Can. Entomol.* **104**: 1505 (1972).
81. M. S. Quraishi, Participating Staff and Graduate Students. *Control of Vectors Through Interference With Normal Processes of Insect Physiology Reproduction and Behavior.* Annual Reports 1–5, 1970–74. Submitted to U.S. Army Medical Research and Development Command, Washington, D. C.
82. B. D. Patterson, S. K. W. Khalil, L. J. Schermeister, and M. S. Quraishi. *Lloydia* **38**: 391 (1975).
83. M. S. Quraishi, L. J. Schermeister, M. Mattson, and P. C. Sandal. Insect-Plant Interactions II: Effects of Selected Plant Extracts on the Developmental Physiology of Diptera. Paper Presented at the Entomological Society of America Meeting, Los Angeles, 1971.

# INSECT CONTROL AGENTS OF INSECT, PLANT AND MICROBIAL ORIGIN

| CHAPTER 14 | Insect Control Agents of Microbial Origin |
|---|---|

Because of the shortcomings of the synthetic organic insecticides (development of resistance, residue problems, environmental pollution, etc.) interest in insect control agents of microbial origin is increasing, though one cannot be sure that the same disadvantages will not accompany their use. Interestingly enough both microbial control and DDT were used at about the same time. The story of DDT is famous, but the use of milky disease for the control of Japanese beetle in the eastern United States (1, 2) is not as wellknown. Several reviews and articles on the control of pests by microbial agents are available (3–7a); in this chapter the subject is dealt with only briefly and mostly with respect to the toxins produced by these organisms. Among the bacteria, two species, namely, *Bacillus thuringiensis* and *B. popilliae*, have acquired considerable importance, and commercial preparations based on these two species are available. Among viruses the *Heliothis* nuclear polyhedrosis virus has been registered for use against cotton bollworm and is being applied in a large area of Missouri.

## BACILLUS THURIGIENSIS

This bacillus is considered to be a pathogenic variety of *B. cereus*. The presence of several toxic agents has been demonstrated in *B. thuringiensis* varieties; of these, four have been generally recognized: $\alpha$-exotoxin, the enzyme phospholipase C; $\beta$-exotoxin, thermostable adenine nucleotide; $\gamma$-exotoxin, unidentified thermolabile phospholipase; $\delta$-endotoxin, proteinaceous parasporal inclusion. The two toxins of major importance in insect control are δ-endotoxin and β-exotoxin. δ-Endotoxin was first discovered by Hannay (8) who reported that the crystal grew to maturity alongside the spore inside the vegetative cell of the bacterium. Angus (9) showed the proteinaceous nature of the crystal and its high toxicity to lepidoterous larvae. This endotoxin is the main ingredient of the commercially available preparations. McConnell and Richards (10) demonstrated that the supernatant from liquid cultures of *Bacillus thurigiensis* was toxic to insects. This toxin (β-exotoxin), which survives autoclaving at 120°C for 15 minutes, is toxic to insects belonging to several orders. Bond et al. (11) further purified this exotoxin, and a detailed review of β-exotoxin has been presented by Bond et al. (12).

The commercially available preparations of *Bacillus thuringiensis* (*B.t.*) (Biotrol®, Dipel®, Microtrol®, Thuricide®) are spore–crystal mixtures for use against lepidopterous larvae. To be effective they must be ingested by the insect; hence their application is recommended when the larvae are feeding actively. This action by ingestion and not by contact limits the impact of the insecticide on nontarget organisms. The structure and the mode of action of δ-endotoxin is not well understood; the gut lining is affected by this toxin, and within 20 minutes after ingestion there is a general breakdown of the epithelium. Some of the insects that have been controlled by *B.t.* preparations include: Cabbage looper *Trichoplusia ni* (Hübner), Codling moth, *Laspeyresia pomonella* (L.), European corn borer, *Ostrinia nubilalis* (Hübner), gypsy moth, *Porthetria dispar* (L.) imported cabbage worm, *Pieris rapae* (L.), Tobacco hornworm, *Heliothis virescens* (Fabricius), and tomato hornworm, *Manduca quinquemaculata* (Haworth). The Food and Drug Administration granted full exemption from tolerance for *B.t.*-based insecticides and in 1965 commercial use of *B.t.* crystals and spores was registered. Fisher and Rosner (13) have discussed the details of the safety of these preparations.

The β-exotoxin of *B.t.* has a wider spectrum of toxicity than the crystal, and it is also toxic to mammals when administered in high dosages through specific routes. Bond et al. (12) have tentatively assigned this exotoxin the following structure:

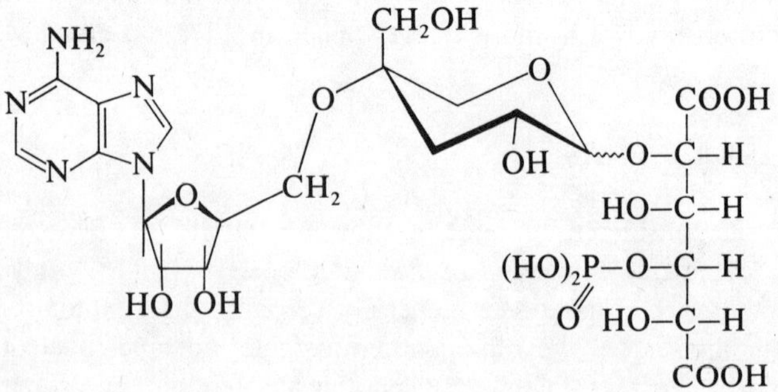

It is an adenine nucleotide that appears to be an ATP analog. Its mode of action is possibly due to competitive inhibition of enzymes that catalyze the splitting of ATP and pyrophosphate. The competitive inhibition of RNA polymerase from *Escherichia coli* and from mammalian systems has been shown for this enzyme (14–16). Toxicity of the purified heatstable toxin to mice by injection and by oral administration was studied by Barjac and Riou (17). By intraperitoneal or subcutaneous injection, doses of 100–800 mg resulted in mortality of the animals. $LD_{50}$ for female mice was calculated as 13 mg/kg by intraperitoneal injection and 16 mg/kg by the subcutaneous route. Considerable necrosis took place in the liver which was the principal organ affected; kidneys, spleen, and adrenal glands also showed signs of damage. Since all strains of *B.t.* do not produce this toxin, for the commercial insecticidal preparation "$\beta$-exotoxin free" strains can be employed.

Unlike the endotoxin, which is toxic mainly to lepidopterous larvae, the $\beta$-exotoxin has a wider spectrum of toxicity and is active against many orders of insects, including Lepidoptera, Diptera, Hymenoptera, Orthoptera, and Coleoptera. The effects develop slowly and manifest themselves during the molt or metamorphosis. Teratogenic effects on the developing insect are evident (18, 19). Treated house fly larvae fail to form normal puparia, or adults fail to emerge from such puparia. The teratogenic effects of this toxin on the immature stages of the house fly have been compared (20, 21) to those reported in the same insect by Quraishi and Thorsteinson (22) as a result of treatment with straight-chain fatty acids.

## BACILLUS POPILLIAE

This bacillus is also being used extensively in the field for the control of Japanese beetle, *Popillia japonica*; a commerial product called Doom® is manufactured from the spores of the bacterium.

## ENTOMOGENOUS FUNGI

Several species of entomogenous fungi are being studied for insect control: *Beauveria bessiana* is being mass produced for the control of leaf-feeding insects; *Coelomomyces* spp. and *Metarrhizium anisopliae* are under investigation for mosquito control. The metabolites produced by some entomogenous fungi are reported to be toxic to insects (4, 5, 23–25). For instance, *Metarrhizium* produces destruxins A and B and other insecticidal toxins (23–25). Roberts (24) has suggested that destruxins are important in producing paroxyms of mycosis. The structure of destruxin B was elucidated by Tamura

$$
\begin{array}{c}
\text{H}_3\text{C} \\
\text{O} \quad \text{H}_3\text{C}{>}\text{CHCH}_2 \quad \text{O} \\
\text{CH}_2-\text{CH}_2-\text{C}-\text{O}-\text{CH}-\text{C}-\text{N}-\text{CH} \\
\text{NH} \qquad\qquad\qquad\qquad \text{C}=\text{O} \\
\text{O}=\text{C} \qquad\qquad\qquad\qquad\quad \text{NH} \\
\text{H}_3\text{C}-\text{C}-\text{H}-\text{N}-\text{C}-\text{CH}-\text{N}-\text{C}-\text{CH}-\text{CH}{<}{\text{CH}_3 \atop \text{C}_2\text{H}_5} \\
\text{CH}_3 \quad \text{O} \quad \text{C}-\text{H} \quad \text{CH}_3 \quad \text{O} \\
\text{H}_3\text{C} \quad \text{CH}_3
\end{array}
$$

Destruxin B (28)

et al. (26), and it was synthesized by Kuyama and Tamura (27). Suzuki and Tamura (28) have reviewed destruxins and insecticides of microbial origin.

Fungi as a group are not completely safe to handle and can infect vertebrates; some mycotoxins are hazardous and therefore extreme care must be taken before any product is used on a large scale. Allergy to *Beauveria bassiana* has been reported in some persons with repeated exposures to massive amounts of powdered preparation of this fungus and the medium on which it was grown.

Tamura et al. (29) discovered that cultured broth from *Streptomyces mobarensis* Nagatsu et Suzuki showed marked insecticidal activity; in 1963 two insecticidal chemicals, namely, piericidin A and piericidin B, were isolated from the mycelia of the microbe (7a). Piericidin A has knockdown properties against insects, but it is also an inhibitor of respiration in mitochondria in insects and mammals. Piericidin B has not been investigated in detail. Both chemicals are toxic to mammals by oral administration, though piericidin B is less toxic than piericidin A. The piericidins, because of their chemical instability and mammalian toxicity, appear to be of limited interest in insect control.

## REFERENCES

1. S. R. Dutky. In *Insect Pathology* (E. A. Steinhaus, ed.), Vol. 2 Academic Press, New York, 1963, p. 75.
2. L. A. Falcon. In *Microbial Control of Insects and Mites* (H. D. Burges and N. W. Hussey, eds.), Academic Press, New York, 1971, p. 67.
3. Y. Tanada. In *Pest Control*, (W. W. Kilgore and R. L. Doutt, eds.), Academic Press, New York, 1967, p. 31.
4. H. D. Burges and N. W. Hussey, eds. *Microbial Control of Insects and Mites*, Academic, New York, 1971, 861 pp.
5. New York Academy of Sciences. Conference on the Regulation of Insect Populations by Microorganisms, Ann. N.Y. Acad. Sci. **217** (June 1973).

6. P. DeBach ed. *Biological Control of Insect Pests and Weeds,* Chapman and Hall, London, 1964.

6a. G. E. Cantwell, ed. *Insect Diseases,* Vol. I, Dekker, New York, 1974, 300 p. plus glossary.

7. J. R. Norris. *Symp. Soc. Gen. Microbiol.* **21**: 197 (1971).

7a. M. Jacobson and D. G. Crosby, eds. *Naturally Occurring Insecticides,* Dekker, New York, 1971.

8. C. L. Hannay. *Nature* **172**: 1004 (1953).

9. T. A. Angus. *Can. J. Microbiol.* **2**: 416 (1956).

10. E. McConnell and A. G. Richards. *Can. J. Microbiol.* **5**: 161 (1969).

11. R. P. M. Bond, C. B. C. Boyce and S. R. French. *Biochem. J.* **114**: 477 (1969).

12. R. P. M. Bond, C. B. C. Boyce, M. H. Rogoff and T. R. Shieh. In *Microbial Control of Insects and Mites* (H. D. Burges and N. W. Hussey, eds.), Academic Press, New York, 1971.

13. R. Fisher and L. Rosner. *J. Agr. Food Chem.* **7**: 686 (1959).

14. K. Šebesta and K. Horská. *Biochem. Biophys. Acta* **169**: 281 (1968).

15. K. Šebesta, K. Horská, and J. Vankova, Abstracts, 5th Meeting FEBS, Czechoslavakia Biochemical Society,1968, p. 250.

16. T. Beebe, A. Korner, and R. P. M. Bond. *Biochem. J.* **127**: 619 (1972).

17. H. de Barjac and J.-Y. Riou. *Rev. Pathol. Comp. Med. Exp.* **6**(6): 19 (1969).

18. J. N. Lillies and P. H. Dunn. *J. Inst. Pathol.* **1**: 309 (1959).

19. A. Burgerjohn and H. de Barjac. *C.r. hebd. Séanc Acad. Sci. Paris* **251**: 911 (1960).

20. A. M. Heimpel. *Annu. Rev. Entomol.* **12**: 287 (1967).

21. M. H. Rogoff and A. A. Yousten. *Annu. Rev. Microbiol.* **28**: 357 (1969).

22. M. S. Quraishi and A. J. Thorsteinson. *J. Econ. Entomol.* **58**: 400 (1965).

23. Y. Kodaira. *Agr. Biol. Chem.* **25**: 261 (1961).

24. D. W. Roberts *J. Invertebr. Pathol.* **8**: 212 (1966).

25. D. W. Roberts. *op. cit.* **14**: 82 (1969).

26. S. Tamura, S. Kuyama, Y. Kodaira, and S. Higashikawa. *Agr. Biol. Chem.* **28**: 137 (1964).

27. S. Kuyama, and S. Tamura. *Agr. Biol. Chem.* **30**: 517 (1965).

28. A. Suzuki and S. Tamura. In *Insecticides,* Proceedings of the 2nd International IUPAC Congress of Pesticide Chemistry, Vol. I Gordon and Breach, New York, 1972, p. 163.

29. S. Tamura, N. Takahashi, S. Miyamoto, R. Mori, S. Suzuki, and J. Nagatsu. *Agr. Biol. Chem.* **27**: 576 (1963).

# DYNAMICS OF INSECT TOXICOLOGY

# INSECT–INSECTICIDE INTERACTIONS

# CHAPTER 15

Sorption, Penetration, Distribution
Inside the Body and Site of Action

When an insecticide comes in contact with an insect the first interaction is evidently the "dissolution" of the chemical in the epicuticular wax (1–3). The affinity of the insecticide for this wax and the physicochemical properties of the wax may initially determine the effectiveness of the chemical as an insecticide. Contact insecticides are lipid soluble and Lord (4) reported that chitin readily absorbed DDT and its analogs from a colloidal suspension. After this initial "dissolution" of the chemical, it evidently spreads over the entire cuticle of the insect. Quraishi and Abdul Matin (5) found that when lindane-$^{14}$C and DDT-$^{14}$C were topically applied to a small area on the prothorax of the American cockroach, they spread all over the insect, including the wings, within a short time after application (Fig. 15.1). Lewis (6) reported that oil films containing diiodooctadecane-1$^{131}$ spread over the whole body in the living fly, *Protophormia terraenovae*, "within a very few minutes of the initial exposure." Gerolt (7) using dieldrin-$^{14}$C and other radioactive insecticides also found that they spread laterally in the insect cuticle.

Thus there seems to be some agreement that contact insecticides spread over the cuticle after they come in contact with the insect, but there seems to be little agreement on how they pass through the cuticular barrier and reach the site of action. Ebeling (3) has very ably summarized the frustration, "How diametrically opposite conclusions can be reached by different investigators doing what appears to be sound, basic research is one of the enigmas in the present status of insect toxicology."

Earlier investigators supported the view that, "the specific toxicity of DDT

**Fig. 15.1.** Print made from the radioautograph of the wings of the American cockroach 1 hour after topical application of DDT-$^{14}$C restricted to a small area on the prothorax. Note that during the process of making the photographic print black and white have been reversed. From M. S. Quraishi and Z. T. Poonawalla, *J. Econ. Entomol.* **62**(5): 988–994 (1969). Reproduced by permission.

to insects is therefore a result of the high permeativity of the insect cuticle to DDT to which the vertebrate skin is an effective barrier" (8). However O'Brien (9), by calculating the penetration of DDT through insect and mammalian integuments showed that times for half penetration are 26 hours for rats and 26.4 hours for the American cockroach. Data obtained to date on the penetration of a topically applied chemical into the insect body are insufficient for a proper understanding of the phenomenon; O'Brien (9) has commented on the deficiency that exists in the field and has summarized the available data. His conclusions are that the half time of penetration ($t_{0.5}$) is a meaningful parameter, that is, the penetration of an insecticide through the insect cuticle is a truly random process and the kinetics of penetration follow first-order form, that is, log of amount penetrating as a function of time is linear. Half time of penetration as a meaningful parameter has been reported in several experiments (10–12). However Olson (13), on the basis of his work, concluded that the penetration of DDT-$^{14}$C into and through the integument of the American cockroach is not a simple first-order function; it may be an aggregate of first-order events. He further surmised that the rate-limiting step in penetration was the partioning of DDT from epicuticular grease into lower phases. When DDT was injected into the hemolymph, about 8% back-diffused into the epicuticular lipids.

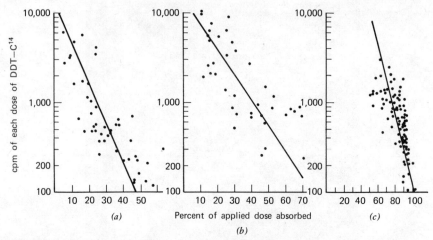

**Fig. 15.2.** Different amounts of DDT-$^{14}$C (in cpm) applied topically on individual house flies, and percent absorbed at each dose level 2 hours (*a*), 6 hours (*b*), and (*c*) 24 hours after application. From Z. T. Poonawalla and M. S. Quraishi, *Can. Entomol.* **102**: 1136–1138 (1970). Reproduced by permission.

Poonawalla and Quraishi (14) studied the dose dependence of the rate of penetration of topically applied DDT-$^{14}$C into the house fly. Individual flies were treated at doses varying between 35 and 2900 ng. Penetration was studied at 2, 6, and 24 hour intervals. The authors showed that the rate of penetration was dose dependent and not exponential even when such small quantities of the insecticide were used (Table 15.1 and Fig. 15.2). The relationship between insect toxicity and rates of penetration and detoxication of insecticides on a mathematical basis has been reported by Sun (15).

**Table 15.1   Dose of DDT-$^{14}$C Applied Topically and Percent Absorption at Different Time Intervals.**

| Time Interval (hr) | Dose (ng) | Mean Percent Absorbed | Significance of Differences |
|---|---|---|---|
| 2 | 35– 285 | 35.78 ± 12.41 | $P < 0.01$ |
|   | 285– 570 | 14.76 ± 6.41 |  |
| 6 | 76– 285 | 50.05 ± 13.44 | $P < 0.01$ |
|   | 285–2900 | 24.19 ± 12.71 |  |
| 24 | 35– 142 | 90.05 ± 5.72 | $P < 0.05$ |
|   | 142– 285 | 84.61 ± 8.88 | $P < 0.01$ |
|   | 285– 804 | 71.00 ± 12.08 |  |

From Z. T. Poonawalla and M. S. Quraishi. *Can. Entomol.* **102:** 1136–1138 (1970). Reproduced by permission.

Further detailed investigations are needed before the controversy can be solved. But we now know that the cuticle is not the type of complicated membranelike inert organ it once was assumed to be. On the contrary, the extracellular cuticle is in dynamic equilibrium with the metabolic pool. Olson's finding that when DDT was injected into the hemolymph of the American cockroach about 8% back-diffused into the epicuticular lipids is important. How the insecticide back-diffuses and how the topically applied insecticide reaches the metabolic pool are the questions for which we do not have definite answers, or even a tacit agreement among all workers. It is generally accepted that the lipid-soluble insecticides pass through the insect integument and reach the site of action (somewhere in the central nervous system) by way of the hemolymph (16). However Gerolt believes that after spreading laterally within the cuticle, probably primarily in the endocuticle, the topically applied

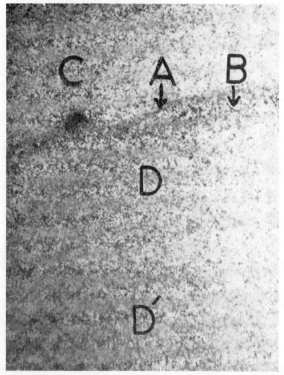

**Fig. 15.3.** Longitudinal section of the mesonotum through the area of topical application 30 minutes after treatment and its radioautograph juxtaposed. The latter has been shifted slightly proximally to show clearly the cuticle (A) through the emulsion and its impression on the radioautograph (B). The diffusion of radioactivity through the intrascutal sutures is clearly visible (C). Dorsal median muscles (D and D') can be seen through the emulsion slide. From M. S. Quraishi and Z. T. Poonawalla, *J. Econ. Entomol.* **62**(5): 988–994 (1969). Reproduced by permission.

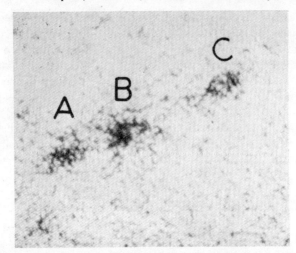

**Fig. 15.4**  Radioautograph of a longitudinal section of the mesonotum through untreated cuticle 30 minutes after topical application. The diffusion of radioactivity through the base of the prescutal seta (A), intrascutal suture (B), and the base of another seta (C) is clearly visible. Note that the cuticle has not left any impression on the radioautograph, indicating the absence of radioactivity from the cuticle itself. From M. S. Quraishi and Z. T. Poonawalla, *J. Econ. Entomol.* **62**(5): 988–994 (1969). Reproduced by permission.

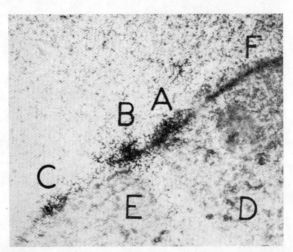

**Fig. 15.5.**  Radioautograph in Fig. 15.4 juxtaposed on the section. Note the inversion of the spots A, B, and C because of the inversion of the radioautograph for juxtaposition. Visible through the emulsion of the radioautograph slide are turgosternal muscles (D), tergal remotors of coxa (E), and cuticle (F). D, E, and F have not produced any impression in the radio-autograph. Compare Fig. 15.4. From M. S. Quraishi and Z. T. Poonawalla, *J. Econ. Entomol.* **62**(5): 988–994 (1969). Reproduced by permission.

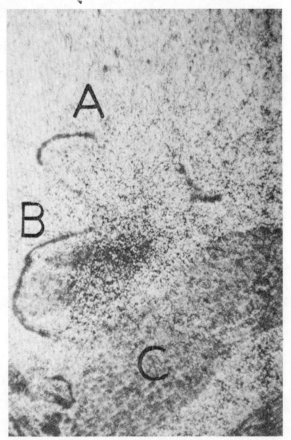

**Fig. 15.6.** Longitudinal section through the mesocutellum and mesopostnotum, 30 minutes after topical application, and its radioautograph juxtaposed. Note the extensive diffusion of radioactivity under the mesopostnotum. There is no trace of radioactivity on the surrounding cuticle that can be clearly seen through the emulsion slide. Visible through the emulsion slide are: mesoscutellum (A), mesopostnotum (B), and lateral oblique dorsal muscle (C). None of these has left an impression on the radioautograph. From M. S. Quraishi and Z. T. Poonawalla, *J. Econ. Entomol.* **62**(5): 988–994 (1969). Reproduced by permission.

insecticide does not penetrate the insect cuticle but reaches the site of action through the integument of the trachae (7, 17). Sellers and Guthrie (18) studied the concentration of dieldrin-$H^3$ in the thoracic ganglion preparation of the house fly by radioautography and did not find any localization of radioactivity in the trachae. Dieldrin localization had occurred mainly in the collagenlike fibers of the neural lamella and the nerve fiber sheaths of the neuropile. Similarly, Burt et al. (19) did not find evidence to suggest that pyrethrin I reached the nervous system through the tracheal system.

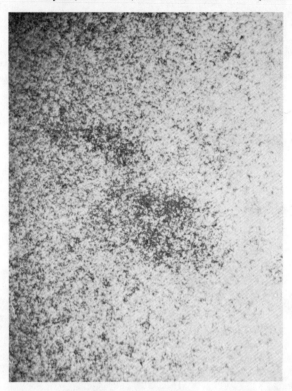

**Fig. 15.7.** Radioautograph of longitudinal section through wing base and the pleural region 30 minutes after topical application. Note the extensive diffusion of radioactivity in the area. None of the cuticular structures have left any trace on the radioautograph. In this case the radioautograph was not juxtaposed on the section. Compare Fig. 15.6. From M. S. Quraishi and Z. T. Poonawalla, *J. Econ. Entomol.* **63**(5): 988–994 (1969). Reproduced by permission.

Quraishi and Poonawalla (20) used freeze-drying and radioautographic techniques to study the diffusion of topically applied DDT-$^{14}$C into the house fly and its distribution in internal organs. They found that the insecticide did not diffuse through the general surface of the cuticle or through the cuticle directly treated. Entry was effected through intrascutal sutures, through bases of setae, under the mesopronotum, through the pleural region, under the noto-pleural suture and through the wing base (Figs. 15.3–15.9). On the basis of their work these authors surmise that at the point of topical application the solvent disrupted the wax monolayer. At this point the insecticide was absorbed onto the "bare" surface of the cuticle. The unadsorbed insecticide interacted with the wax monolayer around it and passed into "solution" in the wax mono-

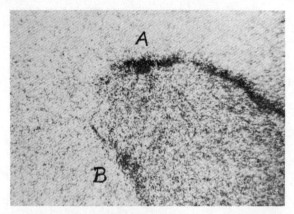

**Fig. 15.8.**   Longitudinal section through the prothorax and part of the mesothorax, 6 hours after topical application, and its radioautograph juxtaposed. Note the adsorption of radioactivity on tergites and sternites. Strong diffusion of radioactivity through the suture between the prothorax and mesothorax (A) and through the membrane between sternites (B) is clearly visible. Note also that there is no selective uptake of radioactivity by any of the thoracic muscles. This section did not cut through the alimentary canal. The emulsion grains in the radioautograph of the cuticle appear darker than their actual color because of their juxtaposition in the background on the dark cuticle which, unlike that of Fig. 15.5 (F), is not clearly visible through the darkened emulsion. Compare emulsion background outside the section with the general darkening of the emulsion exposed to the muscles cut in the section. Compare Fig. 15.9. From M. S. Quraishi and Z. T. Poonawalla, *J. Econ. Entomol.* **62**(5): 988–994 (1969). Reproduced by permission.

layer and through it spread over the entire surface of the insect. Two processes then took place: the rapid diffusion of the chemical into the body "through sutures, membranes, bases of setae, and possibly other portals of entry"; and the "slow interaction of DDT-$^{14}$C with the cuticle at the interphase between the wax monolayer and the cuticle." During the latter process the insecticide was slowly adsorbed onto the surface of the cuticle itself, though "the process of diffusion through various portals of entry was evident throughout the period of this study."

## METABOLISM OF INSECTICIDES IN THE CUTICLE

The insect cuticle is biochemically active and initially interacts with the topically applied toxicants. The conversion of DDT to DDE by the cuticle of grasshopper has been reported by Sternburg and Kearns (21). Quraishi reported similar conversion by the puparial cuticle of the house fly (22). The activation of malathion to malaoxon was surmised by Ahmed and Gardiner (23) in the case of the cuticle of the locust, *Schistocerca gregaria,* because

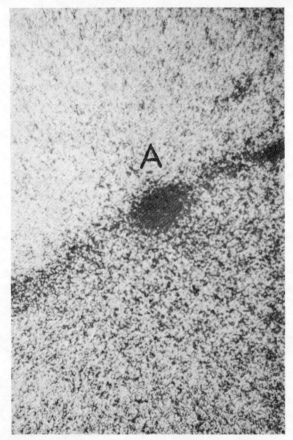

**Fig. 15.9.** Radioautograph of a section through part of the mesothorax 21 hours after topical application. The cuticle shows strong adsorption of radioactivity, but diffusion of radioactivity into the body is taking place through the scutoscutellar suture (A). Note the darkening of the emulsion due to radioactivity in muscles cut in section (below the cuticle); compare with the background (above the cuticle). Compare Fig. 15.8. From M. S. Quraishi and Z. T. Poonawalla, *J. Econ. Entomol.* **62**(5): 988–994. Reproduced by permission.

topically treated insects showed more typical symptoms of malathion poisoning than those in which malathion was injected.

## EFFECT OF TOPICALLY APPLIED CHEMICAL

The effect of the externally applied chemical therefore depends on several factors and represents, to some extent, a series of consecutive reactions:

1. Interaction with the cuticle
   a. Solubility in the epicuticular waxes
   b. Bioactivation and/or degradation in the epicuticular waxes
   c. Loss from the waxes to the integument until a steady state is reached
   d. Bioactivation and/or biodegradation in the integument
   e. Penetration through the integument
2. Distribution inside the body
   a. Bioactivation and/or biodegradation
   b. Travel to the site of action
   c. Excretion

If the amount applied is just sufficient so that the saturation point in the cuticle or epicuticular waxes is reached, penetration continues until a steady state is reached and then ceases. The fate of the insect depends on the sum total of the processes given above.

## DISTRIBUTION INSIDE THE BODY

Little information is available on the internal distribution of a topically applied chemical once it has diffused through the cuticle. Earlier investigations were carried out by several workers (24–26), but histological techniques were rarely used. Quraishi and Poonawalla (20) used two techniques for the study of internal distribution of topically applied DDT-$^{14}$C: (a) At definite time intervals, after topical application of benzene solution of the insecticide on the centre of the mesonotum of lightly anesthetized house flies, the alimentary canal was dissected out, spread over a slide, dried, and radioautographed; (b) transmission of light through radioautographs of histological sections of various organs was measured by a photoelectric cell attached to a microscope with a ×40 objective. "The scale of the photoelectric cell was divided into 15 arbitrary units (0 = no measurable transmission of light, 15 = maximum transmission). The intensity of light was so adjusted that the unexposed emulsion on the slide (background) gave a reading of 15." Transmission of light through the radioautographs of various organs is given in Table 15.2.

"Of all the internal organs, the midgut and hindgut showed the greatest concentration of radioactivity. The alimentary canal thus performed the main role in the absorption and elimination of the topically applied DDT-$^{14}$C and/or its metabolites. Radioautographs of the entire alimentary canal, dissected within 15 minutes after topical application of DDT-$^{14}$C on the mesonotum, showed a uniform acquisition of radioactivity. However the malpighian tubules rarely showed on radioautographs taken earlier than 30 minutes after topical application. On the contrary, the rectal papillae showed a selectively higher

Table 15.2    Transmission of Light Through Radioautographs of Different Organs in Longitudinal Sections of the House Fly Treated Topically on Mesonotum with DDT-$^{14}$C; 21 Hours After Treatment

| Radioautograph of: | Transmission of Light in Arbitrary Units[a] |
|---|---|
| Mesonotum at the site of application | 5 |
| Intrascutal suture at site of application | 3–4 |
| Mesonotum | 6–8 |
| Between gut wall and peritrophic membrane in the thoracic region | 3–4 |
| Midgut cut in transverse section in abdomen, between peritrophic membrane and gut wall | 1–4 |
| Testes | 6–7 |
| Seminal vesicles | 6–7 |
| Muscles of thorax | 11–12 |
| Leg muscles | 14–15 |
| Brain, optic lobes | 11–12 |
| Background (slide emulsion) | 15 |

SOURCE:    From M. S. Quraishi and Z. T. Poonawalla, *J. Econ. Entomol.* **62**(5): 988–994 (1969). Reproduced with permission.

[a] Sections of untreated insects did not leave any impression on the emulsion; transmission of light through background (slide emulsion) is therefore the same as that in controls. A transmission of 15 indicates no radioactivity and a transmission of 1 the maximum concentration of radioactivity.

concentration of radioactivity than the alimentary canal in general, a trend that continued to be discernable until the rectum itself acquired enough radioactivity to make differentiation difficult (about 9 hours after topical application). In the midgut the radioactivity was concentrated only between the inner wall of the gut and the peritrophic membrane" (Figs. 15.10–15.12).

"All other internal organs showed a relatively low and rather uniform uptake of radioactivity. The exceptions were the reproductive organs of both sexes and the seminal vesicles in the males. All of these organs showed a comparatively greater concentration of radioactivity" (Table 15.2)*.

Nakajima et al. (27) used whole-body radioautographic techniques to study the penetration, distribution, and excretion of HCH isomers and nicotine in the American cockroach. Benzene solutions of HCH and ethanol solutions of

* Quotations from M. S. Quraishi and Z. T. Poonawalla, *J. Econ. Entomol.* **62**: 988 (1969) have been made by permission of the Publisher.

**Fig. 15.10.** Longitudinal section through cardia and midgut between the gut wall and the peritrophic membrane of the thorax 21 hours after topical application. From M. S. Quraishi and Z. T. Poonawalla, *J. Econ Entomol.* **62**(5): 988–994 (1969). Reproduced by permission.

nicotine were applied "on the back side of the abdomen" of the insect. They found that lindane reached almost all parts of the central nervous system (CNS), crop, and gizzard within 15 minutes. Later on a "major part of the penetrated $\beta$- and $\gamma$-BHC was found in the midgut, gastric caeca and at the walls of the crop." Nicotine penetration and distribution in the body and CNS was comparatively more rapid.

From the English summary and from the radioautographs in the paper it cannot be discerned whether in the midgut the radioactivity was concentrated only between the inner wall of the gut and the peritrophic membrane as reported by Quraishi and Poonawalla (20) for house fly. Such localized concentration in the case of DDT is interesting and needs further investigation.

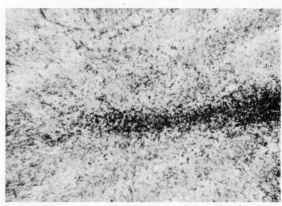

**Fig. 15.11.** Radioautograph of the section in Fig. 15.10 showing concentration of the radioactivity between the peritrophic membrane and the gut wall. From M. S. Quraishi and Z. T. Poonawalla, *J. Econ. Entomol.* **62**(5): 988–994 (1969). Reproduced by permission.

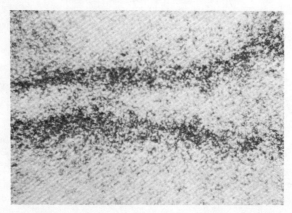

**Fig. 15.12.** Radioautograph of a longitudinal section through the lumen of the midgut 21 hours after topical application. Note the absence of radioactivity from the lumen of the midgut and its concentration between the peritrophic membrane and the gut wall. From M. S. Quraishi and Z. T. Poonawalla, *J. Econ. Entomol.* **62**(5): 988–994 (1969). Reproduced by permission.

## REFERENCES

1. G. Armstrong, F. R. Bradbury, and H. G. Britton. *Ann. Appl. Biol.* **39**: 548 (1952).
2. A. H. McIntosh. *op. cit.* **45**: 189 (1957).
3. W. Ebeling. The Permeability of Insect Cuticle," In *The Physiology of Insecta* (M. Rockstein, ed.), 2nd ed., Academic Press, New York (1974).
4. K. A. Lord. *Biochem. J.* **43**: 72 (1948).

5. M. S. Quraishi and A. S. M. Abdul Matin. CENTO Institute of Nuclear Science. Tehran. Iran, Report No. 9 (a), unpublished. (1962).
6. C. T. Lewis. *Nature* **193**: 904 (1962).
7. P. Gerolt. *Pestic. Sci.* **1**: 209 (1970).
8. H. Martin. *The Scientific Principles of Crop Protection,* 5th ed.. Arnold. London, 1964.
9. R. D. O'Brien. *Insecticides: action and metabolism.* Academic Press. New York, 1967.
10. A. J. Forgash. B. J. Cook. and R. C. Riley. *J. Econ. Entomol.* **55**: 545 (1962).
11. D. A. Lindquist and P. A. Dahm. *op. cit.* **49**: 579 (1956).
12. C. H. Schmidt and P. A. Dahm. *op. cit.* **49**: 729 (1956).
13. W. P. Olson. *Comp. Biochem. Physiol.* **35**: 574 (1970).
14. Z. T. Poonawalla and M. S. Quraishi. *Can. Entomol.* **102**: 1136 (1970).
15. Y.-P. Sun. *J. Econ. Entomol.* **61**: 949 (1968).
16. P. E. Burt and K. A. Lord. *Entomol. Exp. Appl.* **11**: 55 (1968).
17. P. J. Gerolt. *J. Insect Physiol.* **15**: 563 (1969).
18. L. G. Sellers and F. E. Guthrie. *J. Econ. Entomol.* **64**: 352 (1971).
19. P. E. Burt. K. A. Lord. J. M. Forrest. and R. E. Goodchild. *Entomol. Exp. Appl.* **14**: 255 (1971).
20. M. S. Quraishi and Z. T. Poonawalla. *J. Econ. Entomol.* **62**: 988 (1969).
21. J. Sternburg and C. W. Kearns. *J. Econ. Entomol.* **45**: 497 (1952).
22. M. S. Quraishi. *Can. Entomol.* **102**: 1189 (1970).
23. G. A. Ahmed and B. G. Gardiner. *Pestic. Sci.* **1**: 217 (1970).
24. R. W. Fisher. *Can. J. Zool.* **30**: 254 (1952).
25. F. Matsumura. *J. Insect Physiol.* **9**: 207 (1963).
26. W. P. Olson. and R. D. O'Brien. *op. cit.* **9**: 777 (1963).
27. E. Nakajima. H. Shindo. N. Kurihara. and T. Fujita. *Radioisotopes* **18**: 365. In Japanese with English abstract. (1969).

# INSECT–INSECTICIDE INTERACTIONS

# CHAPTER 16

# Mixed-Function Oxidases, Synergism

The detoxication of xenobiotics is brought about by a number of enzymes in a living organism. However investigations carried out in the last 15 years have made it increasingly clear that mixed-function oxidases (MFO) are responsible for initial oxidative enzymatic attack on foreign compounds. These enzymes are not present in all tissues of mammals but are mostly found in the liver and are located exclusively in the endoplasmic reticulum of cells. In insects they are found in various tissues. MFO's are quite versatile and accept most xenobiotics as substrates, and metabolize them by oxidative modifications. They bring about the activation of molecular oxygen by cytochrome P450; nicotinamide adenine dinucleotide in its reduced form (NADPH) delivers the necessary electrons by way of flavoprotein-1 to the cytochrome which is the terminal oxidase of the electron transport chain. The molecular oxygen is activated by the heme moiety of cytochrome P450 and finally oxidizes the xenobiotic. Thus, if $A \cdot X$ represents a foreign compound, the reaction can be written as:

$$NADPH + H + O_2 + A \cdot X \longrightarrow NADP + H_2O + AOX$$

However the picture is much more complicated and all the pathways are not yet known. Since it is not possible to discuss in detail MFO, synergism, and antagonism in a book of this nature, these subjects are dealt with briefly. Those interested in greater details should read the excellent review by Casida (1), which has about 300 references, and Volume II of the *Proceedings of the Second International IPUAC Congress of Pesticide Chemistry*, which deals exclusively with these subjects (2).

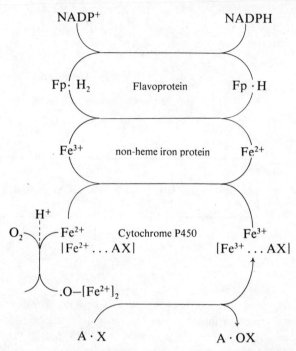

**Fig. 16.1** The mixed function oxidases utilize NADPH to activate molecular oxygen. Cytochrome P450 binds the $O_2$ molecule and activates oxygen. One oxygen atom is transferred to the substrate by the activated oxygen; the other atom is released as $HO^-$. Flavoprotein provides both electrons for the activation of oxygen and is thus oxidized from the dihydro form to the semiquinone and is reduced again by NADPH. The exact mechanism is not fully known.

Figure 16.1 shows some of the possible steps and the roles of cytochrome P450, flavoproteins, and Fe.

MFOs are inducible by lipid-soluble compounds like chlorinated hydrocarbons and barbiturates. By preconditioning of an animal by a chemical that induces MFO, the toxicity of chemicals that are oxidized by these enzymes can be modified toward the animal. These effects are very noticeable in mammals pretreated with chlorinated hydrocarbons, but in insects the level of chlorinated hydrocarbons required for induction of enzymes is high; therefore inductive effects can only be demonstrated in the insects already resistant to the inducing compound. However benzylic hydroxylation of DDT was found in *Drosophila melanogaster* in 1959 and soon afterwards microsomal oxidation came to be recognized as an important mechanism for the metabolism of insecticides in insects. Sun and Johnson (3) were the first to report that synergism was manifested only when oxidative metabolism was involved in the detoxication

Conversion of phosphorothionates to phosphates

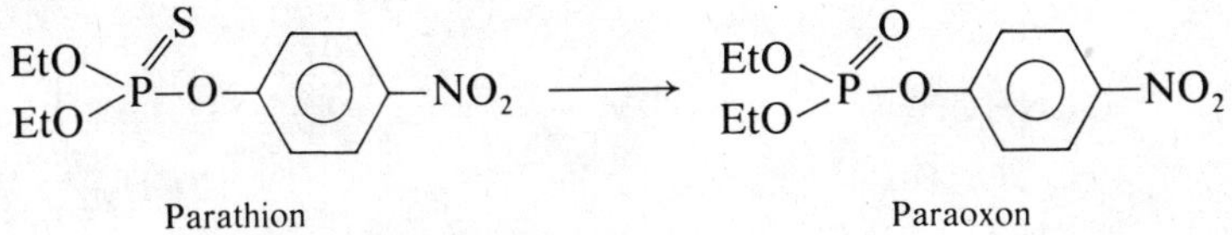

$$\underset{\text{EtO}}{\overset{\text{EtO}}{>}}\overset{\text{S}}{\underset{\|}{P}}-O-\text{⟨○⟩}-NO_2 \longrightarrow \underset{\text{EtO}}{\overset{\text{EtO}}{>}}\overset{\text{O}}{\underset{\|}{P}}-O-\text{⟨○⟩}-NO_2$$

Parathion                Paraoxon

(a)

Sulfur oxidations

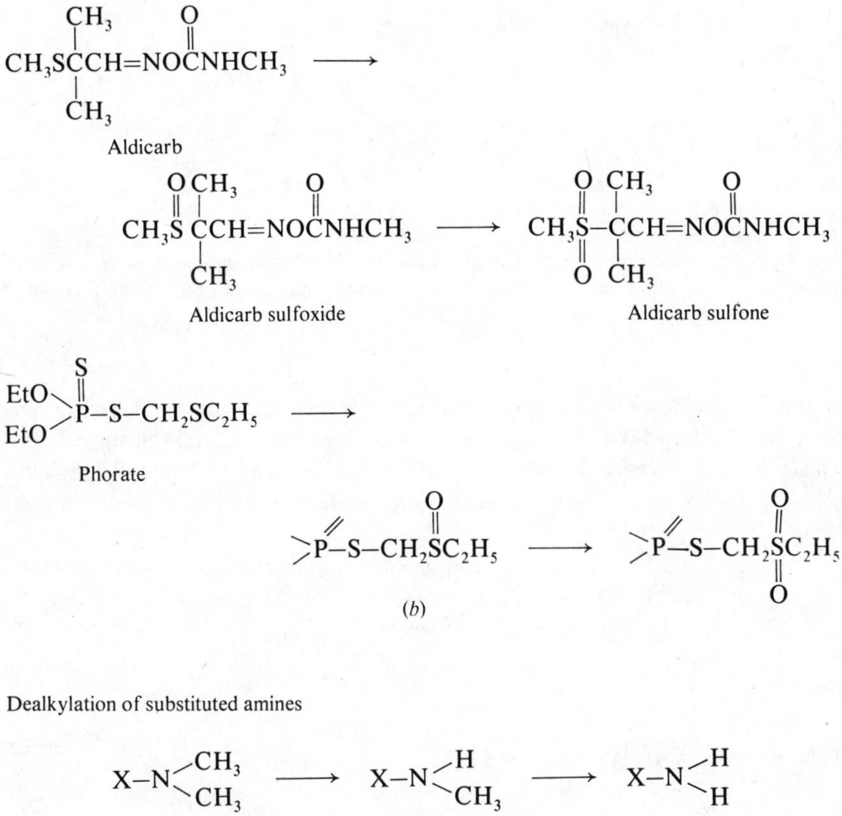

$$\underset{\overset{|}{CH_3}}{\overset{\overset{CH_3}{|}}{CH_3SC}}CH=N\overset{O}{\overset{\|}{O}}CNHCH_3 \longrightarrow$$

Aldicarb

$$\underset{\overset{|}{CH_3}}{\overset{O\;CH_3}{\overset{\|\;|}{CH_3S\;C}}}CH=N\overset{O}{\overset{\|}{O}}CNHCH_3 \longrightarrow \underset{\overset{\|}{O}\;CH_3}{\overset{O\;CH_3}{\overset{\|\;|}{CH_3S-C}}}CH=N\overset{O}{\overset{\|}{O}}CNHCH_3$$

Aldicarb sulfoxide              Aldicarb sulfone

$$\underset{\text{EtO}}{\overset{\text{EtO}}{>}}\overset{\text{S}}{\underset{\|}{P}}-S-CH_2SC_2H_5 \longrightarrow$$

Phorate

$$>\overset{O}{\overset{/\!\!/}{P}}-S-CH_2\overset{O}{\overset{\|}{S}}C_2H_5 \longrightarrow >\overset{O}{\overset{/\!\!/}{P}}-S-CH_2\overset{O}{\underset{\|}{\overset{\|}{S}}}C_2H_5$$

(b)

Dealkylation of substituted amines

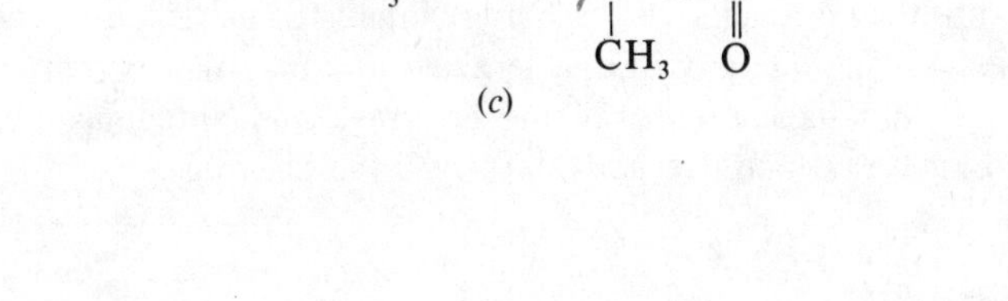

$$X-N\overset{CH_3}{\underset{CH_3}{<}} \longrightarrow X-N\overset{H}{\underset{CH_3}{<}} \longrightarrow X-N\overset{H}{\underset{H}{<}}$$

Bidrin                Azodrin

$$X= \underset{CH_3O}{\overset{CH_3O}{>}}\overset{O}{\overset{/\!\!/}{P}}O-\underset{\overset{|}{CH_3}}{C}=CH\overset{}{\underset{O}{\overset{\|}{C}}}$$

(c)

Hydroxylation of aromatic and alicyclic rings

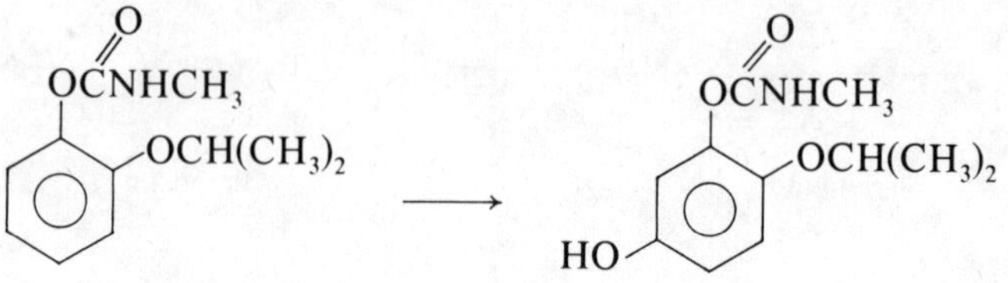

Baygon

(d)

Epoxidation of double bonds

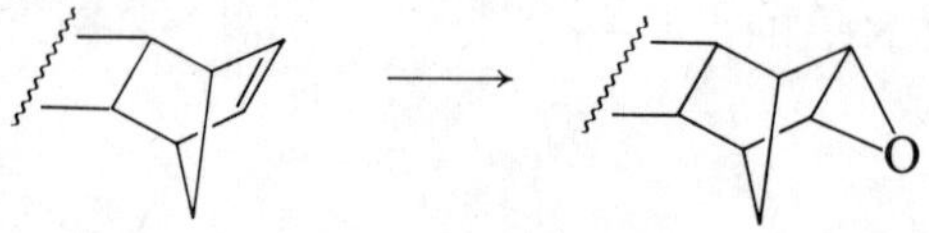

Aldrin                                              Dieldrin

(e)

**Fig. 16.2** (a) Conversion of phosphorothionates to phosphates; (b) sulfur oxidations; (c) dealkylation of substituted amines (d) hydroxylation of aromatic and alicyclic rings; (e) epoxidation of double bonds.

of the insecticide. This explained the synergistic action of methylenedioxy-phenyl (1,3-benzodioxole) compounds toward pyrethroids, and the suggestion was made that synergists interfere with oxidative detoxication. Sun and Johnson (3) also showed that sesamex inhibited epoxidation of aldrin *in vivo* in house flies. Hodgson and Plapp (4) have reviewed the work on insect microsomes, and effects of synergists on the metabolism and toxicity of anticholinesterases have been discussed by Wilkinson (5).

Some of the reactions in which the MFO play an important role are shown in Fig. 16.2.

## SYNERGISTS AND ANTAGONISTS

Synergists are those chemicals that, when administered with an insecticide, enhance its toxicity to insects. Synergists are generally of two types: they may be close structural analogs of insecticides and synergize them by competing for detoxifying enzymes; or they may inhibit MFO and thereby minimize the detoxication of the insecticide. We are more interested in the latter type of synergists. These, by inhibiting MFO, synergize those insecticides that are detoxified by microsomal drug-metabolizing enzymes and antagonize those like phosphorothionates which are activated by these enzymes.

Benzo-($d$)-1,3-dioxole (methylenedioxyphenyl) compounds, which were originally developed for use with pyrethroids, were, until recently the only important group of synergists known. These include sesamin, sesmolin, sesamex, safrole, and so forth. Recently several new compounds that synergize insec-

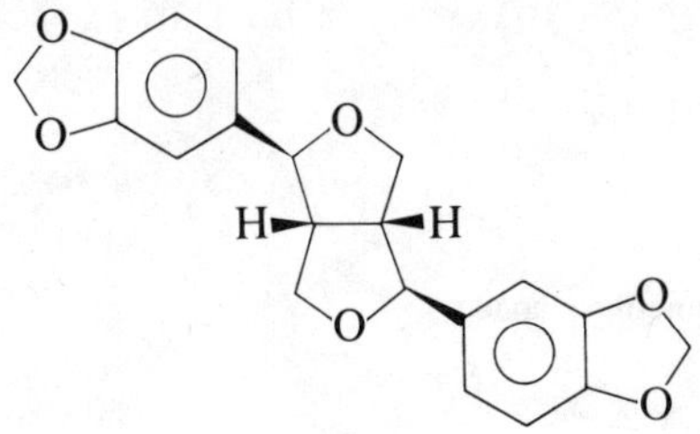

Sesamin 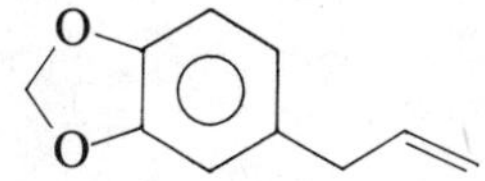                                                    Safrole

ticides have been discovered (Fig. 16.3). Among these are included aryloxyl-amines, such as SKF 525 A, MGK 264, Lilly 18947, WARF antiresistant, and pentynyl phthalimide; compounds like aryl-2-propynyl that contain the

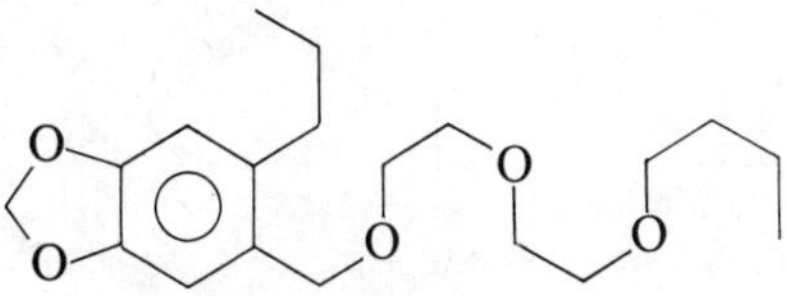

Piperonyl butoxide 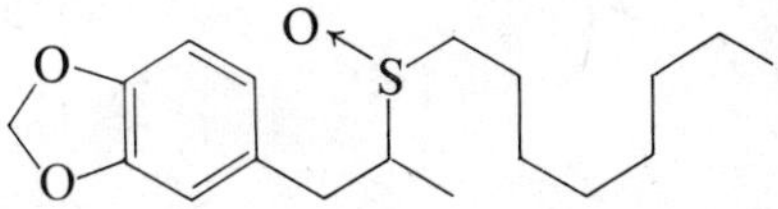                          Sulfoxide

(a)

Aryloxylamines (N-Alkyl compounds)

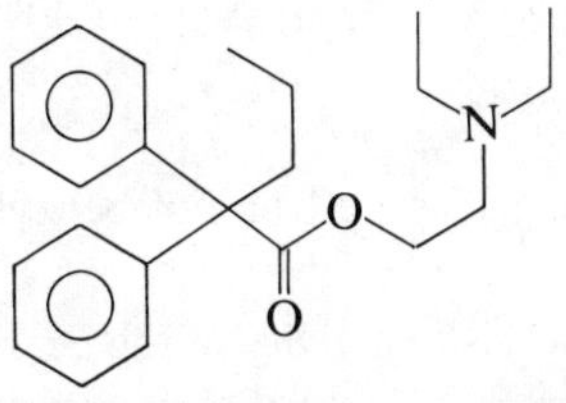

SKF 525A 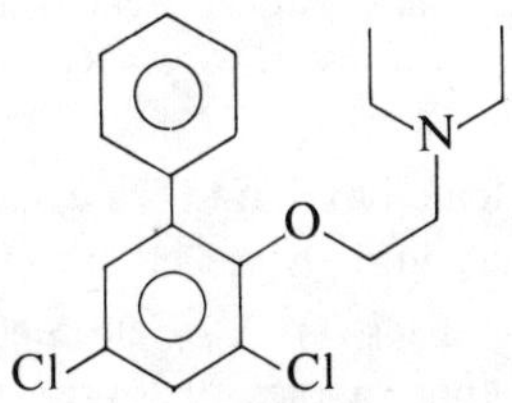                                        Lilly 18947

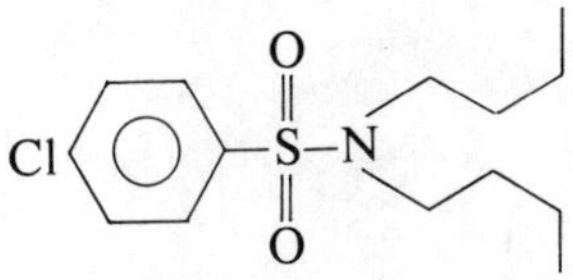

WARF antiresistant

(b)

Aryl-2-propynyl ethers and esters

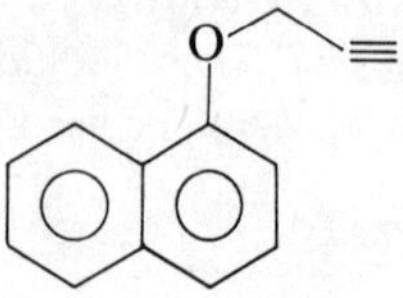

Naphthyl propynyl ether

(c)

Organothiocyanates

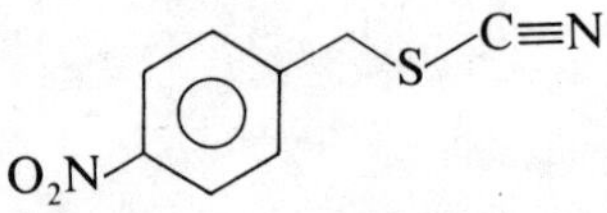

$O_2N$—

Nitrobenzyl thiocyanate

(d)

Oxime ethers

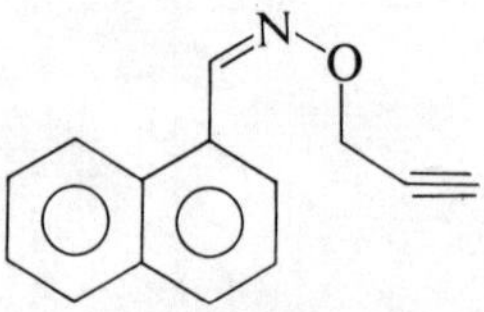

Naphthaldoximino propynyl ether

(e)

**Fig. 16.3**  Some recently discovered compounds that synergize insecticides. (a) Methylenedioxy-phenyl compounds (b) aryloxylamines (N-alkyl compounds) (c) Naphyl propynyl ether (d) Organothiocyanates (e) oxime ethers.

acetylenic bond, and the organothiocyanates like Thanite® or nitrobenzyl thio-cyanate and oxime ethers (1, 2).

The discovery by Felton et al. (6) of the synergistic activity of the benzo-(d)-1,2,3-thiadiazoles for pyrethroids and carbamates has added another group of materials of interest in synergism.

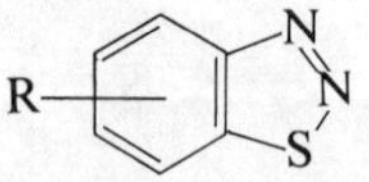

$R$—

Recently materials like metyrapone (2-methyl-1,2-bis [3-pyridyl]propanone), belonging to the group (1-naphthylvinyl) pyridines, that are known to block the steroid 11-$\beta$-hydroxylase of adrenal cortical mitochondria (which is cytochrome P450 mediated) are being investigated. Such compounds have been shown to inhibit the microsomal oxidation of several substrates (7). A series of 1-arylimidazoles and 4(5)-substituted imidazoles have been found to be the most powerful microsomal enzyme inhibitors tested. Such compounds are under investigation (8–12).

The spectra of synergists interacting with cytochrome P450 vary with the type of synergist used. Piperonyl butoxide, NIA 16824, and MGK 264 give type I spectra at 0.1 or 1.0 m$M$, that is, they resemble hexobarbital substrate. Type II spectra (aniline type) are obtained with metyrapone. This indicates that synergists do not all bind at the same site (13–15).

The piperonyl butoxide and other methylenedioxyphenyl (MDP) type synergists persist longer in house flies than in mammals. In the latter case, after an initial inhibition of MFO, there is an induction phase that leads to the metabolism of these synergists themselves by the mammals (1, 16).

It is generally accepted that synergists inhibit the MFO system, but the exact mechanisms involved have not been worked out. The synergists could compete for the activated oxygen, could interfere with the activation of oxygen, or could interfere with microsomal electron transport chain. For a more detailed discussion of the subject see refs. 1, 2, 4, 5, 16.

Among the structurally similar (analog) synergists are DMC, the [1,1-bis ($p$-chlorophenyl) ethanol], the noninsecticidal analog of DDT, and F-DMC [1,1-bis ($p$-chlorophenyl)-2,2,2,-trifluoroethanol]. They are inhibitors of DDT-dehydrochlorinase. Some noninsecticidal analogs of organophosphate and carbamate insecticides synergize these against resistant strains of insects (17–20). Malathion is synergized by EPN because the latter inhibits the enzyme carboxylesterase which hydrolyses malathion and malaxon in mammals. In those insects like *Culex* spp. which use a phosphatase route for the detoxification of malathion this synergism does not take place.

Besides synergizing insecticides, synergists have other uses as well. There are several chemicals that are effectively detoxicated by insects before they reach the site of action; if their detoxication is blocked by synergists, could we have a wider variety of insecticides? Additionally, there are several insecticides with a narrow spectrum or selective toxicity; could the use of synergists result in their becoming wide-spectrum pesticides, thereby killing pests and other organisms alike? Also, could synergists help overcome resistance?

## REFERENCES

1. J. E. Casida. *J. Agr. Food Chem.* **18**: 753 (1970).
2. A. S. Tahori, ed. *Insecticides: Resistance, Synergism, Enzyme Induction* Vol. II, Gordon and Breach, New York, 1972.

3. Y. P. Sun and E. R. Johnson. *J. Agr. Food Chem.* **8**: 261 (1960).

4. E. Hodgson and F. W. Plapp. *J. Agr. Food Chem.* **18**: 1048 (1970).

5. C. F. Wilkinson. *Bull. WHO* **44**: 171 (1971).

6. J. C. Felton, D. W. Jenner, and P. Kirby. *J. Agr. Food Chem.* **18**: 671 (1970).

7. K. C. Leibman. *Mol. Pharmacol.* **5**: 1 (1969).

8. K. C. Leibman and E. Ortiz. *Drug. Metab. Disposition* **1**: 184 (1973).

9. A. L. Johnson, J. C. Kauer, D. C. Sharung, and R. I. Dorfman. *J. Med. Chem.* **12**: 1024 (1969).

10. K. C. Leibman and E. Ortiz. *Pharmacologist* **13**: 223 (1971).

11. K. C. Leibman and E. Ortiz. *Drug Metab. Disposition* **1**: 775 (1973).

12. C. F. Wilkinson, K. Hetnarski, and L. J. Hicks. *Pestic. Biochem. Physiol.* **4**: 299 (1974).

13. R. B. Mailman, A. P. Kulkarni, R. C. Baker, and E. Hodgson. *Drug Metab. Disposition* **2**: 301 (1974).

14. R. B. Mailman and E. Hodgson. *Bull. Environ. Contam. Toxicol.* **8**: 186 (1973).

15. A. P. Kulkarni, R. B. Mailman, R. C. Baker, and E. Hodgson. *Drug Metab. Disp.* **2**: 309 (1974).

16. C. F. Wilkinson and L. J. Hicks. *J. Agr. Food Chem.* **17**: 829 (1969).

17. H. H. Moorefield and C. W. Kearns. *J. Econ. Entomol.* **48**: 403 (1955).

18. A. S. Perry, A. M. Mattson, and A. J. Buckner. *Biol. Bull.* **104**: 426 (1953).

19. S. Cohen and A. Tahori. *J. Agr. Food Chem.* **5**: 519 (1957).

20. F. W. Plapp. In *Biochemical Toxicology of Insecticides* (R. D. O'Brien and I. Yamamoto, eds.), Academic Press, New York, 1970, p. 179.

# INSECT–INSECTICIDE INTERACTIONS

# CHAPTER 17

# Activation, Degradation, and Excretion

Upon entering the body of an organism a xenobiotic is usually converted into more polar substances and excreted from the body. These are generally detoxifying reactions because the transformation products are less toxic than the parent compound. However some chemicals are converted to more toxic compounds than the parent xenobiotic. Such conversions in the case of insecticides have been called "bioactivation," and are brought about in most cases by MFOs that require NADPH and oxygen *in vitro*. Examples of bioactivation have been discussed in detail where individual classes of insecticides have been treated; they are summarized below.

## CHLORINATED HYDROCARBONS

Among the chlorinated hydrocarbons the bioactivation mechanism most commonly found is the epoxidation of cyclodienes; it was first reported in 1953 in the case of heptachlor in dog by Davidow and Rodomski (1), and subsequently confirmed in farm animals (2, 3). Conversion of aldrin to dieldrin in animals was reported in 1956 (4); this conversion also takes place in insects. Similarly, isodrin is converted to endrin. Since these reactions are mediated by MFOs, they are inhibited by structurally dissimilar synergists like piperonyl butoxide or SKF 525-A.

Aldrin and dieldrin also are converted to photoaldrin and photodieldrin in sunlight, and dieldrin is converted to photodieldrin by different microorganisms obtained from soil, water, and intestines of animals (5–12). Thus the conversion of aldrin and dieldrin to photodieldrin in living organisms is conceivable. Both photoaldrin and photodieldrin are more toxic than their precursors. Photoaldrin is remarkably toxic to mosquito larvae and is converted to photodieldrin, which in living organisms can form the pentachloro ketone of dieldrin. This ketone appears to be more toxic to house flies and mosquitoes than its parent compound (13, 14).

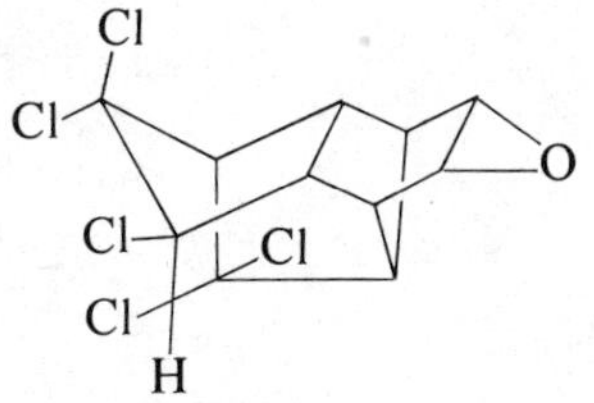

Photodieldrin

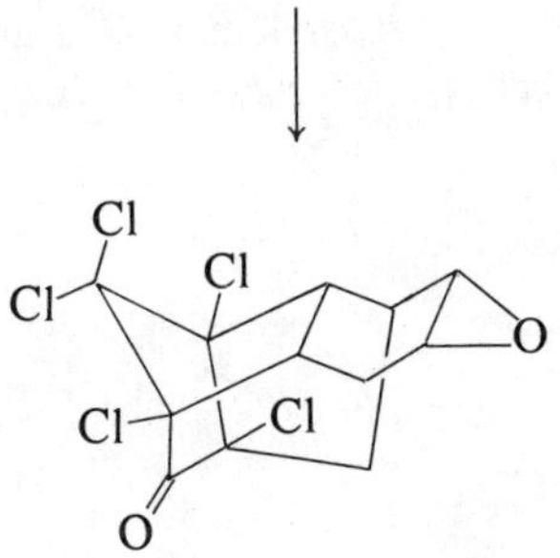

Pentachloroketone of dieldrin

The conversion of DDT to dicofol is an example of the transformation of an insecticide to a miticide in an insect; this conversion is inhibited by DMC, piperonyl butoxide, and sesamex, and evidently it does not take place in red spider mites. The reductive dechlorination of DDT to DDD may enhance the toxicity of DDT to some species of insects.

The activation of organophosphorus compounds has been discussed in detail in Chapter 3 and is summarized below.

1. Desulfuration: conversion of P=S to P=O
2. Hydroxylation of N-methyl groups followed by N-dealkylation
3. Thioether oxidation
4. Side-group oxidation

Bioactivation does take place in some cases with carbamates; aldicarb is converted to its sulfone which is 17 times as toxic as aldicarb; the sulfone is further oxidized, albeit, very slowly, to sulfoxide, which is more than four times as toxic as its precursor (15–17).

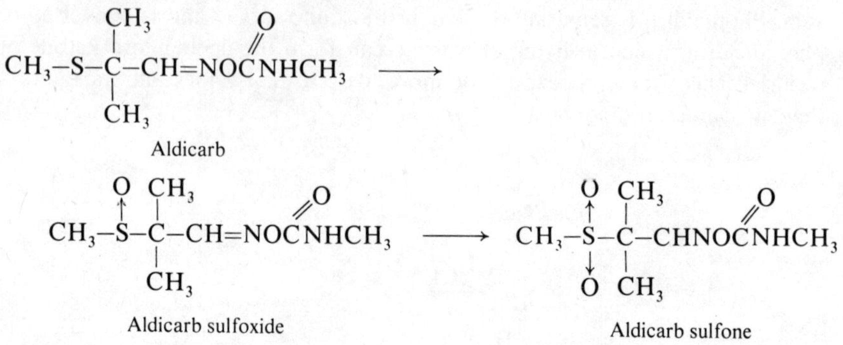

$$CH_3-\underset{\underset{CH_3}{|}}{\overset{\overset{CH_3}{|}}{C}}-CH=NO\overset{O}{\overset{\|}{C}}NHCH_3 \longrightarrow$$

Aldicarb

$$CH_3-\underset{\underset{CH_3}{|}}{\overset{\uparrow}{\underset{}{\overset{O}{S}}}}-\underset{\underset{CH_3}{|}}{\overset{CH_3}{C}}-CH=NO\overset{O}{\overset{\|}{C}}NHCH_3 \longrightarrow CH_3-\underset{\underset{O}{\downarrow}}{\overset{\overset{O}{\uparrow}}{S}}-\underset{\underset{CH_3}{|}}{\overset{CH_3}{C}}-CHNO\overset{O}{\overset{\|}{C}}NHCH_3$$

Aldicarb sulfoxide                                      Aldicarb sulfone

The dialkylamino oxidations in the case of Metacil and Zectran result in the formation of their corresponding methylamino and amino analogs which are more toxic than their respective precursors (18–20).

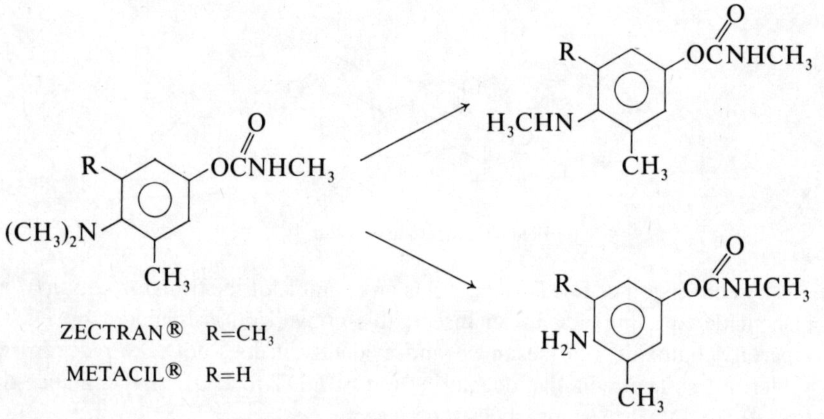

ZECTRAN® R=CH₃

METACIL® R=H

At least one of the metabolites of dimetilan that contains *N*-hydroxymethyl groups with intact dimetilan ring structure is stable and as toxic as dimetilan (21). However in insect control the only important activation reaction among carbamates appears to be in the case of aldicarb.

The degradation, inactivation, and metabolism of various classes of insecticides have been discussed and metabolic pathways have been shown where necessary. Among organochlorines the degradation of cyclodienes is very complex because of the possibilities afforded by the stereochemical three-dimensional molecules. Hydrolytic opening of the epoxide ring, as in the case of dieldrin, takes place to give *trans*-diol. The toxicity of this compound to mice is about one-twelfth that of its precursor (22). The methano bridge in dieldrin is hydroxylated to yield 9-*syn*-hydroxy dieldrin which is readily conjugated with glucuronic acid. Characterization of the metabolites of cyclodienes is a tedious and time-consuming process because of the stereochemistry and the complexities of the molecules involved. Isodrin is converted by epoxidation to endrin which is nonpersistent in mammals; possibly because of its stereochemistry it is more readily attacked by enzymes than dieldrin. The hexachloronorborene nucleus in cyclodienes is rather stable and remains intact in most of the metabolites (13, 14).

Epoxide rings in many chemicals are hydrated by epoxide hydrases in living organisms (23–25) to form the corresponding *trans*-dihydroxy compounds, thus changing the bioactivity of the toxicants. In the case of dieldrin the epoxide hydrases are not very efficient in attacking the epoxide ring. However the detoxication of the analog HEOM of dieldrin is species specific. It is toxic to

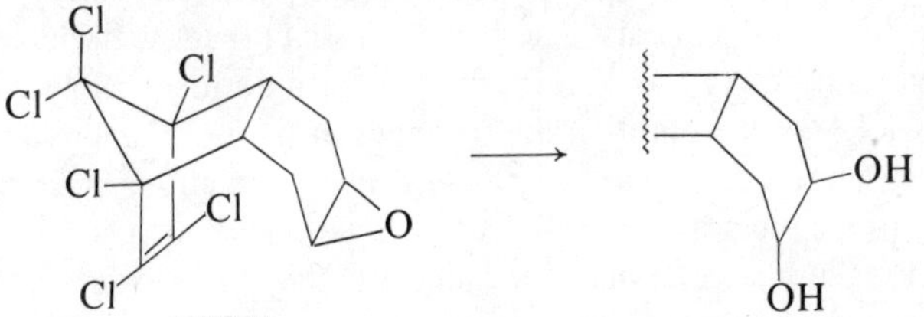

HEOM

HEOM *trans*-diol

the tsetse fly because of the poor activity of epoxide hydrases in this insect, but it is inactive against other insects (26). Phenols and dihydrodiols are formed through an intermediate arene epoxide (27) in living organisms.

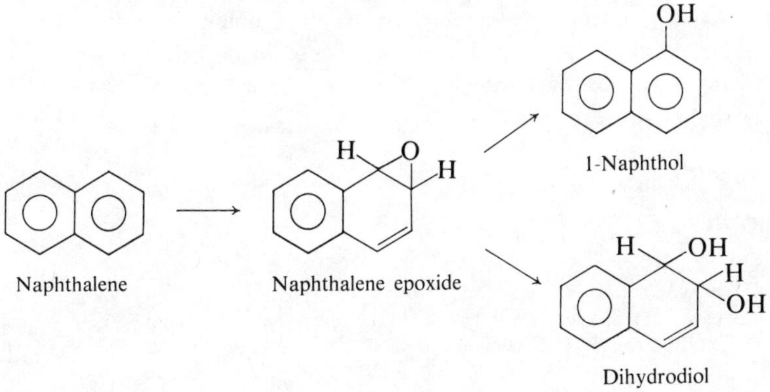

Naphthalene

Naphthalene epoxide

1-Naphthol

Dihydrodiol

Inactivation of JH analogs in some strains of flour beetles (26) and house flies may be due to epoxide hydrase(s) attack on the epoxide ring, epoxidation and hydration of double bond, and carboxylesterase attack on the $COOCH_3$ group.

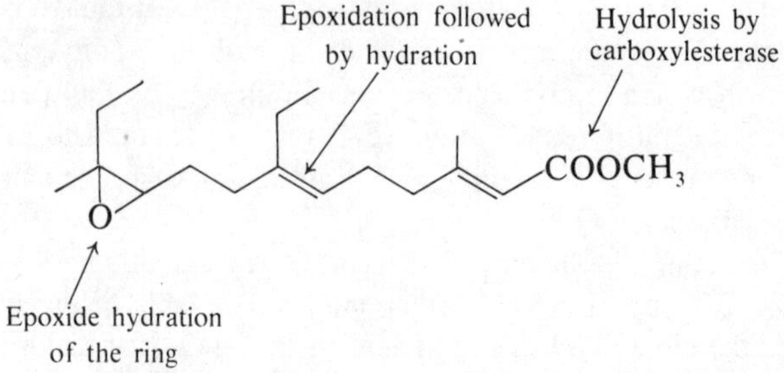

Inhibitors of these enzymes may prevent inactivation of these compounds (26).

The formation of intermediate arene epoxides may be responsible for some chronic toxic effects of chemicals. These epoxides can react with nucleophilic groups like amino, hydroxyl, and sulfhydryl to form an irreversible complex. This can ultimately result in undesirable changes in tissues. Continued insults to tissues may result in necrosis or ultimately cancer formation. More research is needed to better understand metabolism in relation to chronic toxicity. In chemical carcinogenesis prolonged feeding of the carcinogen is usually necessary for the induction of tumors.

The metabolism of carbamates, organophosphorus compounds, and other insecticides have been discussed previously, and the reader should refer to relevant chapters for details.

If the xenobiotic or its metabolites have a functional group, such as amino, carboxyl, epoxide, hydroxyl, and sulfhydryl, then one of the endogenous substrates reacts with it to form ethereal sulfates, glucuronides, mercapturic acids, amino acid conjugates, and so forth. These conjugates are excreted. Glutathione plays an important role in the formation of water-soluble derivatives that are eliminated by excretion.

## REFERENCES

1. B. Davidow and J. L. Radomski. *J. Pharmacol. Exp. Ther.* **107**: 259 (1953).
2. B. Davidow, J. L. Radomski, and R. E. Ely. *Science* **118**: 383 (1953).
3. R. E. Ely, L. A. Moore, P. E. Hubanks, R. H. Carter, and F. W. Poos. *J. Dairy Sci.* **38**: 669 (1955).
4. J. M. Bann, T. J. DeCino, N. W. Earle, and Y. P. Sun. *J. Agr. Food Chem.* **4**: 937 (1956).
5. G. W. Ivie and J. E. Casida. *J. Agr. Food Chem.* **19**: 405 (1971).
6. G. W. Ivie and J. E. Casida. *op. cit.* **19**: 410 (1971).
7. M. A. Q. Khan, R. H. Stanton, D. J. Sutherland. *Science* **168**: 318 (1969).

8. M. A. Q. Khan, R. H. Stanton, D. J. Sutherland, J. D. Rosen, and N. Maitra. *Arch. Environ. Contam. Toxicol.* **1**: 159 (1973).

9. W. Klein, J. Kohli, I. Weisgerber, W. Klein, and F. Korte. *J. Agr. Food Chem.* **21**: 152 (1973).

10. J. Kohli, S. Zarif, I. Weisgerber, W. Klein, and F. Korte. *op. cit.* **21**: 855 (1973).

11. F. Matsumura, K. C. Patil, and G. M. Boush. *Science* **170**: 1208 (1970).

12. A. Robinson, A. Richardson, and B. Bush. *Bull. Environ. Contam. Toxicol.* **1**: 127 (1966).

13. G. T. Brooks. Chlorinated Insecticides, Vol. I, *Technology and application*, CRC Press, Cleveland, 1974, 249 pp.

14. G. T. Brooks. Chlorinated Insecticides, Vol. II, *Biological and Environmental Aspects*, CRC Press, Cleveland, 1974, 197 pp.

15. N. R. Andrawes, H. W. Dorough, and D. A. Lindquist. *J. Econ. Entomol.* **60**: 979 (1967).

16. J. B. Knaak, M. J. Tallant, and L. J. Sullivan. *J. Agr. Food Chem.* **14**: 573 (1966).

17. R. L. Metcalf, T. R. Fukuto, C. Collins, K. Bork, J. Burk, H. T. Reynolds, and M. F. Osman. *op. cit.* **14**: 579 (1966).

18. A. M. Abdel-Wahab and J. E. Casida. *op. cit.* **15**: 479 (1967).

19. A. M. Abdel-Wahab, A. M. Kuhr, and J. E. Casida. *op. cit.* **14**: 290 (1966).

20. E. S. Oonnithan and J. E. Casida. *Bull. Environ. Contam. Toxicol.* **1**: 59 (1966).

21. M. Y. Zubairi and J. E. Casida. *J. Econ. Entomol.* **58**: 403 (1965).

22. F. Korte and H. Arent. *Life Sci.* **4**: 2017 (1965).

23. G. T. Brooks. *World Rev. Pest Control* **5**: 62 (1966).

24. P. L. Grover, A. Hewer, and P. Sims. *Biochem Pharm.* **21**: 2713 (1972).

25. D. M. Jerina, J. W. Daly, B. Witkop, P. Zaltmann-Nirenberg, and S. Udenfriend. *Biochem.* **9**: 147 (1970).

26. G. T. Brooks. *Nature* **245**: 382 (1973).

27. D. M. Jerina, J. W. Daly, B. Witkop, P. Zaltmann-Nirenberg, and S. Udenfriend. *J. Amer. Chem. Soc.* **96**: 6525 (1968).

# RESISTANCE AND GENETICS OF RESISTANCE

| CHAPTER 18 | Natural versus Artificial Selection and the Impact of Insecticide Pressure |

Since individual variations in species are always present, it is highly improbable that any normal treatment by an insecticide will kill 100% of the target insects; some individuals always survive insecticide pressure. These are usually those insects that, by virtue of their genetic makeup, can effectively render the chemical nontoxic or are irritated by the insecticide and thus avoid taking a lethal dose. The former type of insects possess physiological resistance and the latter, behavioristic resistance. In some cases both types of resistance mechanisms may be present.

In nature perhaps there are certain factors that favor the preponderance of the susceptible individuals among insects, but once the environment is changed and insecticide pressure applied, artificial factors favoring the inherently resistant individuals are created. Such individuals then start taking over, and the longer the insecticide pressure, the higher the rate of the reproduction of the insect, and the more rapid the succession of generations, the sooner is the takeover by the resistant individuals. Insecticide resistance has been well documented and has been the subject of several reviews and articles (1–5a). The World Health Organization periodically issues circulars and monographs on the status of resistance in insects. By development of resistance we mean cases in which ability to resist high doses of insecticides has developed within a normally susceptible species, and not insects like the Mexican bean beetle, *Epilachna varivestis* Mulsant, and grasshoppers which even initially were tolerant to DDT. Tolerance is often nonspecific; on the contrary, resistance is usually specific to one insecticide or a group of insecticides.

Standardization of resistance has been the aim of the World Health Organization (WHO), and several tests have been developed for the purpose. Kits for these tests are available from WHO. The Food and Agriculture Organization (FAO) also has a working party on Pest Resistance to Pesticides. The WHO Expert Committee on Insecticides gave the following definition for the phenomenon of resistance (1).

Resistance to insecticides is the development of an ability in a strain of insects to tolerate doses of toxicants which would prove lethal to the majority of individuals in a normal population of the same species. The term "behavioristic resistance" describes the development of an ability to avoid a dose which would prove lethal.

Resistance did not originate with the use of synthetic organic insecticides, but their widespread use brought into focus a phenomenon that had been reported earlier. The first documented cases of resistance were those of San Jose scale to lime sulfur in 1908 and California red scale to HCN in 1916 (3).

If a dosage–mortality curve is plotted on a linear scale a sigmoid curve is obtained (Fig. 18.1), because initially there is a small increase in mortality with in-

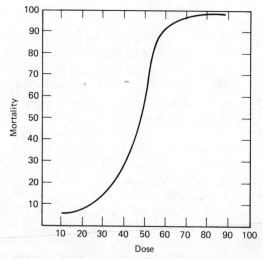

**Fig. 18.1.** Hypothetical dose–mortality relationship showing percent mortality as a result of treatment with hypothetical doses of a toxicant. Mortality and dose plotted on an arithmetic scale.

creasing dose (this part of the curve represents the most susceptible insects in population), and this is followed by a rapid increase in mortality with increasing dose (this represents a majority of individuals in a randomly chosen population). After this, large increments in doses result only in a small increase in mortality (this represents the more tolerant or even the resistant individuals in the population). If dosage mortality data are transformed such that mortality is expressed in probits (6) and dose is expressed on a log scale (Fig. 18.2) a straight line, the regression line, can be calculated from the data or fitted by eye. If these data are available for a population before insecticide treatment is started or before the insects start showing signs of resistance, it is called the "base line." The changes in the base line then give the trends in the develop ment of resistance in insects. To start with, the slope of the regression line may change, indicating the elimination of the more susceptible individuals from the population; but as resistant populations start building up and taking over, a heterogeneous population may result (Fig. 18.2, curve C). Finally the entire regression line shifts as resistant individuals almost completely take over (Fig. 18.2, line D and E).

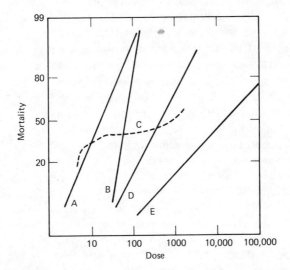

**Fig. 18.2.** Mortality in probits and dose in log scale in a hypothetical insect population subjected to heavy insecticide pressure. Regression line A represents the sigmoid curve in Fig. 18.1. B represents the "susceptible" population from which the very susceptible individuals have been eliminated. C represents heterogeneous population. D and E show the development of resistance to the toxicant.

One of the advantages of the regression line for dose–mortality response is that the $LD_{50}$ and $LD_{90}$ values can be read directly from the line.

Considerable research has been done on the genetics of resistance, but even a reasonably brief discussion of the subject is beyond the scope of this book; however we do know that resistance is due to preexisting alleles in a population. Frequently it is due to a single gene allele (monofactorial). In such cases there is a strong development of resistance to an insecticide. At times resistance is due to a combination of genes (polyfactorial); such resistance usually takes a long time to develop. In some cases, like DDT-resistance in some strains of the house fly, the individuals possessing the DDT-resistant gene are not favored by the environment; also, for normal survival and propagation of the strain, ancillary genes have to be combined with DDT-resistant genes, and though monofactorial in origin, it takes longer for the resistance to develop.

Though in genetic literature dominant or recessive characters of the genes are emphasized, in the case of insecticide resistance this is not always as clear-cut and intermediate types are often found. Georghiou (7) has suggested classifying recessiveness and dominance as "completely recessive," "incompletely recessive," "intermediate," "incompletely dominant," and "completely dominant," to describe the various types of resistance in $F_1$ heterozygotes. A simple case of Mandelian inheritance of resistance to dieldrin in the case of a Nigerian strain of *Anopheles gambiae* Giles was demonstrated by Davidson (8). Dieldrin resistance is, however, usually intermediate. The homozygotes RR and rr are highly resistant and highly susceptible, respectively, and Rr are almost exactly intermediate. Exceptions are found in the cases of the cattle tick, *Boophilus microplus*, and the Ivory Coast strain of *Anopheles gambiae*; in both cases resistance is dominant.

DDT resistance in the house fly is dominant, as it is in the case of *Anopheles stephensi* List., but in other *Anopheles* species it is usually recessive. Some of the major genes for resistance have been located on chromosomes in the house fly, mosquitoes and other insects (9, 10). For instance, the DDT-ase gene is the case of the house fly is of incomplete dominance and is located on chromosome 2. This gene is often designated as Deh because it dehydrochlorinates DDT. Genes for resistance to organophosphorus and carbamate insecticides are also located on chromosome 2.

Chlorinated hydrocarbon insecticides present two different types of resistance mechanisms: one to DDT and its analogs like DDD, methoxychlor, and Perthane, and the other to cyclodiene derivatives and lindane (11). Thus insects subjected to DDT pressure usually acquire resistance not only to DDT, but also to its analogs. Similarly, those reared under lindane pressure also show resistance to cyclodienes. Maximum resistance often develops to the inducing compound itself and cross resistance to other compounds is less marked. Thus insects reared under lindane pressure show greater resistance to lindane than to dieldrin. This type of resistance is known as cross resistance and it differs from poly- or multiresistance which is developed as a result of sequential or concomitant pressure with two or more unrelated insecticides on the same strain of the insect (12). However rearing of house flies under organophosphate pressure results in high DDT and cyclodiene resistance in these insects (13).

Carbamate resistance develops more rapidly in insects that are initially tolerant to the carbamate.

Several species have developed resistance to more than one insecticide, and resistance appears most frequently in Diptera where more than 80 species are involved. In the house fly DDT resistance first appeared in 1946, followed by cyclodiene resistance in 1949 and organophosphorus resistance in 1955. Among mosquitoes, *Culex pipiens fatigans* Wied. (*C.p. quinquefasciatus* Say) is initially somewhat tolerant to DDT and readily develops resistance to this insecticide. When DDT is replaced by cyclodienes it acquires an even higher order of resistance to lindane and cyclodienes (3). *Anopheles* (*M*) *stephensi mysorensis* Sweet and Rao, which is initially DDT susceptible, acquired DDT resistance in Iran, and in countries around the Persian Gulf, in the early 1960s, when DDT was replaced by dieldrin, within 2 years it acquired an even higher order of resistance to dieldrin and lindane (14). Next in importance to Diptera, in the development of resistance, are the Hemiptera, Lepidoptera, and Acarina. The bedbug, the German cockroach, and the house fly are among those insects that have developed resistance to more than one (usually three or four) group of insecticides. DDT-resistant house flies and codling moth populations are present in all regions of the world where this insecticide has been used. *Tetranychus urticae* Koch is resistant to most organophosphate insecticides.

Among insects of agricultural importance more than 119 species have developed resistance to insecticides; the highest number is found among Hemiptera, 33; followed by Lepidoptera (caterpillars), 29; Coleoptera, 23; mites, 16; and others (2). Principal crops threatened by resistance in various countries include cotton in the United States by boll weevil as a result of endrin resistance and by *Heliothis* as a result of resistance to DDT, toxaphene, and methyl parathion; cotton in Egypt as a result of resistance to toxaphene in cotton leaf worm (*Spodoptera littoralis*); and cocoa in Ghana by Cocoa capsid as a result of lindane and aldrin resistance (5). The European red mite, *Panonychus ulmi* (Koch), is now resistant to almost all available acaricides used in orchards.

Among insects of medical importance, 15 species of *Anopheles* that carry malaria are resistant to DDT and 36 to dieldrin and lindane (3). This development of resistance has seriously affected malaria eradication programs in Central America, Indonesia, the Middle East, and Pakistan. Physiological resistance is another factor to be reckoned with among mosquitoes. The floodwater mosquitoes, *Aedes* spp., in California have acquired resistance to almost all available insecticides (3a). Similarly, the human louse is becoming more and more difficult to control. Resistance to DDT was first reported in Korea and soon was found in all areas where DDT was being used. The human louse is now resistant to 1% lindane as well and is acquiring resistance to malathion in Africa.

# REFERENCES

1. A. W. A. Brown. "Insecticide resistance in arthropods", *WHO Monogr. Ser.* **38**: (1958).
2. A. W. A. Brown and R. Pal. *WHO Monogr. Ser.* **443**: 491 pp. (1971).
3. A. W. A. Brown. In *Pesticides in the Environment*, Vol. I, Part II (R. White-Stevens, ed.), Dekker, New York, 1971, p. 457.
3a. National Academy of Sciences. Mosquito Control. Washington, D.C. 1973.
4. G. T. Brooks. Chlorinated insecticides, Vol. I, *Technology and Application*. CRC Press, Cleveland, 1974.
5. G. T. Brooks. Chlorinated Insecticides, Vol. II, Biological and Environmental Aspects, CRC Press, Cleveland, 1974.
5a. FAO report, Pest Resistance to Pesticides, Food and Agriculture Organization, Rome, 1970.
6. C. I. Bliss. *Ann. Appl. Biol.* **22**: 134 (1935).
7. G. P. Georghiou. *Exp. Parasitol.* **26**: 224 (1969).
8. G. Davidson. *Nature* **178**: 861 (1956).
9. A. W. A. Brown. *World Rev. Pest Control* **6**: 104 (1967).
10. M. Tsukamoto. *Residue Rev.* **25**: 289 (1969).
11. J. R. Busvine. *Nature* **174**: 783 (1954).
12 J. M. Grayson and D. G. Cochran. *World Rev. Pest Control* **7**: 172 (1968).
13. A. W. A. Brown and Z. H. Abedi. *Mosq. News* **20**: 118 (1960).
14. Ch. M. H. Mofidi. In *The Role of Science in the Development of Natural Resources with Particular Reference to Iran, Pakistan and Turkey*. (M. L. Smith, ed.), Pergamon Press, London. 1963.

# RESISTANCE AND GENETICS OF RESISTANCE

# CHAPTER 19

Different Types of
Resistance Mechanism

## DDT RESISTANCE

DDT resistance in the house fly was the first to be reported and this insect has provided a model for the investigation of resistance in general. This resistance is believed to be due to a high titer of DDT-ase. Its mechanism was first elucidated by Sternburg et al. in house fly strain L at the University of Illinois and at about the same time by Perry and Hoskins (1, 2). The enzyme requires glutathione for activation *in vitro* (3). As stated earlier this enzyme was purified and its molecualr weight was found to be 36,000 and an optimum pH for its activity was 7.4. It dehydrochlorinates methoxychlor and DDD even more rapidly than DDT (4). It is inhibited by the DDT analog DMC (chlorfenethol). DDT-ase is synthesized during larval growth; thus late-emerging flies contain more of the enzyme and are tolerant to DDT (5, 6). The DDT-detoxifying enzyme in the body louse, *Pediculus humanus humanus*, is different from that in the house fly in that the former is heat stable, resists the action of alcohol and protease (7), and is activated not only by glutathione but also by cysteine or ascorbic acid. Lice metabolize DDT to DDA, DBP, and small amounts of DDE. The resistance mechanism to DDT in the case of mosquitoes is also due to conversion of DDT to DDE; this has been demonstrated in, among other species, *Aedes aegypti, Aedes taeniorhynchus* (Wied), and *Culex pipiens fatigans* (8, 9). The DDT resistance of *Culex pipiens fatigans* also extends to the *o,p'* isomer (the house fly DDT-ase cannot dehydrochlorinate *o,p'*-DDT, nor the *ortho*-chloro derivative of DDT). Besides, the resistant strain of *Culex pipiens fatigans* absorbs less DDT than the susceptible ones, and its enzyme is less active on DDD than DDT (10) (compare this with the house fly enzyme).

Interestingly, the microsomal fractions from house fly hydroxylate naphthalene in direct proportion to the level of resistance to DDT (11). And a Danish strain of house fly (Fc) raised under diazinon pressure was also resistant to DDT, but it metabolized about 50% of the applied dose of DDT to water-soluble metabolites, indicating involvement of microsomal enzymes. Microsomal preparations gave four unknown metabolites (12). Sesamex [sesoxane; 2-(3,4-methylenedioxyphenoxy)-3,6,9-trioxaundecane] was a powerful synergist for DDT and diazinon in this (Fc) strain, but not in the case of Illinois strain L in which the resistance mechanism is due to the conversion of DDT to DDE. This indicated the involvement of microsomal enzymes in strain Fc. The gene responsible for this resistance was found on chromosome 5 and was designated DDT-*md*, this gene has been found in house fly strains Fc and SKA only.

DDT resistance in the German cockroach is due to an oxidation process brought about by mixed-function oxidases, and the metabolic products are dicofol and other polar metabolites (13). Similar resistance mechanism has been reported for the American cockroach.

## CYCLODIENE AND LINDANE RESISTANCE

We are still very much in the dark regarding the mechanism of resistance to cyclodiene insecticides. This ignorance is further complicated by the fact that we have no knowledge of the mode of their action. Yamasaki and Narahashi (14, 15) found that dieldrin-poisoned nerve of the American cockroach showed spontaneous bursts of action potential, but in the case of nerve of resistant house fly there was a much longer latent period between the application of dieldrin and the appearance of discharge. These findings relate the action of dieldrin to the central nervous system (CNS). Biochemical differences between resistant and susceptible house flies have not been found. Similarly, Ray (16) found no difference between resistant and susceptible German cockroaches so far as the absorption of dieldrin through the cuticle and its concentration in the nerve cord are concerned. However, in the same insect, Matsumura (17) has suggested some relationship between ATP-ase activity and the resistance mechanism and found that the resistant nerve cord bound less DDT-$^{14}$C than the susceptible one; possibly there are alterations at the site of action.

Insects show cross resistance between cyclodienes and lindane, but lindane-resistant strains of house flies produce several lindane metabolites (18). House fly homogenates degrade lindane *in vitro* in the presence of glutathione (19).

The enzyme was isolated by Ishida and Dahm (20), but the titer of the enzyme bore no relation to the extent of lindane resistance and it acted faster on DDT than on lindane.

Even in the case of DDT all cases of resistance cannot be explained on the basis of detoxication alone. Oppenoorth (21) compared three DDT-resistant strains of house flies with respect to their ability to metabolize DDT; one of the strains carrying the gene kdr (knockdown resistance) metabolized only 50% of an applied dose of DDT (1 $\mu$g) and the other two metabolized 98% 18 hours after application. Metcalf (22) opines that the failure of synergists in enhancing toxicity of DDT in the field may be due to the selection of the kdr gene.

## ORGANOPHOSPHATE RESISTANCE

Hydrolysis of the organophosphate by the insect is the most important mechanism of resistance in this group of insecticides. However thionophosphates are first activated to phosphates. Therefore in such cases we have two processes taking place: oxidation that activates the insecticide, which is also called the activation process, and hydrolysis that renders it nontoxic, which is also called the detoxication process. The toxicity of the insecticide therefore depends on the speeds with which these processes proceed, or

$$\text{Toxicity} = \frac{\text{activation}}{\text{detoxication}}$$

Thus the low toxicity of isopropyl parathion for bees as compared to house flies is possibly due to slow oxidation of the chemical in bees because of isopropyl groups (23). Oxidation in house flies is 5–10 times as fast. When both insects are treated with isopropyl paraoxon there is no difference in toxicity.

Since hydrolases have low substrate specificity, cross resistance is common among organophosphates metabolized by them; thus, if there is more cross resistance confined to a group of compounds, the probability is that there is a common enzymatic resistance mechanism. Another case of low substrate specificity is attack by phosphatases that metabolize chemicals containing phosphoryl radicals. Malathion, among the organophosphorus insecticides, stands apart in the sense that it often evokes no cross resistance to other insecticides of this group. It is metabolized by the attack of carboxylesterases on the carbethoxy group to form malathionic acid. Esters without this selectophore group are not attacked by these specific enzymes, though the phosphoryl radicals in malathion are attacked by phosphatases as well.

Considerable evidence has accumulated during the last decade to show that the increase in certain hydrolases is genetically controlled. One of the earlier theories of organophosphate resistance was proposed by van Asperen and Oppenoorth (24–28). They hypothesized that the mutation of a gene that normally causes aliesterase synthesis results in the synthesis of a phosphatase that can attack and degrade many organophosphates. This hypothesis was based on the discovery that the carboxylesterase content of organophosphate-resistant house flies was significantly reduced and low aliesterase and high phosphatase are genetically linked. For a detailed discussion and criticism of this theory see O'Brien (29, 30).

The physiological resistance to organophosphates is now considered to be dependent on a single gene that is usually dominant. Sometimes it can be semidominant or recessive. In the last two cases resistance is slow in developing.

## CARBAMATE RESISTANCE

Resistance to carbamates involves detoxication by nonspecific esterases; resistant house flies metabolize carbaryl to 1-naphthol and this hydrolysis is inhibited by sesamex (31). House fly strains resistant to carbaryl do not show cross resistance to other carbamate insecticides (32). However, a propoxur-resistant strain of *Culex fatigans* was highly resistant to closely related chemicals but not to structurally unrelated carbamates. It metabolized propoxur at twice the rate of the susceptible strain (33).

## PYRETHRUM RESISTANCE

At present pyrethrum resistance is more of academic rather than practical interest; nevertheless, some investigations have been carried out and detoxication of pyrethrins is only partially understood. Pyrethrin I is converted to alcohol, aldehyde, and carboxylic acid derivatives (34, 35). Allethrin resistance in the house fly, at least in one strain (R-Baygon), appears to be due to oxidase(s) controlled by a gene on autosome-2 (36). Possibly two detoxication factors are involved, one an attack on the acid moiety and the other on the alcohol moiety, depending on the compound used. Other factors in resistance may include rate of absorption or penetration and reduced nerve sensitivity (37). Plapp and Casida (38) found that DDT- and dieldrin-fed house fly strains contain an increased level of allethrin detoxifying oxidases.

## PENETRATION THROUGH THE CUTICLE AS A RESISTANCE FACTOR

The rate of penetration of a toxicant through the cuticle in determining insecticide resistance has not been clearly established. Forgash et al. (39) found that in the diazinon-resistant (Rutgers A strain) house flies the penetration of diazinon was slower compared to the susceptible strains. Farnham et al. (40) reported that in the diazinon-R SKA strain of house flies as compared to a Rothamstead normal one there was a definite indication of decreased penetration. Continuing their work with the SKA strain Sawicki and Lord (41) isolated the penetration factor by crossing experiments and found that delayed penetration was responsible for slower knockdown and was found only in strain SKA or those crosses that carried SKA's third chromosome. They also confirmed that the penetration-delaying factor alone was not the major reason for resistance in diazinon-selected strains of the house fly; but when this is combined with desethylating factor, resistance to many organophosphorus compounds is greatly increased.

## BEHAVIORISTIC RESISTANCE OR THE AVOIDANCE OF TOXIC CHEMICALS

Some strains of insects instinctively avoid toxic chemicals. Strains of house flies have been discovered that avoid malathion in sugar baits (42) while others avoid DDT-treated surfaces. Similarly, spotted root fly, *Euxesta notata* (Weidemann), avoids residual deposits of malathion and parathion (43). In the case of mosquitoes this type of behavioristic resistance becomes important because treatment by insecticides of residences is the most common method of controlling mosquito-borne diseases. DDT-resistant strains of *Anopheles sacharovi* Farr, *Anopheles culicifacies* Giles, *Anopheles albimanus* Wied, *Culex p. quinquefasciatus*, and *Aedes aegypti* are less irritable to DDT-treated surfaces than susceptible ones (44–47). By laboratory selection strains of *Anopheles atroparvus* Van Thiel that show hyperirritability to DDT deposits have been developed (48). *Anopheles gambiae* in Africa are reported to be irritated by DDT deposits and thus escape lethal dosages (49).

## REFERENCES

1. J. Sternburg, C. W. Kearns, and W. N. Bruce. *J. Econ. Entomol.* **43**: 214 (1950).
2. A. S. Perry and W. M. Hoskins. *Science* **111**: 600 (1950).
3. J. Sternburg, E. B. Vinson, and C. W. Kearns. *J. Econ. Entomol.* **46**: 513 (1953).
4. H. Lipke and C. W. Kearns. *J. Biol. Chem.* **234**: 2123 (1959).
5. D. Pimentel, J. E. Dewey, and H. H. Schwardt. *J. Econ. Entomol.* **44**: 477 (1951).
6. F. F. Sanchez and M. Sherman. *op. cit.* **59**: 272 (1966).
7. A. S. Perry and A. J. Buckner. *Amer. J. Trop. Med. Hyg.* **7**: 620 (1958).
8. Z. H. Abedi, J. R. Duffy, and A. W. A. Brown, *J. Econ. Entomol.* **56**: 511 (1963).
9. A. W. A. Brown. *Annu. Rev. Entomol.* **5**: 301 (1960).

10. T. Kimura, J. R. Duffy, and A. W. A. Brown. *Bull. WHO* **32**: 557 (1965).
11. R. D. Schonbrod, W. W. Philleo, and L. C. Terriere. *J. Econ. Entomol.* **58**: 74 (1965).
12. F. J. Oppenoorth and K. van Asperen. *Entomol. Exp. Appl.* **11**: 81 (1968).
13. M. Agosin, B. C. Fine, N. Scaramelli, J. Ilivicky, and L. Aravena. *Comp. Biochem. Physiol.* **18**: 101 (1966).
14. T. Yamasaki and T. Narahashi. *J. Econ. Entomol.* **51**: 146 (1958).
15. T. Yamasaki and T. Narahashi. *J. Insect Physiol.* **3**: 230 (1959).
16. J. W. Ray. *Nature* **197**: 1226 (1963).
17. F. Matsumura. In *Insecticides* (A. S. Tahori, ed.), Vol. II, 1971, p. 95.
18. F. R. Bradbury. *J. Sci. Food Agr.* **8**: 90 (1957).
19. M. Ishida and P. A. Dahm. *J. Econ. Entomol.* **58**: 383 (1965).
20. M. Ishida and P. A. Dahm. *op. cit.* **58**: 602 (1965).
21. F. J. Oppenoorth. *J. Entomol. Exp. Appl.* **10**: 75 (1967).
22. R. L. Metcalf. *Annu. Rev. Entomol.* **12**: 229 (1967).
23. R. L. Metcalf and M. Frederickson. *op. cit.* **58**: 143 (1965).
24. K. van Asperen and F. J. Oppenoorth. *Entomol. Exp. Appl.* **2**: 48 (1959).
25. K. van Asperen and F. J. Oppenoorth. *op. cit.* **3**: 68 (1960).
26. F. J. Oppenoorth and K. van Asperen. *Science* **132**: 298 (1960).
27. F. J. Oppenoorth and K. van Asperen. *Entomol. Exp. Appl.* **4**: 311 (1961).
28. F. J. Oppenoorth and K. van Asperen. *Annu. Rev. Entomol.* **10**: 185 (1965).
29. R. D. O'Brien. *Insecticides: Action and Metabolism*, Academic Press, New York, 1967.
30 R. D. O'Brien. *Annu. Rev. Entomol.* **11**: 369 (1966).
31. M. E. Eldefrawi and M. W. Hoskins. *J. Econ. Entomol.* **54**: 501 (1961).
32. H. Ikemoto. *Botyu-Kagaku* **29**: 68 (1964).
33. G. P. Georghiou, R. B. March, and G. E. Printy. *Bull. WHO* **35**: 691 (1966).
34. I. Yamamoto and J. E. Casida. *J. Econ. Entomol.* **59**: 1542 (1966).
35. I. Yamamoto and J. E. Casida. *J. Agr. Food Chem.* **17**: 1227 (1969).
36. F. W. Plapp and J. E. Casida. *J. Econ. Entomol.* **62**: 1174 (1969).
37. A. W. Farnham. *Pestic. Sci.* **2**: 138 (1971).
38. F. W. Plapp and J. E. Casida. *J. Econ. Entomol.* **63**: 1091 (1971).
39. A. J. Forgash, B. J. Cook, and R. C. Riley. *op. cit.* **55**: 544 (1962).
40. A. W. Farnham, K. A. Lord, and R. M. Sawicki. *J. Insect Physiol.* **11**: 1475 (1965).
41. R. M. Sawicki and K. A. Lord. *Pestic. Sci.* **1**: 213 (1970).
42. J. W. Kilpatrick and H. F. Schoof. *J. Econ. Entomol.* **51**: 18 (1958).
43. G. H. S. Hooper and A. W. A. Brown. *Entomol. Exptl. Appl.* **8**: 263 (1965).
44. S. C. Bhatia and R. B. Deobhankar. *Indian J. Entomol.* **24**: 36 (1962).
45. A. W. A. Brown. *Bull. WHO* **30**: 97 (1964).
46. J. R. Busvine. *op. cit.* **31**: 645 (1964).
47. J. de Zulueta. *op. cit.* **20**: 797 (1959).
48. J. L. Gerold and J. J. Laarman. *Nature* **204**: 500 (1964).
49. A. W. A. Brown. In *Pesticides in the Environment*, Vol. I, Part II, (R. White-Stevens, ed.) Dekker, New York, 1971, p. 457.

# INSECTICIDE–ENVIRONMENT INTERACTIONS

# CHAPTER 20

# Effects of Insecticides on Nontarget Organisms

Pesticides are primarily used to kill pest species and if they accomplished only this and then dissipated or decomposed there would be no reason to worry. However, once a chemical is released in the environment, meteorological, chemical, physical, biological, and other allied factors determine its fate and distribution in the ecosystem where it interacts with nontarget species and materials. Our knowledge of these interactions is far from satisfactory. A review of the ecological effects of pesticides on nontarget species was done by Pimentel (1) and deals comprehensively with the effects of pesticides on various organisms. In the present chapter a brief review of the effects of persistent pesticides on nontarget species is attempted.

In the laboratory, chlorinated hydrocarbons have been reported to cause, at very low concentration (10 ppb), decreased growth rate and developmental failure of phytoplankton and molluscan larvae that feed on them (2, 3). Laboratory experiments have also indicated that chlorinated hydrocarbons interfere with photosynthesis in single-celled plants at concentrations of a few parts per billion (4, 5). The lower organisms may concentrate chlorinated hydrocarbons. Diatoms that have a high oil concentration pick them up. These algae are ingested by invertebrate organisms that serve as food for small fish; larger fish feed on smaller fish. If one takes a concentration factor of 10 at each trophic level, then a theoretical concentration of several hundredfold is possible. Since phytoplankton are at the base of the food chain any concentration of stable lipophilic chemicals in them is passed on the food chain. Cox (6) has reported that the DDT content of phytoplankton in the sea has been steadily increasing since 1955 even though the amount of DDT being used has been declining since about 1965. Nimmo et al. (7) found that continuous exposure of shrimp to a DDT concentration of 0.2 ppb resulted in 100% mortality in 18 days and at 0.12 ppb there was 100% mortality in 28 days. Oyster growth is affected by many pesticides at concentrations as low as 0.1 ppb (8). Molluscs are known to concentrate chlorinated hydrocarbons; eastern oysters were kept for 7 days in water containing 10 ppb of DDT. They had concentrated 151 parts per million (ppm) of DDT in their tissue (9). Oysters are known to filter 1–4 litres of water per hour. If an oyster weighing 20 grams were to filter about 2 liters/hour containing 10 ppb of DDT, in 1 week it would have extracted more than 85% of DDT present in the water.

The lipid tissues of all fish examined contain chlorinated hydrocarbons and since ovaries contain lipids, the concentration of insecticides in these organs is high, which can cause adverse effects on reproduction; a concentration of 5 ppm in fresh water trout causes 100% failure in the development of sac fry (10). The potential hazards of pesticide residues on the reproduction of fish have been well documented (11–14).

Because fishery resources in the lakes were at a low ebb in 1965, coho salmon from the Pacific Coast were introduced into Lake Michigan and Lake Superior. This introduction was very successful, but soon scientists from the U.S. Bureau of Commercial Fisheries found DDT and dieldrin residues in all species of fish in the Great Lakes. The highest concentration was reported from major game species in Lake Michigan (11). At that time no detrimental effects on fish were noted, but in 1967 with the first return of mature coho salmon and the development of their eggs in the hatcheries, there was significant mortality of fry. DDT residues were related to these mortalities. Some of the surplus adult salmon were sold to mink ranchers. Rations containing up to 15% of salmon were fed to mink. Ranchers reported very high reproductive losses when these rations were fed during reproductive period. No adult mortalities were reported (15).

One of the best recorded examples of trophic concentration of pesticides of the DDT family is Clear Lake, California. This lake was treated three times during 9 years (1949, 1954, 1957) with DDD (about 120,000 lb in all was used) for the control of a small midge, the clear lake gnat, *Chaoborus asticopus* (Chaoboridae). In addition, approximately 500,000 lb of DDT were applied to the surrounding watershed during the years 1949–1964. The final concentration of DDT alone in the lake was calculated as 14 ppb. Three months following the second treatment of the lake (December 1954), 100 western grebes were reported dead and in March of the following year and again in December 1957 additional mortalities of grebes were reported. Samples of fat were analyzed from two "sick" grebes in 1958 and were found to contain 1600 ppm of DDD on a wet basis. All other organisms that were analyzed from the lake contained varying amounts of DDD and related chemicals. This indicated the transferral of DDD through the food chain (16).

A general decline in the thickness of shells has been reported in the eggs of many species of birds since 1946 (17, 18). In some cases changes between $-15$ and $-25\%$ have been reported. Maximum changes have occurred in great blue herons followed by red-breasted mergansers ($-23\%$) and common mergansers and double crested cormorants ($-15\%$ each). Since DDT also came into use in the late 1940s and since residues of DDT and its analogs were found in these eggs, the thinning was associated with chlorinated hydrocarbons. Evidence of poor hatchability in herring gull, *Larus argentatus*, eggs contaminated with DDT was also found (19). The entire subject has been reviewed in detail by Anderson and Hickey (20).

Several investigations have been carried out in the laboratory on the effect of feeding DDT, DDD, DDE, and dieldrin on egg-shell thickness of birds (21–24). In all cases some thinning of egg shell was observed as a result of these chemicals; decrease in egg shell calcium was also noted (24). But the egg-shell thinning in experimental birds has never been very dramatic.

Along with chlorinated hydrocarbon insecticides there is another group of chemicals, the polychlorinated biphenyls (PCB), that are equally ubiquitous. For quite some time scientists had been finding peaks on gas chromatograms that came close to or looked similar to those of chlorinated hydrocarbon insecticides. They remained unidentified until 1967 when reports from Sweden and Great Britain identified these as PCBs (25, 26). Recently techniques have been developed to separate PCB from chlorinated hydrocarbons and papers have reported separately on these chemicals.

Longcore and Mulhern have recently examined 61 black duck eggs from the United States and Canada. PCBs were found in 57; mean PCB residues ranged from 0.05 ppm in samples from Nova Scotia to 3.30 ppm in those from Massachusetts, and all of them contained DDE. Eggs collected prior to 1940 were compared in shell thickness to those collected in 1964 and in 1971; these had shell thicknesses of 0.348, 0.321, and 0.343 mm, respectively (27). In the years 1969 and 1970, Faber and Hickey (28) collected and examined 117 eggs from 19 species of aquatic-feeding birds and compared their shell thicknesses and a thickness index to those of eggs collected prior to 1947. They found that DDE was a prominent factor in egg-shell thinning, especially in herons. PCBs were important in mergansers and "mercury was positively correlated with thickness in grebes and negatively correlated in mergansers."

Aldrin and other chlorinated hydrocarbons interrupt estrus in rats (29) when they are fed at several parts per million. This is possibly due to increased MFO activity that hydroxylates estradiol-17$\beta$, estrone, and other estrogens. On the other hand, estrogenic activity of DDT in low dosages (1 ppm) as measured by increased uterine wet weight has been reported (30). Probably these differences in action are due to different dose levels and experimental conditions. Also, technical DDT consists of a mixture of several chemicals, including $o,p'$-DDT which has been shown to be estrogenic (31). Subnormal reproduction in beagle dogs, when fed DDT and aldrin, has been reported by Deichmann et al. (32). Endrin, when fed for 120 days at 5 ppm in the diet, increased mouse parent mortality and the animals produced significantly small litters (33).

In the case of organophosphorus compounds, Johnson (34) has reported delayed neurotoxic action in animals. Lesions in axons in the spinal cord, sciatic nerve, and medulla oblongata were found. The brain, however, did not show signs of damage.

## REFERENCES

1. D. Pimentel. Ecological Effects of Pesticides on Non-target Species, Executive Office of the President, Office of Science and Technology, 1971.
2. R. Ukeles. *Appl. Microbiol.* **10**: 352 (1962).
3. H. C. Davis. *Fish Rev.* **23**: 8 (1961).
4. D. W. Menzel, J. Anderson, and A. Randtke. *Science* **167**: 1724 (1970).
5. C. F. Wurster. *Science* **159**: 1474 (1968).
6. J. L. Cox. *Science* **170**: 1974 (1970).
7. D. R. Nimmo, A. R. Wilson, Jr., and R. R. Blackman. *Bull. Environ. Contam. Toxicol.* **5**: 333 (1970).
8. P. A. Butler. *J. Appl. Ecol.* **3** (suppl.): 253 (1966).
9. S. Udall. Testimony before U.S. Senate Sub-committee on Reorganization and International Organizations, pt. I: 71, May 1963.
10. P. A. Butler. *Bioscience* **19**: 889 (1969).
11. S. H. Smith. *J. Fish Res. Board Can.* **25**: 667 (1968).
12. K. J. Macek. *op. cit.* **25**: 1787 (1968).
13. H. E. Johnson. *Bull. Ecol. Soc. Amer.* **48**: 72 (1967).
14. G. E. Burdick, E. J. Harris, H. J. Dean, T. M. Walker, J. Skea, and D. Colby. *Trans. Amer. Fish Soc.* **93**: 127 (1964).
15. R. J. Auerlich, R. K. Ringer, H. L. Seagran, and W. G. Youatt. *Can. J. Zool.* **49**: 611 (1971).
16. M. W. Miller and G. G. Berg, eds. *Chemical Fallout*, Thomas, Springfield Ill. 1969, 531 pp.
17. D. A. Ratcliffe. *Nature* **215**: 208 (1967).
18. J. J. Hickey and D. W. Anderson. *Science* **162**: 271 (1968).
19. J. O. Keith. *J. Appl. Ecol.* **71**: 85; **3** Suppl. 57 (1966).
20. D. W. Anderson and J. J. Hickey. In *Proceedings XVIth International Ornithological Congress* (K. H. Voous, ed.), E. J. Brill, Leiden, 1972.
21. S. N. Wiemeyer and R. D. Porter. *Nature* **227**: 737 (1970).
22. J. H. Enderson and D. D. Berger. *Bioscience* **20**: 335 (1970).
23. R. K. Tucker and H. A. Haegle. *Bull. Environ. Contam. Toxicol.* **20**: 335 (1970).
24. J. Bitman, H. C. Cecil, S. J. Harris, and R. M. Prouty. *Nature* **224**: 44 (1969).
25. G. Widmark. *J. Assoc. Off. Anal. Chem.* **50**: 1069 (1967).
26. D. C. Holmes, J. H. Simmons, and J. O'G. Tatton. *Nature* **216**: 227 (1967).
27. J. R. Longcore and B. M. Mulhern. *Pestic. Monit. J.* **7**: 62 (1973).
28. R. A. Faber and J. J. Hickey. *op. cit.* **7**: 27 (1973).

29. W. L. Ball, K. Kay, and J. W. Sinclair. *A.M.A. Arch. Ind. Hyg. Occup. Med.* **7**: 292 (1953).
30. R. M. Welch, A. Loh, and A. H. Conney. *Toxicol. Appl. Pharm.* **14**: 358 (1969).
31. J. Bitman, H. C. Cecil. *J. Agr. Food Chem.* **18**: 1108 (1970).
32. W. B. Deichmann, W. E. McDonald, A. G. Beasley, and D. Cubit. *Ind. Med. Surg.* **40**: 10 (1971).
33. E. E. Good and G. H. Ware. *Toxicol. Appl. Pharmacol.* **14**: 201 (1969).
34. M. K. Johnson. *Br. Med. Bull.* **25**: 231 (1969).

# INSECTICIDE–ENVIRONMENT INTERACTIONS

# CHAPTER 21

## Effects of Insecticides on Man

## CHLORINATED HYDROCARBONS

In a book of this sort this subject can be discussed only very briefly. Literature concerning the effects of pesticides on man has been summarized and some of the recent reviews include those by Hayes and Kay (1–3). Considerable work has been done on chlorinated hydrocarbons, and because they are chemically stable and persist in the environment for a long time they are discussed first. Except for methoxychlor, the chlorinated hydrocarbons represent a family of highly stable chemical compounds. They are inducers of MFO, and DDT is one of the best inducers known. They are both hepatotoxic and neurotoxic. Hayes (4) first reported the neurotoxicity of chlorinated hydrocarbons in personnel involved in indoor spraying of dieldrin in the tropics. Since then electroencephalographic abnormalities have been found in factory workers exposed to cyclodiene insecticides (5, 6). Some workers in an aldrin factory suffered convulsions, and the epoxide of aldrin, dieldrin, was present in high concentration in body fat. Convulsions in rats given 100 ppm aldrin in diet are common (7). In addition, laboratory animals on high dosages of chlorinated hydrocarbons exhibit hyperexcitability and irritability.

Laws et al. (8) studied 35 workers of a plant producing DDT. They had worked for 11–19 years in the plant and may have had as much as 17.5–18 mg of DDT per person per day. Their fat had mean values of $p,p'$-DDT from 39 to 128 times those reported by Hayes et al. (9) for the general population. DDT-related materials were 12 to 32 times those found in the general population. These factory workers showed no signs of blood dyscrasias or cancer.

However, in an earlier study, Birks (10) reported 19 cases of proven or probable malignancies in 400 industrial workers he had examined.

A group of 79 subjects with varying degrees (non-production workers and production workers) and various duration (several weeks to more than 15 years) of exposure to lindane were examined by Samuels and Milby (11). They found no clinical symptomatology or physical evidence of disease clearly attributable to lindane. However, as reported earlier, lindane is suspect in cases of serious blood dyscrasias, primarily aplastic anemia. Exposure to vaporizers of lindane was reported in most of these cases.

Kay (12) reports that metabolites of DDT and malathion can form antigens with tissue proteins. This immune response raises the possibility of allergic tissue damage. But this damage has not been proved experimentally, though the possibility exists that genetic factors may predispose rare individuals to such damage.

DDT-derived materials in the blood of the general population have been reported; these materials are proportional to DDT stored in fat (13, 14). A ratio of 1:140 has been found. It therefore appears that storage in fat is a prophylactic mechanism that prevents the chemical from circulating into the organs and thus affords protection to the individual. However, under conditions of stress when reserve fat is used more rapidly, the blood concentration of DDT may increase; but evidence for this is lacking in man. There has been a report that persons taking anticonvulsants do not accumulate DDT in their bodies even if they are heavily exposed (15).

DDT has been reported in the milk of nursing mothers from several countries, including Czechoslovakia and Russia (16, 17). Kroger (18), on the basis of 53 samples, found that human milk contained an average of 2.4 ppm $p,p'$-DDT and its metabolites. DDT residues have been reported in human fetuses in concentrations as high as 26 ppm (19).

Hayes et al. (20) studied the effect of ingested DDT on 24 volunteers who were given daily oral doses from 3.5 to 35 mg/per man for 21.5 months, and they were observed for an additional 25.5 months. Sixteen of these volunteers were followed up to 5 years. Those receiving the highest dosages contained between 105 and 619 ppm of DDT in their fat when feeding was ceased. No definite clinical or laboratory evidence of injury as a result of DDT intake was found; excretion of DDA and storage of DDT and DDE were proportional to DDT intake.

A group of 233 persons was chosen by Jager from among 800 workers who had been exposed from 4 to 14 years to aldrin, dieldrin, endrin, and telodrin in an insecticide plant. The blood of this group contained dieldrin at an average level of 0.035 $\mu$g/ml. This represents more than 50 times the level in the general population. No adverse effects were found among these workers as a result of this high level of dieldrin. Hunter and Robinson (21) and Hunter et al. (22)

studied effects of oral administration of dieldrin (up to a maximum dosage of 0.21 mg/kg/man/day in volunteers and found no adverse effects in them either during the course of investigations or at their conclusion after 24 months.

Investigations on chlorinated hydrocarbons have shown that the levels of these chemicals in blood and in adipose tissue are proportional to the daily dosage. An equilibrium is reached and there is a finite upper limit to storage that is related to daily intake. This leads to a steady state and no higher levels of buildup in the body take place if the daily intake remains constant.

Duggan (23) calculated the daily intake of 19 year old men in the United States as 0.028 mg/man/day for DDT; for DDT-related compounds the figure was 0.063.

Since chlorinated hydrocarbons are inducers of microsomal enzymes (MFO) they increase the levels of MFO in man and bring about the metabolism of endogenous steroids, fatty acids, thyroxin, tryptamine, and so forth. Also, they compete with these endogenous chemicals for active sites of enzymes, because chlorinated hydrocarbons in their turn also are metabolized by the very same enzymes that they help induce. When organochlorines are administered in sufficient dosages to induce enzymes, proliferation of the smooth endoplasmic reticulum in the liver takes place. This change is reversible and is evidently adaptive because the body is usually exposed to a large variety of different compounds. Nevertheless, the effect of these changes on hormone regulation and drug therapy is worth considering, especially in the cases of people occupationally exposed to chlorinated hydrocarbons.

Rat liver is sensitive to enzyme inducers, and a dietary level of 1 ppm cyclodiene epoxides has been regarded as threshold for enzyme induction; for DDT this level is about 5 ppm. No threshold value for man has been established, but the present daily intake of these insecticides by man through his food is far below the threshold level for rats (24). Also, see Chapter 22.

## ORGANOPHOSPHATES AND CARBAMATES

These chemicals, as indicated earlier, inhibit acetylcholinesterase, which is responsible for hydrolyzing acetylcholine at synaptic sites. They exhibit a wide range of toxicity to mammals. Parathion and aldicarb, for instance, are highly toxic. Occupational fatalities have therefore resulted from the use of the highly toxic chemicals. In the case of toxic organophosphates the enzyme is irreversibly phosphorylated, whereas in the case of carbamates the inhibition is reversible. Clinical signs of carbamate poisoning as compared to those of organophosphate poisoning develop much more rapidly; hence fatalities or overexposures are less likely to occur. In the case of organophosphate poisoning certain oximes act as antidotes because they accelerate reactivation of the phosphorylated enzyme. Workers chronically exposed to organophosphates show low levels of cholinesterase. A value of less than $0.5^\Delta$ pH/hour should be considered significant. In some cases values of $0.2^\Delta$ pH/hour have been found in workers exposed to organophosphates without any apparent evidence of sickness. The adaptation to this low level of enzyme activity and its mechanisms are not fully understood (25).

Jager et al. (26) conducted electromyographic (EMG) studies of 66 workers engaged in the manufacture and formulations of pesticides. They had varying degrees of exposure to organophosphorus and organochlorine compounds. Precise exposure times were not recorded. Signs of impaired nerve and muscle function in about 50% of those working with organophosphorus compounds (*cis*-isomer of Bidrin and $\beta$-isomer of Gardona) were discovered. Only 4% of those working with organochlorine pesticides showed these signs. These authors concluded that blood cholinesterase activity does not provide a sensitive index of functional impairment of nerve and muscle.

Chlorphos inhalation tests in mice have resulted in abnormalities in the development of embryos and changes in placenta (27). There have been other reports suggestive of teratogenic effects of organophosphates and carbamates, but no conclusive evidence has been presented.

## REFERENCES

1. W. J. Hayes, Jr. *Ann. N.Y. Acad. Sci.* **160**, art. 1: 40 (1969).
2. K. Kay. *Environ. Res.* **6**: 202 (1973).
3. K. Kay. *op. cit.* **7**: 243 (1974).
4. W. J. Hayes, Jr. *Bull WHO* **20**: 891 (1959).
5. I. Hoogendam, J. P. Versteeg, and M. DeVlieger. *Arch. Environ. Health* **4**: 86 (1962).
6. G. Kazantzis, A. I. G. McLaughlin, and P. F. Prior. *Br. J. Ind. Med.* **21**: 46 (1964).
7. W. L. Ball, K. Kay, and J. W. Sinclair. *A.M.A. Arch. Ind. Hyg. Occup. Med.* **7**: 292 (1953).
8. E. R. Laws, Jr., A. Curly, and F. J. Biros. *Arch. Environ. Health* **15**: 766 (1967).
9. W. J. Hayes, jr., W. E. Dale, and V. W. Burse. *Life Sci.* **4**: 1611 (1965).
10. R. E. Birks. *Arch. Environ. Health* **1**: 291 (1960).
11. A. J. Samuels, T. H. Milby. *J. Occup. Med.* **13**: 147 (1971).
12. K. Kay. *Ind. Med.* **38**: 52 (1969).
13. J. L. Radomski, W. B. Deichmann, A. A. Rey, and T. Markin. *Toxicol. Appl. Pharmacol.* **20**: 175 (1971).
14. M. Wasserman, M. Gon, D. Wasserman, and L. Zellermeyer. *Arch. Environ. Health* **11**: 375 (1965).
15. D. S. Kwalick. *J. Amer. Med. Assoc.* **215**: 120 (1971).
16. L. N. Suvak. *Zdravookhr. Kishinev.* **13**: 19 (1970).
17. G. V. Grechava. *Vopr. Pitan.* **29**: 75 (1970).
18. M. Kroger. *J. Pediatrics* **80**: 401 (1972).
19. R. E. Rappolt and W. E. Hale. *Clin. Toxicol.* **1**: 57 (1968).
20. W. J. Hayes, Jr., W. E. Dale, and C. I. Pirkle. *Arch. Environ. Health* **22**: 169 (1971).
21. C. G. Hunter and J. Robinson. *op. cit.* **15**: 614 (1967).

22. C. G. Hunter, J. Robinson, and M. Roberts. *op. cit.* **18**: 12 (1969).
23. R. E. Duggan *Pestic. Monit. J.* **2**: 2 (1968).
24. G. T. Brooks. Chlorinated Insecticides, Vol. II. *Biological and Environmental Aspects*, CRC Press, Cleveland, 1974, 197 pp.
25. O. C. Beasley and K. R. Long. *J. Iowa Med. Soc.* **56**: 45 (1966).
26. K. W. Jager, D. V. Roberts, and A. Wilson. *Br. J. Ind. Med.* **27**: 273 (1970).
27. V. A. Gofmekler and S. A. Tabakova. *Farmacol. Toksikol.* **33**: 737 (1970), English abstract.

# INSECTICIDE–ENVIRONMENT INTERACTIONS

# CHAPTER 22

# Carcinogenicity, Mutagenicity, and Teratogenicity

Recent advances in the fields of analytical chemistry, molecular biology, biochemistry, and pharmacology have drawn attention to the subtle effects of apparently subliminal dosages of chemicals on the individual organism. That chronic exposures to chemicals may result in cellular insults is possible. Research in the field of carcinogenesis has demonstrated the ability of various stresses to change cellular protein or deoxyribonucleic acid (DNA). Even when each of these stress-producing agents is individually far below the threshold level of physiological effects, it is possible that when more than one such agent is present they can have a synergistic effect or act in tandem to produce changes at the cellular level. It is therefore imperative that we make careful investigations on the potentialities of pesticidal chemicals as carcinogenic, mutagenic, and teratogenic agents.

Very limited data are available on the carcinogenic effects of insecticides on man. However information on laboratory animals has been gathered over a number of years and some insecticides have been suspected of carcinogenic activity. But species and strains of animals differ in their response to a carcinogen and the number and types of tumors induced. Even when careful studies have been made, and despite the observation that in many cases tumorigenic responses for many chemicals are similar in some animals and man, extrapolation of results from experimental animals to man presents many difficulties and many uncertainties. Among the animals most frequently used for such studies are the rat and the mouse, and strains of these animals with different sensitivities to spontaneous incidence of various types of tumors are available.

## DDD, DDT AND MIREX

Fitzhugh and Nelson (1) first reported the mildy tumorigenic action of DDT on Osborne-Mendel rats. Hepatic cell tumors were suspected; however the same laboratory could not reproduce these results and other workers also failed to reproduce them. The Bionetics Research Laboratories (2) under a contract with the National Cancer Institute, screened 130 pesticides and industrial chemicals and found that DDT at 46.4 mg/kg/day and 10 other registered pesticides, including mirex, chlorbenzilate, and strobane, produced a significant increase in the incidence of tumors in hybrid mice. Aldrin and dieldrin were not included among the chemicals tested. The data of the Bionetics Research Laboratories has been received with mixed feelings, because very high dosages were used and feeding was started when the animals were 7 days old. Two hybrid strains of mice were used and tumors were found especially in liver, lung, and lymphoid organs; they were mostly benign, though the possibility of their being malignant was later pointed out. However it should be noted that enzyme inducers appear to greatly affect rat liver; a dietary level of 1 ppm of cyclodiene epoxides has been established as a threshold. Hayes (1969) also expressed the opinion that the changes produced in the liver of rats involved the cellular tissue that produces microsomal enzymes. These changes are reversible and are characteristic of rodents; many chemicals that induce MFO can produce such changes. Agathe et al. (3) did not find DDT tumorigenic when fed for almost 2 years to Syrian golden hamsters at 500 and 1000 ppm.

Multigeneration studies extending over five generations were undertaken by Tarjan and Kemeny in inbred BALB/c mouse strain (4). Mice were given a daily intake of 0.4–0.7 mg/kg of DDT in their diet for 6 months and the experimental animals were observed for 26 months. Tumor and lukemia incidence was higher in the DDT group than in controls, and according to the authors the incidence increased in later generations. However in a letter to the editor, Terracini (5) pointed out that after analyzing the data he did not find an increase in the proportion of tumor and leukemia incidence in succeeding generations. Kay (6) calculated tumor incidence in the succeeding generations from the data of Tarjan and Kemeny and arrived at the same conclusion. Another multigenerational study is being conducted by Tomatis et al. (7), who have reported that the liver tumors appeared earlier in the $F_1$ mice fed 50–250 ppm DDT. This work is continuing and further investigations when published will add to our knowledge regarding the multigenerational effect. Ovarian tumors in the $F_1$ generation showed an inverse relationship to dose (compare aldrin and dieldrin), possibly because of increased induction of mixed-function oxidases resulting in metabolism of steroids.

DDD has been used in the chemotherapy of tumors of adrenal cortex (8), and Okey (9) has shown that pretreatment with DDT substantially reduces mammary tumor incidence due to dimethylbenzanthracene (DMBA), probably because the induced MFOs metabolize DMBA.

## ALDRIN, DIELDRIN, AND ENDRIN

The picture regarding these insecticides appears a little confusing. Some studies have concluded that aldrin and dieldrin were tumorigenic (10–12), others that they were inactive (13–15), and one concluded that endrin was inactive, but aldrin and dieldrin were antitumorigenic (16).

Davis and Fitzhugh (10) reported that feeding of both aldrin and dieldrin to mice resulted in a statistically significant increase in the incidence of hepatic tumors. However the tumors were morphologically benign; these authors considered the two insecticides to be tumorigenic but not carcinogenic. In a later study Fitzhugh et al. (11) did not find any closely dose-related relationship between incidence of tumors and aldrin and dieldrin; nevertheless there was an increased incidence of tumors in animals fed aldrin and dieldrin.

Deichmann et al. (16) conducted lifetime tests with aldrin and dieldrin in Osborne-Mendel strain of rats and found a decrease in the incidence of all tumors, particularly tumors of the mammary and lymphatic tissue when both sexes were fed aldrin or dieldrin. The dose levels were 20, 30, and 50 ppm. Walker et al. (13) fed dieldrin at dietary levels up to 10 ppm to CFE rats for 2 years and did not find any increase in the incidence of tumors. However microscopic nodules were found in three of the animals receiving the highest dose. They were also present in one of the controls.

Hunter and Robinson (14) and Hunter et al. (15) studied volunteers who ingested dieldrin on a daily basis for 2 years. The maximum dosage used was 0.21 mg/kg/man/day. No adverse effects were found either during the course of the study or at its conclusion. Jager (17) studied 233 workers occupationally exposed to cyclodiene insecticides and found no adverse effects of the exposure on these men.

However the careful investigations of Walker et al (13) and Thorpe and Walker (18) have clearly shown that in the CFl strain of mice dieldrin is carcinogenic and produces liver cancer with metastases to the lung in some cases. The response is strongly dosage related. Cancer formation occurred even at the lowest dosage fed (0.1 ppm) and was the principal reason for the aldrin/dieldrin ban by the Environmental Protection Agency. Liver lesions were also observed in controls, but increased incidence of liver lesions was evident in the treated groups.

## HEXACHLOROCYCLOHEXANES

Kay (6) has reported two studies by Nagasaki et al. (1971 and 1972) in which α, β, γ, and δ isomers of HCH have produced tumors in mice.

## HEMPA

The carcinogenic action of hempa (19) has already been mentioned.

## ARAMITE

The carcinogenic properties of Aramite, 2-(p-tertiary-butylphenoxy)isopropyl-2-chloroethyl sulfite, an acaricide, have been confirmed by several workers (2, 8).

A major difficulty in evaluating the carcinogenic effect on animals in long-term studies is the exact nature of hepatic lesions. Also, spontaneous tumors do occur in animals under laboratory conditions. Some strains are particularly prone to these spontaneous tumors; the sites of these tumors differ. If they are in the liver they are usually called "hepatoma," a word that has been used for both benign and malignant types. Unless the type of hepatoma is exactly described, or photomicrographs given, precise evaluation of the results is not possible.

Because of the thalidomide disaster, investigations in the field of teratology have gained great importance; insecticides have been tested for teratogenic effects, but positive results in literature are not many. Carbaryl has been found to be teratogenic in guinea pigs (20), in beagle dogs (21), and in mice (2). Meiniel et al. (22) found a strong teratogenic effect of parathion on the Japanese quail, *Coturnix coturnix japonica*, when eggs were immersed in a dilute solution of the insecticide. The teratogenic action of β-toxin of *Bacillus thuringiensis thuringiensis* has already been discussed.

The mutagenic properties of tepa and metapa have been reported by a number of investigators. Epstein found a tentative single-dose threshold value of 0.04 mg/kg for tepa and 1.4 mg/kg for metepa (8).

## REFERENCES

1. O. G. Fitzhugh and A. A. Nelson. *J. Pharm. Exp. Ther.* **89**: 18 (1947).
2. J. R. M. Innes, B. M. Ulland, M. G. Valerio, L. Petrucelli, L. Fishbein, E. R. Hart, A. J. Pallota, R. R. Bales, H. L. Falk, J. J. Gart, M. Klein, I. Mitchell, and S. Peters. *J. Nat. Cancer Inst.* **42**: 1101 (1969).

References 241

3. C. Agathe, H. Garcia, P. Shubik, L. Tomatis, and E. Wenyon. *Proc. Soc. Exp. Biol. Med.* **134**: 113 (1970).
4. R. Tarjan and T. Kemeny. *Food Cosmet. Toxicol.* **8**: 478 (1969).
5. B. Terracini. Letters to the Editor, *op. cit.* **8**: 478 (1970).
6. K. Kay. *Environ. Res.* **7**: 243 (1974).
7. L. Tomatis, V. Turusov, N. Day, and R. T. Charles. *Int. J. Cancer* **10**: 489 (1972).
8. W. F. Durham and C. H. Williams. *Annu. Rev. Entomol.* **17**: 123 (1972).
9. A. B. Okey. *Life Sci.* pt 1. **11**: 833 (1972).
10. K. J. Davis and O. G. Fitzhugh. *Toxicol. Appl. Pharmacol.* **4**: 187 (1962).
11. O. G. Fitzhugh, A. A. Nelson, and M. L. Quaife. *Food Cosmet. Toxicol.* **2**: 551 (1964).
12. A. I. T. Walker, E. Thorpe, and D. E. Stevenson. *Food Cosmet. Toxicol.* **11**: 415 (1973).
13. A. I. T. Walker, D. E. Stevenson, J. Robinson, E. Thorpe, and M. Roberts. *Toxicol. Appl. Pharmacol.* **15**: 345 (1969).
14. C. G. Hunter and J. Robinson. *Arch. Environ. Health* **15**: 614 (1967).
15. C. G. Hunter, J. Robinson, and M. Roberts. *op. cit.* **18**: 12 (1969).
16. W. B. Deichmann, W. E. McDonald, E. Blum, M. Bevilacqua, J. Radomski, M. Keplinger, and M. Balkus. *Ind. Med.* **39**: 426 (1970).
17. K. W. Jager. *Aldrin, Dieldrin, Endrin, and Telodrin. An Epidemiological and Toxicological Study of Long-term Occupational Exposure.* Elsevier Publishing Co., Amsterdam, 1970, 234 pp.
18. E. Thorpe and A. I. T. Walker. *Fd. Cosmet. Toxicol.* **11**: 433 (1973).
19. J. A. Zapp, Jr. "HMPA: a Possible Carcinogen." Letters to the Editor, *Science* **190**: 422 (1975).
20. J. F. Rubens. *Toxicol. Appl. Pharmacol.* **15**: 152 (1969).
21. H. E. Smalley, J. M. Curtis, and F. M. Earl. *Toxicol. Appl. Pharmacol.* **13**: 392 (1968).
22. R. Meiniel, Y. Lutz-Ostertag, and H. Lutz. *Arch. Anat. Microsc. Morphol. Exp.* **59**: 167 (1970).

# INSECTICIDES, FUNGICIDES, AND WORLD ECONOMY

| CHAPTER 23 | Role of Insecticides and Fungicides in Food and Agriculture, and Public Health |

We are living in what may be described as "explosive times." Everything seems to be expanding very rapidly; our knowledge is doubling at least every 10 years and our population is nearing the four billion mark. Unfortunately, more than two-thirds of these people are denied the satisfaction of two square meals a day, and each day about 10,000 people—most of them children—die as a direct or indirect result of malnutrition (1). In the United States about 14.3 billion dollars worth of agricultural commodities were lost to pests annually between 1951 and 1960 (2).* An additional $1 billion dollars are lost in storage and forest trees. More than 200 insects attack our main crops, animals, and forests; only a few, like cottonycushion scale of citrus, *Icerya purchasi* Maskall, the spotted alfalfa aphid, *Therioaphis maculata* (Buckton), and the

---

* Recent figures for such losses are not available, and, to the best of my knowledge, have not been compiled. In view of the food shortages plaguing so many countries it is unfortunate that strenuous efforts are not being directed by various governments concerned in this direction and that our knowledge remains woefully inadequate. The differing views cited in this chapter bear testimony to my premise. Before reliable solutions to a problem are sought, precise definition of the problem is important, and in order to obtain optimum benefit from our pest control efforts, it seems logical that countries establish mechanisms for the evaluation of long term losses due to pests so that priorities for research may be reassigned if necessary. The specialized agencies of the United Nations can perform a valuable service by providing help for the member states, especially the developing countries.

screwworm fly, *Cochliomyia hominivorax* Coq., have been dramatically controlled by nonchemical methods. Conventional insecticides control between 80 and 90% of these pests (3). Whenever pest populations reach economic levels, recourse to pesticides appears to be the only possible protection. However pesticides are not selective; they do not actively search the pest and are not self-propagating. They often eliminate parasites and predators as well. The use of lindane or DDT for the control of the codling moth which results in increased population of mites, and secondary pest problems in the case of cotton as a result of excessive use of insecticides are good examples of these complex interactions.

After World War II many developing countries learned by experience that emphasis on industrialization without first building a sound agricultural base results in the failure of both. The judicious use of fertilizers, pesticides, and farm machinery in the United States since 1950 has resulted in 35% greater yield with 45% less farm labor and use of 11% less land. The national average yield on field corn has doubled; similarly, the production of rice and wheat per acre has nearly doubled. The use of modern technology has completely transformed our agriculture; one farmer could feed and clothe only 4 persons in 1850 and 25 in 1960, and the figure today is 46.

In 1964 it was estimated that about 85% of the total volume of insecticides used by United States farmers consisted of 12 insecticides. Of these, toxaphene occupied first place, with DDT a close second; both of these accounted for about 46% of the total insecticides used in 1964. Cotton, corn, and apples were treated with two thirds of the total quantity of the insecticides used. Of these, cotton occupied first place; 80% of methyl parathion, 86% of endrin, 70% of DDT, and 69% of toxaphene were used on cotton. Ten percent of the total was used on corn (96% of all aldrin and 84% of heptachlor) (4, 5).

Rice and wheat together form the principal source of the world's food supply; one half of the world's population depends on rice as the staple food, and grains form an important part of diet in a major portion of the world. In the 1960s these crops were treated with aldrin, dieldrin, and endrin throughout the world. Endrin was used in India and Pakistan for the control of rice stem borers. In the United States the use of aldrin was mainly limited to field crops, especially corn. The use of endrin in this country was mainly limited to non-agricultural commodities. It is estimated that the total investment in 1966 by Corn Belt farmers for aldrin was approximately $22 million. This investment resulted in increased production estimated at $198 million (6).

In 1966 pesticide materials worth 561 million (353 million lb) were used by farmers in the United States; this comprised about 2% of the total farm production expense for that year. The cost of capital investment in application machinery, labour, and management time devoted to inspection is not included in this 2% (5). Not much progress has been made in the economic analysis of the benefits of pesticides. Headly (7, 8) estimates that for every dollar spent by the farmer on pesticides the return is between $4 and $5. Apart from this return there are also public and regulatory pressures on the farmer to farm premium quality produce without infestation by pests.

For most developing countries agricultural products form a large portion of their exports. This is especially true in Africa, where in Sudan, Republic of Zaire, and Tchad in the late 1960s agricultural exports represented close to 100% of their total exports (9). Pests are of great economic importance on export crops, especially cotton, cocoa, coffee, banana, and peanuts (groundnuts). Cotton forms the backbone of the economy of three African countries, namely, Egypt, Sudan, and Tchad. This crop cannot be profitably grown without insect control. The Food and Agriculture Organization of the United Nations estimates that the whitefly, *Bemisia tabaci* Genn., can reduce irrigated seed cotton yield by 44% in the Gezira area of the Sudan (10); and *Heliothis armigera* (Hubner) (cotton bollworm), if not controlled, can make cotton production uneconomical.

World cotton production has not kept pace with the increasing demand, and increased production per acre is one of the best solutions available. During the 1960s India had the largest area of any country under cotton production, but it ranked fourth in total production and seventh in yield per hectare (11). In the four countries, namely, Dahomey, Ivory Coast, Upper Volta and Cameroun, there has been an increase of 767, 550, 300, and 85%/hectare in cotton production between the mid 1950s and 1968 (12). This is due to improved techniques, use of higher yielding varieties, fertilizers, and insecticides. Experiments conducted in French-speaking West African countries indicate that when the use of insecticides is coordinated with the use of fertilizers, insecticidal spray can result in 13–17% increase in the yield of cotton. An optimum increase of 300% in yield was obtained after 10 sprays. However under field conditions the results may not be as dramatic, but if insecticide use is abandoned, it is estimated that Dahomey and the Ivory Coast would lose 50 and 60% respectively, of their total cotton production; loss in export earnings would amount to approximately $6 million (9).

At the turn of the last century tropical America was a leader in cocoa production and grew 85% of the world's supply. Fungus pests reduced cocoa yield and in the absence of effective pesticides, production in America gradually dropped, but Africa, which was free of the fungus disease, started increasing its cultivation of the crop. By 1964 the American share of the world market had dropped to 20%, whereas that of Africa rose to 78% (12, 13). At present Ghana is the largest producer of cocoa and has about 4 million acres (1.5 million hectares) under cultivation; Nigeria occupies second place. Among insect pests, the capsid bugs, *Sahlbergella singularis* Haglund, which is distributed throughout West Africa's cocoa growing areas, and *Distantiella theobroma* Dist., which is restricted in distribution from the Ivory Coast to the Central African Republic and Cameroun, are important. Lindane is effective in controlling these pests. In the mid 1950s replanting of trees and spraying was started in Ghana and the increase in crop yield was beyond all expectations. On some farms production rose as high as 244% after 3 years of treatment. The overall increase was more than 230% during the years 1964–1965; 581,000 tons as compared with 250,000 tons annually for the years 1947–1951. Later the annual average fell to 370,000 tons. The exact contribution of insecticidal sprays alone cannot be accurately assessed, but it is estimated that it saved between 70,000 and 140,000 tons of cocoa worth more than $30 million at the prevailing world price. Similarly, in Nigera it is estimated that spraying against capsids and black pod disease of cocoa caused by the fungus *Phytophthora palmivora* (Butl.) saved Nigera between $30 and $60 million. Thus in these two countries an investment of $6 million in pesticides resulted in additional earnings of between $66 and $132 million, a cost/benefit ratio of 1 : 11–22 (9).

In Sri Lanka (Ceylon) in the 1870s the coffee crop was seriously damaged by the coffee rust *Hamileia vastatrix* B. and Br.; in the absence of fungicides no remedy was available and coffee had to be replaced by tea. In 1970 coffee rust made its appearance in Brazil, but effective fungicides have saved the industry from the fate it met in Sri Lanka 100 years ago (9). Protection of coffee requires precision in the application of insecticides and fungicides. This recently has been demonstrated in the case of Arabica coffee in Kenya which commands premium prices. In 1967 and 1968 coffee berry disease (brown blight) caused by *Colletotrichum coffeanum* Noack almost ruined the entire crop. Treatment with newer organic fungicides and changes in agricultural practices resulted in Kenya's recuperating from the loss. Among insect pests of Arabica coffee in Kenya are the mealy bugs, *Pseudococcus* spp., and the leaf miner *Leucoptera coffeina* Meyricki Ghesq. Improved agricultural practices, careful surveillance, and precision in the application of protective pesticides have prevented the recurrence of pest problems on a large scale.

In Latin America more than 33% of the potential crop estimated at $4.5 billion is lost annually; losses due to insects amount to 10% of the potential crop. Since agricultural practices differ from primitive small farms to modern large-scale operations, it is difficult to estimate the part played by pesticides in total agriculture, but losses would be far greater if no pesticides were used (9).

More than 244 million cattle are raised in South America and the grasshopper problem is important on grazing lands. Normally natural grasslands support 3 sheep or 0.8 cattle per hectare; however if grasshoppers are present, competition for available food is greater, and a density of 8 grasshoppers per square meter results in the animals not getting enough food. Use of insecticides keeps the grasshopper population under control; without the use of insecticides there would be disasterous consequences for the cattle industry (9). Insecticides have greatly helped in breaking locust cycles in Africa, and with their judicious use considerable success in reducing locust cycles has been achieved. On the same continent the cattle industry also faces another insect problem of a slighly different kind—the tsetse fly; control of this fly by insecticides and the manipulation of its breeding and resting grounds have opened up new grazing grounds (9).

## COST/BENEFIT AND BENEFIT/RISK EQUATIONS

As indicated earlier, definite figures on the economic benefits of the use of pesticides are not available and indeed would be extremely difficult to calculate accurately. Almost always the use of pesticides is accompanied by the planting of better varieties, use of fertilizers, and better soil management. In fact, some of the varieties that have brought about the so-called green revolution would not be successful unless the changes mentioned above are made; and some, like IR-8 rice, are attacked by some pests in preference to older varieties. In the succeeding paragraphs the role of insecticides in some important crops is discussed.

Corn occupies first place among the crops of the United States; its total value in 1966 was close to $5 billion and it was grown on 66 million acres. The insecticides mostly used to protect corn were aldrin and heptachlor among the organochlorines and diazinon and parathion among the organophosphates. In that year no alternatives were available for 16% of the United States cropland. Organophosphates and carbamates could have been substituted for organochlorines to control insect infestation on the remainder, but this would have increased the cost by $7.3 million (14). If left untreated, the insect damage could have exceeded 20%.

Cotton is the fifth most valuable crop in the United States and it is a crop of international importance with problems of international magnitude. Newly irrigated areas in developing countries are usually planted with cotton, and soon insect problems arise, especially with varieties of better quality and higher yield. The use of insecticides usually gives good control in the first few years; later more frequent applications of insecticides are needed to obtain economic control. Often resistance develops and new insecticides that are usually more costly are substituted; but pests soon develop resistance to these insecticides as well. This is followed by a resurgence of the original pests accompanied by secondary pests that acquire an important role. This was interestingly demonstrated in Turkey which is an important cotton producer. Since 1950, because of better varieties, use of pesticides, and better farming and irrigation, the yield per acre has more than doubled. However the Egyptian leafworm

*Spodoptera littoralis* (Boisd.) has become resistant to methyl parathion and the spiny bollworm *Earis insulana* (Boisd.) to endrin. In the Adana area the spider mites that were once of no economic importance have acquired pest status.

In the United States in the 1960s cotton was grown on 10.3 million acres and the total value of the crop was $1.3 billion dollars. Only about 54% of the total cotton acreage receives insecticide treatment. The southeast and delta states account for about 79% of the treated acreage, whereas in the southern plains only 37% is treated. In the year 1966 application of insecticides to cotton represented 44% of all insecticides and 60% of all chlorinated hydrocarbons used on all crops (88% toxaphene; 73% DDT; and 91% parathion). If organochlorines were selectively restricted, the increase in cost would have been $15.4 million (14). Without the use of insecticides it would not have been profitable to grow cotton in those areas where insecticides are regularly employed. In fact, where a major pest has developed resistance to available insecticides, cotton can no longer be profitably grown. This is the case in Tampico, Mexico, where, because of tobacco budworm resistance, cotton production has been stopped. The same insect, because of development of resistance, is threatening cotton production in Arkansas, Louisiana, Mississippi and Texas (15).

In some cases pests are responsible for the destruction of an entire industry. For example, because of the boll weevil, production of Sea Island cotton in the southeastern United States is no longer profitable. (15).

A provisional regional study conducted by the Food and Agriculture Organization (16) has calculated the cost/benefit relationship with respect to rice in India. Their findings indicate that because of insect and fungus control there were 18 and 14% increases in production; this gave cost/benefit relationships of 1:4 and 1:3, respectively.

In Nigeria nearly all cotton seed is treated for the control of black-arm disease caused by the bacterium *Xanthomonas malvacearum* (E. F. Smith) Dowson. Losses due to this disease were estimated at 10%; the small investment in the chemical used for seed dressing brings the country an additional income of $1.3 million (9).

Cost/benefit ratios have been calculated for various crops by scientists (17–22). Metcalf (21) calculated a ratio of 1:29 for potatoes when DDT was used for the control of Colorado potato beetle and the potato leafhopper, *Empoasca fabae* (Harris); Reynolds et al. (22) reported a return of $18 for every dollar spent on phorate for sugar beets in California for the control of the green peach aphid, *Myzus persicae* (Sulzer), the southern garden leafhopper, *Empoasa solana* De-Long, and the mite, *Tetranychus urticae* Koch. Use of aldrin for the control of the susceptible corn rootworms, *Diabrotica* spp., returned $4 in profits for every dollar spent on the insecticide (23). A cost-benefit ratio of 1:9

for corn (6), and 1 : 11–22 for cocoa (9) have already been reported. The use of insecticides on the sugar crop in Pakistan resulted in an increased yield of over 30%. The amount spent on insecticides was $77,000, but the increase in yield gave a profit of $7.2 million, a cost benefit ratio of 1 : 63.5 (9).

There is a feeling that "naturally grown" food is cleaner and better for human health, but this is not necessarily true; several fungi and other organisms attack food and food products and some produce toxins that are injurious. Similarly, weevils |*Sitophilus granarius* (L.)| that infest grain contain several quinones that are carcinogenic. In Germany analysis of a large sample of flour showed 225 weevil fragments in 250 grams of flour (9).

According to the latest (1951–1960) estimates of crop losses, 33.6% of the total production is lost to all pests (this includes loss in yield and quality). The annual cost of this loss is estimated as $14.3 billion dollars (2, 17). Pimentel (18) estimated crop losses due to insects if no insecticides were used to be about 16.3% of the total production. The total loss to pests and diseases is estimated at about 40.7% of potential crop production. Thus, according to Pimentel, if no pesticides were used the overall crop loss would be 40.7–33.6 or 7.1% (18). There is an annual surplus of 10% in quantity in the United States (5), and Pimentel argues that if no pesticides were used, the supply of food for the nation would be ample, but quantities of certain fruits and vegetables would be significantly reduced. However one point to remember is that agricultural exports would almost be eliminated. These accounted for $7.4 billion in 1967, about 24% of our total export income; moreover, countries that depend on food imported from the United States would face starvation. Pimentel (18) estimates the cost of the loss of 7.1% as $2.1 billion. The cost of pesticides used in agriculture is about $0.56 billion (18, 19). Assuming application cost to be about $0.19 billion, the return per dollar invested for pest control, according to Pimentel, is $2.82 dollars. This is less than that calculated by the President's Scientific Advisory Committee (PSAC) (20) and Headley (7, 8). They estimate returns of between $4 and $5 dollars for every dollar invested in pesticides.

Headley has made the proposal that planting an additional 12% acreage could compensate for this loss and reduce insecticide use to 20–30% of the present level (8), but standards set by the public are so high that either treatment with insecticides at the present level or education of the public is essential.

According to Borlaug (3), if pesticide use in the United States were to be completely banned, crop losses would probably be about 50% of the total yield, and increase in food prices would be four- to fivefold. If the United States were still using 1940 technology, not only 50 million more acres now held under reserve would have to be cultivated, but about 242 million acres of marginal land also would have to be brought under the plough if the nation were to be fed adequately. The ecological consequences of cultivating these

292 million acres ("an area roughly equivalent to the total land area of the United States east of the Mississippi river and south to the Ohio river"), not only on man but also on wildlife, would be obvious (3). Ennis et al. (15), referring to Uphchurch and Heisig, and T. R. Eichers (private communication) have presented a slightly different view. According to their estimate if pesticides were not used, "the total combined output of crops; livestock, and forests would be reduced by at least 25 per cent;" the increase in the prices of farm products would be at least 50%. Our expenditure on food would be 25% of our income.

## ROLE OF INSECTICIDES IN PUBLIC HEALTH

Nowhere has the success of DDT been so dramatic as in the control of insect-borne diseases, especially malaria, plague, and typhus. DDT also has been used successfully against vectors of filariasis and cutaneous and visceral leishmaniasis (kala-azar). Thanks to this insecticide approximately one billion persons have been freed from malaria since the late 1940s. Malaria has almost been eliminated from 19 countries and about 85 countries are on their way to getting rid of this disease. Sri Lanka (Ceylon) presents a graphic example of the importance of DDT. Before the malaria control program (as they were called then, as compared to "malaria eradication program," a term adopted later) was started in the early 1950s there was an annual incidence of two million malaria cases in Sri Lanka; two-thirds of the island was neglected and lived a marginal existence. In 1963 there were only 17 cases of malaria and because of financial stringency, the malaria eradication program was discontinued. Within four years the number of malaria cases had risen to 3000 and the next year more than five times as many cases were reported; in late 1969 two million persons were sick with malaria. The program was restarted in 1969 (24).

One aspect of malaria and kala-azar control that the author has especially noted is the expression on the faces of children and youth. In 1949 when I was associated with starting the United Nations World Health Organization Malaria Control Demonstration Program in Bangladesh, in our prespraying survey we visited village after village with a high incidence of malaria (between 80 and almost 100%). There was a strange listless appearance on the faces of children and youth. After 2 years of malaria control the entire expression on the faces of young persons had changed. They were alert and alive and indeed one could discern that *joie de vivre* that was missing when malaria was present.

In India in the early 1950s there were close to 75 million cases of malaria and about a 10% mortality as a direct or indirect result of this disease. After about 10 years of malaria control there were only about 100,000 cases of this fever, and life expectancy during this period rose from 32 to 47 years (24–26).

As a result of malaria control agricultural productivity almost always shows an increase. In Sri Lanka malaria control resulted in the agricultural development of the neglected two-thirds of the island that previously had very scant to marginal agricultural production. Similarly, in Nepal land that was abandoned before is being brought under plough as a result of malaria control. In Tabasco state of Mexico agricultural development could take place only after malaria was "eradicated." A 10% increase in agricultural production as a result of malaria control alone, has already been reported in Bangladesh (Chapter 5).

## REFERENCES

1. H. J. Waters. Asst. Administrator for War on Hunger. Agency for International Development, Dept. of State. Address before the Pesticide Chemical Industry. A.I.D. Conference, May 9, 1967, Washington, D.C.
2. The Council on Environmental Quality. Integrated Pest Management, U.S. Government Printing Office. Washington, D.C., 411 pp.
3. N. E. Borlaug. *Bioscience* **22**: 41 (1972).
4. USDA Agr. Econ. Report No. 147, 1968.
5. USDA. Agricultural Statics. U.S. Government Printing Office, Washington, D.C., 1970, 627 pp.
6. Shell Chemical Company. "Aldrin, Dieldrin, Endrin" (a status report), 1967, 59 pp.
7. J. C. Headley. *Annu. Rev. Entomol.* **17**: 273 (1972).
8. J. C. Headley. "Productivity of Agricultural Pesticides." In *Economic Research on Pesticides for Policy Decisionmaking.* Proc. Symp. Econ. Res. Ser., USDA, 1971, 172 pp.
9. Cooperative Program of Agro-Allied Industries with FAO and other UN Organizations. GIFAP, Brussels, Belgium, 1972.
10. Food and Agricultural Organization: Crop Assessment Methods, 1970.
11. Scan (Shell Chem. Agr. News). San Ramon, CA. Vol. 18, No. 2, 1972.
12. Food and Agriculture Organization. Production Yearbook 23, 1969.
13. H. H. Cramer. *Plant Protection and World Crop Production,* Pflanzenschutz-Nachrichten, Bayer, 20, 1967.
14. USDA. "Economic Consequences of Restricting the use of Organochlorine insecticides on Cotton, Corn, Peanuts and Tobacco." *Agr. Econ.* Report No. 178, 1971, 52 pp.
15. W. B. Ennis, Jr., W. M. Dowler, and W. Klassen. *Science* **188**: 593 (1975).
16. Indicative World Plan for Agricultural Development to 1975 and 1985. Provisional Regional Study No. 4, FAO, Rome, 1968.
17. USDA. *Losses in Agriculture,* Agr. Handbook No. 291. Agr. Res. Ser. U.S. Government Printing Office, Washington, D.C., 1965, 120 pp.
18. D. Pimentel. *J. N.Y. Entomol. Soc.* **81**: 13 (1973).
19. USDA. "Farmer's Pesticide Expenditures in 1966," Agr. Econ. Rep. No. 192, Econ. Res. Ser., 1970, 43 pp.
20. PSAC (the President's Scientific Advisory Committee). "Restoring the Quality of our Environment," Report of environmental pollution panel, PSAC. The White House, 1965, 317 pp.

21. R. L. Metcalf. Methods of Estimating Effects," In *Research on Pesticides* (C. O. Chichester, ed.), Academic Press, New York, 1968, pp. 17–29.
22. H. T. Reynolds, R. C. Dickson, R. M. Hannibal, and E. F. Laird, Jr. *J. Econ. Entomol.* **60**: 1 (1967).
23. H. B. Petty. "DDT, Other Persistent Insecticides, and Our Environment," Proc. 22nd. Ill. Custom Spray Oper. Training School, 1970, pp. 169–189.
24. WHO. "Vector Control" *WHO Chron.* **25**: 5 (1971).
25. L. J. Bruce-Chwatt. *Misc. Publ. Entomol. Soc. Amer.* **7**(1): 7 (1970).
26. L. J. Bruce-Chwatt. *Bull. WHO* **44**: 419 (1971).

# INTEGRATED CONTROL
# AND PEST MANAGEMENT

<table>
<tr><td>**CHAPTER 24**</td><td>Economics, Efficacy, and Ethics<br>of Insect Control</td></tr>
</table>

Of all the necessities of man, food is of prime importance, without adequate food man's progress would have been greatly hampered. As population increases, more and more food is needed; in many densely populated countries large tracts of land suitable for cultivation are just not available and reliance has to be made on increased yields per acre of land. In the 1970s remarkable progress has been made in this direction. The so-called green revolution has been possible because of high-yielding varieties of wheat, rice, maize, and other grains and cereals. Countries that have shown spectacular progress include Pakistan, Philippines, and India. But there is a general misconception that green revolution means only substituting high-yielding variety for the low-yielding one. This is only partially true, because the most important aspect of the green revolution is the care required by the new varieties. Their yield depends on certain optimum conditions, that is, more fertilizers, proper soil conditioning, and better protection against pests. Because of vigorous and more luxuriant growth and other complex factors depending on the crop, its geographical location, its microecology, and its environment, pest problems are intensified. Resort to pesticides appears, at first sight, to be the best solution, but past experience has shown that insects subjected to intensive insecticide pressure soon become tolerant to the pesticide; this necessitates higher doses and more frequent application of the chemical, and ultimately development of resistance is most likely. Also, frequent use of insecticides results in almost complete suppression of the populations of parasites and predators, thereby further increasing the pest population. This sequence of events has occurred in

254

many countries and in relation to various crops; cotton is a good example because chemicals have been heavily relied upon for the control of cotton pests. The Canete Valley of Peru presents an interesting case (1). In the early 1950s heavy reliance was placed on planting better and higher-yielding varieties of cotton and protecting them with organochlorines (DDT, HCH, and toxaphene). Cotton yields rose from 440 lb/acre in 1950 to 648 lb/acre in 1954, but resistance problems started early; aphids could not be controlled by HCH in 1952, and in 1954 the leafworm, *Anomis texana*, acquired resistance to toxaphene. A year later the weevil, *Anthonomus vestitus*, appeared in large numbers early in the growing season. A hitherto unimportant insect, the tortrix, *Argyrotaenia sphaleropa*, acquired pest status. The tobacco budworm, *Heliothis virescens*, heavily infested the cotton fields and could not be controlled by DDT. Several secondary pests appeared and resort to organophosphorus insecticides was made; these were applied at very short intervals. Despite this heavy application of pesticides, the yield of cotton dropped to 296 lb/acre. In neighboring valleys where insecticides were either not used or used judiciously, similar losses did not occur. Finally integrated pest control was started in Canete Valley in 1957 and the yield of cotton per acre showed gradual improvement; the highest yield recorded ever in the valley, 922 lb/acre, was reached in some places after resort to this type of control (1).

In integrated pest management use is made of all factors to maximize control of pests with the minimum use of extraneous materials and chemicals (2–5). Efforts are made to optimize existing natural controls, mostly by cultural methods. Moreover, rather than apply pesticides on a fixed schedule, as is done in chemical control, regular scouting of the fields is done to determine the need for control measures; these are initiated as soon as it is discovered that the pest is approaching an economically damaging level (threshhold level). Control measures consist of a combination of pest suppression techniques. Extraneous chemicals in large quantities are used only when essential, especially when the pest buildup is very rapid and soon reaches economic threshold, and other combinations of control are not expected to yield desired results. Or if a pest population rapidly reaches economic threshold it may be necessary to reduce its population by using short-lived insecticides and then releasing parasites and predators and other control agents when manageable population density has been reached. Some of the techniques that can be employed in integrated control are as follows.

1. Cultural methods or environmental manipulation:
   a. Change of farming practices in such a way that the pest's environment is adversely affected.
   b. Avoidance of 100% plantation by monoculture and cultivars. In this way natural selection of strains of pests that can adapt themselves to improved agricultural varieties is greatly delayed.

The southern corn leaf blight that reduced United States corn production by 15% is an example of the dangers involved in planting monocultures. About 800 million bushels of corn were lost, resulting in a loss of more than $2 billion (6).

    c. Use of disease-free seed.

    d. Use of resistant varieties or incorporation of resistant genes in cultivars.

  2. Control measures:

    a. Use of alternative means of pest control such as pheromones, attractants, repellents, and sterilization techniques.

    b. Mass rearing of parasites and predators and their release at appropriate time.

    c. Use of insect pathogens and their release at the opportune moment.

    d. Use insecticides only when necessary.

Select nonpersistent insecticides, such as nicotine, mevinphos, and trichlorfon, apply at the minimum effective dose. There is a general feeling that the use of insecticides is recommended at a higher dose than necessary and Adkisson (7) believes that in some cases effective control can be obtained by using only 50% of the recommended dose. Of course application would have to coincide with the occurrence of the pest(s), but effort should be made also to see if beneficial insects can in some way be saved. For instance, alfalfa aphid, *Therioaphis maculata* (Buckton), can be killed by mevinphos without killing all the parasites in the aphid. Malathion and parathion do not show this selectivity. Mevinphos could therefore be the insecticide of choice in this case (8). Batiste et al. (9) and Madsen and Vakenti (10) have used the principle of timed application and reduced dosage of azinphosmethyl on pears and apples, respectively, for the control of the codling moth; insecticide application was supplemented with baited pheromone traps. Wilson and Armbrust (11) took advantage of the differences in the life cycles of alfalfa weevil, *Hypera postica* (Gyllenhal), and its parasite *Bathyplectes curculionis* Thoms; the overwintering eggs of the weevil hatch in March, but the parasite is still in its pupal stage, one application of the insecticide methyl parathion in late March killed the weevil larvae but left the parasites unharmed.

Two cooperative Federal–State projects have been initiated by the USDA in 1971 to further demonstrate the benefits of field scouting in pest control programs (5). The crops concerned are cotton in Arizona and the Southeastern United States, and tobacco in North Carolina and South Carolina. Each scout collects data from 1000–2000 acres and reports to an agriculture extension agent who compiles the results and recommends the control measures, if necessary, to the farmer. In the cotton program there was a saving of about $12.50/acre as compared to chemical control (5).

In 1972 a concerted effort was begun in the United States to more effectively use a variety of control principles; several agencies cooperated and 15 crops in 29 states were subjected to integrated control. The California Agricultural Extension service, in cooperation with private pest management consultant firms, is evolving for cotton a complete insect pest management project. A system for data handling and retrieval has been developed and the project covers almost all facets involved in raising cotton. Increased grower participation to the extent that they provide most of the finances is a testimony to the success of the project. Michigan is using a similar program in the apple ecosystem which was started in 1972 (12).

Integrated pest management services have developed in California, Arizona, and Texas (5, 6). In California the Farm Bureau has found that cotton, grape, and citrus farmers have increased their net profits by 22% with the aid of private integrated pest management firms (5). An increase of 7.3% in net income as a result of integrated pest management has also been reported by grape farmers in the Delano area of California. The main pests of grapes in this area are the leafhopper, pacific mite, the grape mealy bug, and omnivorous leaf roller. For the chemical control of leafhopper and pacific mite two applications of Zolone® (phosalone) were necessary, and for the other two pests one application of parathion was made (5).

Skilled manpower is very important for this type of control, and the training of the public to accept the importance of the principle of live and let live is essential. In other words, our cosmetic standards ought to be such that we should accept a wormy apple or an insect in our produce once in a while. Nothing is more important for the success of integrated control than the change in the attitudes of all concerned. To start with, the farmer finds it difficult to change cultivation and pest control practices. He goes more by the calendar than by the subtle changes in the complex environment and field conditions that surround his crops. He plants, sprays, and harvests his crop according to a routine and is generally satisfied with the results. He is afraid of crop loss resulting from untried changes and is aware of the public demand for cosmetics; this makes him skeptical even if the pest population is below the economic threshold. To repeat, there has got to be a change in the attitude of the public; we should accept a worm in a can of corn once in a while as a fact of life.

There is also a lack of trained personnel; it is important that our colleges and universities place a great emphasis on training persons in integrated control. This training should not be limited to undergraduate and graduate students, but special courses should be started for professional entomologists at all age levels. The Federal Government should make lucrative funds available for senior scientists to spend time in different parts of the country and train themselves in integrated means of control. Federal aid should be made available to senior entomologists who intend to specialize in integrated control. Insurance companies should be encouraged to provide protection both to farmers against crop failures by use of integrated control and to specialists making pest control recommendations. No matter how considered an advice for integrated control is, there are bound to be some failures and accidents, as indeed they are in all forms of control; malpractice and liability insurance for the specialist is therefore essential.

A recent book, *Introduction to Insect Pest Management*, edited by Robert L. Metcalf and William Luckmann, has treated the various aspects of integrated control in detail (13).

# REFERENCES

1. R. F. Smith. "Patterns of Crop Protection in Cotton Ecosystems." Talk given at Cotton Symposium on Insect and Mite Control. Problems and Research in California. Berkley. 1969.
2. V. M. Stern. R. F. Smith. R. van den Bosch, and K. S. Hagen. *Hilgardia* **29**: 81 (1959).
3. FAO. Report of the 2nd Session of the FAO Panel of Experts on Integrated Pest Control. Rome. PL/1968/M/3. 1968.
4. National Academy of Sciences. Insect–Pest Management and Control. *Principles of Plant and Animal Pest Control*. Vol. 3, 1969, 508 pp.
5. The Council on Environmental Quality. *Integrated Pest Management*. U.S. Government Printing Office. Washington. D.C., 1972, 41 pp.
6. W. Shaw. In *Agricultural Chemicals* (J. E. Swift, ed.), Univ. of California Press. Berkley. 1971, p. 18.
7. P. L. Adkisson. "Objective Uses of Insecticides in Agriculture," *Agr. Chem. Symp.* **1971** pp. 43–51.
8. V. M. Stern and R. van den Bosch. *Hilgardia*. **29**: 103 (1959).
9. W. C. Batiste. A. Berlowitz. W. H. Olson, J. E. DeTar, and J. L. Loos. *Environ. Entomol.* **2**: 387 (1973).
10. H. F. Madsen and J. M. Vakenti. *Environ. Entomol.* **2**: 677 (1973).
11. M. C. Wilson and E. J. Armbrust. *J. Econ. Entomol.* **63**: 554 (1970).
12. W. B. Ennis. Jr.. W. M. Dowler, and W. Klassen. *Science* **188**: 593 (1975).
13. R. L. Metcalf and W. Luckman. eds. *Introduction to Insect Pest Management*. Wiley. New York. 1975, 587 pp.

# AUTHOR INDEX

Abdel-Hamid, F. M., 37, 63
Abdellatif, M. A., 90, 96
Abdel Wahab, A. M., 87, 90, 96, 206, 209
Abe, K., 9, 19
Abedi, Z. H., 214, 218
Acree, J., 98, 112
Adkisson, P. L., 256, 258
Aeschlimann, J. A., 68, 94
Agathe, C., 238, 241
Agosin, M., 219, 223
Ahmed, M. K., 44, 63
Albone, E. S., 107, 113
Aldridge, W. N., 32, 55, 59, 63, 64
Alley, E. G., 122, 123
Anderson, D. W., 226, 228
Anderson, J., 225, 228
Andrawes, N. R., 58, 64, 90, 96, 206, 209
Andrews, A. K., 107, 113
Andrews, T. L., 77, 78, 95
Angus, T. A., 174, 177
Aravena, L., 219, 223
Arent, H., 122, 123, 207, 209
Armbrust, E. J., 256, 258
Armstrong, G., 180, 192
Arthur, B. W., 37, 58, 63
Ascher, K. R. S., 142, 147, 152, 154
Ashworth, R. J., 90, 96
Auerbach, C., 142, 147
Auerlich, R. J., 226, 228
Augustinsson, K. B., 59, 62, 64

Bagley, W. P., 90, 96
Baker, R. C., 201, 202
Balcus, M., 239, 241
Bales, R. R., 238, 240
Ball, W. L., 227, 229, 231, 234
Bann, J. M., 204, 208
Barlow, F., 77, 95
Barnes, J. R., 145, 147
Barnett, J. R., 116, 122
Baron, R. L., 58, 64, 122, 123
Barron, J. R., 58, 64
Barthel, W. F., 7, 19
Bartley, W. J., 87, 90, 96
Batiste, W. C., 256, 258
Beasley, A. G., 228, 229
Beasley, O. C., 234, 235
Beck, S. D., 153, 154
Beck, V., 122, 123
Beckman, R., 157
Beebe, T., 175, 177
Benskin, J., 167, 171
Benyon, K. I., 62, 64
Berg, G. G., 227, 228
Berger, D. D., 227, 228
Bergot, B. J., 165, 177
Berlowitz, A., 256, 258
Beroza, M., 149, 151, 153, 156, 157, 158, 160, 163, 170
Bevilacqua, M., 239, 241
Bhatia, S. C., 222, 223

Bickley, W. E., 164, 170
Bierl, B. A., 158
Bigger, J. H., 121, 123
Birch, N. C., 156, 158
Birks, R. E., 232, 234
Biros, F. J., 231, 234
Bishop, J. L., 87, 89, 96
Bitman, J., 98, 99, 109, 112, 113, 227, 228, 229
Blaakmeer, P., 62, 65
Blackman, R. R., 226, 228
Blair, D. P., 107, 113
Blickenstaff, C. C., 163, 164, 170
Blinn, R. C., 104, 113
Bliss, C. I., 213, 216
Blum, E., 239, 241
Blum, M. S., 13, 19
Bond, R. P. M., 174, 175, 177
Booth, G., 71, 95
Borck, K., 79, 90, 95, 96
Bork, K., 206, 209
Borkovec, A. B., 142, 144, 145, 146, 147
Borlaug, N. E., 244, 249, 250, 251
Bousch, G. M., 205, 209
Bowers, W. S., 161, 162, 163, 170
Bowman, M. C., 98, 108, 112, 152, 154
Boyland, E., 55, 64
Bradbury, F. R., 127, 129, 180, 192, 219, 223
Braid, P. E., 50, 63
Braun, B. H., 146, 147
Bridges, R. G., 127, 129
Bridges, P. M., 15, 20
Brindley, T. A., 153, 154
Brindley, W. A., 56, 64
Britton, H. G., 180, 192
Brodie, B. B., 81, 95
Brooks, G. T., 115, 121, 123, 131, 134, 205, 207, 208, 209, 211, 215, 216, 233, 235
Brown, A. W. A., 98, 108, 112, 167, 171, 211, 212, 214, 215, 216, 218, 219, 220, 221, 222, 223
Brown, H. D., 101, 112
Brown, T. M., 167, 171
Brownlee, R. G., 157, 158
Bruce-Chwatt, L. J., 251, 252
Bryant, C., 13, 20
Buchner, A. J., 99, 107, 112, 121, 123, 201, 202, 222

Burdick, G. E., 226, 228
Burgerjohn, A., 175, 177
Burges, H. D., 173, 175, 176
Burk, J., 79, 90, 95, 206, 209
Burse, V. W., 231, 234
Burt, P. E., 183, 185, 193
Burton, R. L., 152, 154
Bush, B., 118, 122, 205, 209
Busvine, J. R., 214, 216, 222, 223
Butenandt, A., 156, 157, 158, 168, 171
Butler, P. A., 226, 228
Butterworth, J. H., 152, 153
Byrne, K. J., 151, 153

Cairns, K. G., 122, 123
Camougis, C., 13, 14, 19
Cantwell, G. E., 173, 177
Cardé, R. T., 157, 158
Carlson, D. A., 158
Carman, G. E., 104, 113
Carpenter, C. P., 87, 96
Carter, R. H., 204, 208
Casida, J. E., 7, 15, 19, 24, 25, 37, 39, 40, 44, 46, 56, 58, 59, 60, 62, 63, 64, 72, 85, 87, 88, 90, 95, 96, 107, 113, 195, 201, 205, 206, 221, 223
Cassil, C. C., 94, 96
Cecil, H. C., 109, 113, 227, 228, 229
Cerf, D. C., 167, 171
Černy, V., 162, 170
Chadwick, R. W., 129
Chamberlain, R. W., 14, 19
Chamberlain, W. F., 166, 171
Chambers, D. L., 151, 153
Chancey, E. L., 90, 96
Chang, M. L., 169, 171
Chang, S. C., 8, 15, 19, 20, 44, 145, 146, 147
Charles, R. T., 238, 241
Chasseud, L. F., 55, 64
Chen, P. R., 50, 63
Cheng, H.-M., 24, 25
Chikamoto, T., 7, 19
Clemens, G. P., 56, 64
Coats, J. R., 107, 110, 113
Cochran, D. G., 214, 216
Cohen, J. A., 30, 63
Cohen, S., 201, 202
Colby, D., 226, 228
Collier, C. W., 157, 158

Collins, C., 78, 95, 206, 209
Conin, R. V., 163, 170
Conney, A. H., 227, 229
Cooperative Program of Agro-Allied
  Industries, 245, 246, 247, 248, 249, 251
Cook, B. J., 181, 193, 221, 223
Cook, J. W., 50, 63
Coppedge, J. R., 80, 90, 95
Coppel, H. C., 156, 158
Council on Environmental Quality, The,
  243, 249, 251, 255, 256, 257, 258
Cox, J. L., 225, 228
Cramer, H. H., 245, 251
Cresswell, K. M., 165, 171
Crombie, L., 7, 19, 24, 25
Crosby, D. G., 173, 176, 177
Cross, A. D., 168, 171
Cubit, D., 228, 229
Cueto, C., Jr., 122, 123
Cunningham, R. T., 151, 153
Cupp, E. W., 165, 170
Curly, A., 231, 234
Curtiss, J. M., 240, 241
Cutkomp, L. K., 127, 129

Dahm, L. K., 161, 170
Dahm, P. A., 15, 19, 39, 56, 60, 62, 63, 64,
  65, 127, 129, 181, 193, 219, 220, 223
Dale, W. E., 231, 234
Daly, J. W., 207, 209
Darrow, D. I., 37, 63
da Silva, R. F. P., 167, 171
Datta, P. R., 107, 113, 122, 123
Dauterman, W. C., 55, 56, 62, 64, 65
David, J., 152, 154
Davidow, B., 121, 123, 204, 208
Davidson, A. N., 32, 63
Davidson, G., 214, 216
Davies, J. D., 70, 95
Davis, H. C., 225, 228
Davis, K. J., 239, 241
Davis, R. B., 86, 96
Davison, K. L., 122, 123
Day, N., 238, 241
Dean, H. J., 226, 228
DeBach, P., 173, 177
DeCino, T. J., 204, 208
Decker, G. C., 121, 123
Deichmann, W. B., 228, 229, 232, 234,
  239, 241

DeMeo, G. M., 146, 147
De Milo, A. B., 145, 146, 147
Deobhankar, R. B., 222, 223
De Tar, J. E., 256, 258
Deutch, E. W., 81, 82, 95
De Vlieger, M., 231, 234
Dewey, J. E., 218, 222
Dickson, R. C., 248, 251
Dicowsky, L., 40, 63
Diekman, J., 158
Djerasi, C., 158
Do, F. M., 127, 129
Doane, C. C., 18, 20, 156, 158
Dolejš, L., 162, 170
Donninger, C., 42, 61, 63, 64
Dorfman, R. I., 201, 202
Dorough, H. W., 58, 64, 85, 86, 87, 88, 90,
  96, 116, 122, 123, 206, 209
Dowler, W. M., 248, 250, 251, 257, 258
DuBois, K. P., 39, 56, 63, 64
Duckles, C. R., 78, 95
Duffy, J. R., 218, 219, 222, 223
Duggan, R. E., 233, 235
Dunbar, D. M., 18, 20
Dunham, L. J., 165, 171
Dunn, P. H., 175, 177
Dupras, E. F., 165, 170
Durham, W. F., 239, 241
Dutky, S. R., 173, 176
Dyte, C. E., 167, 171
Dzuik, L., 165, 170

Eagan, H. J., 119, 123
Earl, F. M., 240, 241
Earle, N. W., 204, 208
Ebeling, W., 139, 140, 180, 192
Eddy, G. W., 37, 63
ElBashir, S., 56, 64
Eldefrawi, A. T., 23, 25
Eldefrawi, M. E., 23, 25, 87, 96, 221, 223
Elgar, K. E., 118, 122, 205, 209
Elington, G., 107, 113
Elliott, J. W., 107, 113
Elliott, M., 7, 8, 11, 16, 17, 18, 19, 20
Ely, R. E., 204, 208
End, C. S., 35
Enderson, J. H., 227, 228
Engelman, F., 160, 169
Ennis, W. B., 248, 250, 257
Environmental Protection Agency, U. S.,

133, 134, 135
Eto, M., 54, 58, 59
Evans, N. C., 107, 113

Faber, R. A., 227, 228
Fahmy, M. A. H., 77, 78, 95, 105, 113
Falcon, L. A., 173, 176
Fales, H. M., 162, 170
Falk, H. L., 238, 240
Farm Chemicals Handbook, 35
Farnham, A. W., 11, 18, 19, 20, 221, 222
Feil, V. J., 85, 86, 88, 96, 122, 123
Felig, J., 145, 147
Felton, J. C., 78, 79, 95, 200, 202
Fenwick, M. L., 58, 64
Findlay, J. B. R., 152, 154
Fine, B. C., 219, 223
Fishbein, L., 238, 240
Fisher, R., 174, 177
Fisher, R. W., 189, 193
Fitzhugh, O. G., 131, 134, 238, 239, 240,
    241
Flowers, L. S., 77, 95
Food and Agriculture Organization, United
    Nations, 119, 122, 211, 216, 245, 248,
    251
Ford, I. M., 39, 63
Forgash, A. J., 128, 129, 181, 193, 221,
    223
Forrest, J. M., 185, 193
Fraser, E. D., 165, 170
Fraser, J., 77, 78, 95
Freal, J. J., 129
Frederickson, M., 220, 223
Fried, J. R., 168, 171
Fries, G. R., 107, 113
Friori, B., 151, 153
Fuhremann, T. W., 118, 122
Fujita, T., 81, 95, 190, 193
Fujitani, J., 7, 19
Fujitomo, F. S., 149, 153
Fukami, H., 24, 25
Fukuto, T. R., 30, 31, 32, 37, 49, 56, 58,
    62, 63, 64, 65, 69, 71, 72, 73, 75, 77,
    78, 79, 81, 82, 94, 95, 105, 113, 206,
    209
Furleinmeier, A., 168, 171
Furst, A., 168, 171
Furst, C. I., 55, 58, 64
Fye, R. L., 146, 147

Gadallah, A. J., 146, 147
Gage, J. C., 56, 64
Gahan, J. B., 99, 112
Gaines, T. B., 144, 147
Gannon, N., 121, 123
Garcia, B. A., 163, 170
Garcia, H., 238, 241
Gardiner, B. G., 187, 193
Gart, J. J., 238, 241
Gaston, L. K., 151, 153
Gaudette, L. E., 81, 95
Gaughan, L. C., 18, 20
Geigy Company, 98
Georghiou, G. P., 167, 171, 214, 216, 221,
    223
Gerold, J. L., 222, 223
Gerolt, P., 122, 123, 180, 183, 185
Gersdorff, W. A., 8, 19
Gianotti, O., 121, 123
Gibson, J. R., 122, 123
Gijswijt, M. J., 168, 171
Gilbert, B. N., 166, 171
Gilmore, L., 70, 95
Gingrich, A. R., 166, 171
Ginsburg, S., 70, 82, 94, 96
Gnadinger, C. B., 7, 19
Goda, M., 23, 25
Godin, P. J., 7, 19
Gofmekler, V. A., 234, 235
Gon, M., 232, 234
Good, E. E., 228, 229
Goodchild, R. E., 185, 193
Gordon, C. H., 107, 113
Göthe, R., 107, 113
Grayson, J. M., 214, 216
Grechava, G. V., 232, 234
Green, N., 8, 19
Greenberg, H. W., 8, 19
Grose, J. E. M., 77, 95
Grover, P. L., 128, 129, 207, 209
Gueldner, R. C., 157, 158
Gunther, F. A., 49, 62, 63, 104, 113
Guthrie, F. E., 185, 193
Gysin, H., 68, 94

Hadaway, A. B., 77, 95
Haegle, H. A., 227, 228
Hagen, K. S., 255, 258
Hahn, F., 145, 147
Hale, W. E., 232, 234

Hammett, L. P., 81, 95
Hampshire, F., 168, 171
Hanken, R. W., 116, 122
Hannay, C. L., 174, 177
Hannibal, R. M., 248, 251
Hansch, C., 81, 82, 95
Hanzmann, E., 161, 170
Hardee, D. D., 157, 158
Harper, 7, 19
Harrett, R. A., 90, 96
Harris, E. J., 149, 153, 226, 228
Harris, R. L., 165, 170
Harris, S. J., 109, 113
Harris, S. T., 227, 228
Harrison, A., 15, 20
Harrison, I. R., 77, 78, 95
Hart, E. R., 238, 240
Hartley, C. S., 58, 64
Haskell, P. T., 152, 153
Hassan, A., 37, 38, 63, 90, 96
Hastings, F. L., 71, 82, 95, 96
Hatch, M. A., 70, 94
Hayes, D. K., 13, 20
Hayes, W. J., Jr., 107, 113, 122, 123, 231, 232, 234
Haynes, H. L., 69, 94
Headley, J. C., 245, 249, 251
Heath, D. F., 30, 56, 63, 122, 123
Hecker, E., 157, 158
Hedde, R. D., 122, 123
Hedin, P. A., 157, 158
Heimburger, G., 59, 64
Heimpel, A. M., 175, 177
Hein, R. E., 15, 20
Heinricks, E. A., 167, 171
Hellenbrand, K., 80, 95
Henrick, C. A., 162, 163, 164, 170
Hermanson, H. P., 90, 96
Hess, C., 107, 113
Hetnarski, K., 201, 202
Hewer, A., 207, 209
Hickey, J. J., 227, 228
Hicks, B. W., 90, 96
Hicks, L. J., 201, 202
Higashikawa, S., 176, 177
Hill, L., 160, 170
Hillman, R. C., 139, 140
Hilton, B. D., 70, 95
Hingman, K. C., 160, 170
Hirwe, A. S., 103, 107, 110, 112, 113

Hocks, P., 168, 171
Hodgson, E., 40, 62, 63, 65, 87, 88, 96, 198, 201, 202
Hoffman, L. J., 39, 63
Hogendijk, C. J., 59, 64
Holan, G., 100, 104, 105, 112
Hollingworth, R. M., 55, 56, 60, 61, 62, 64, 85, 94, 96
Holloway, W. J., 164, 170
Holmes, D. C., 227, 228
Hoogendam, I., 231, 234
Hooper, G. H. S., 222, 223
Hopkins, D. E., 166, 171
Hopkins, L. O., 4, 19
Hopkins, T. L., 15, 20, 37, 63
Hoppe, W., 168, 171
Horn, D. H. S., 168, 171
Horowitz, S. B., 145, 147
Horská, K., 175, 177
Hoskins, M. W., 87, 96, 218, 221, 222, 223
Hsu, H. Y., 169, 171
Hubanks, P. E., 204, 208
Huber, R., 168, 171
Hummel, H. E., 151, 153
Hunt, LaW. M., 166, 171
Hunter, C. G., 232, 234, 235, 239, 241
Huppi, G., 168, 171
Hussey, N. W., 173, 175, 176
Hutchinson, C., 120, 122
Hutson, D. H., 42, 61, 62, 63, 64
Hyman, J., 115

Ignoffo, C. M., 164, 170
Ikemoto, H., 221, 223
Ilvicky, J., 219, 223
Inamasu, S., 9, 19
Incho, H. H., 8, 19
Ingangi, J. C., 151, 153
Ingle, L., 116, 122
Innes, J. R. M., 238, 240
Inouye, Y., 7, 19
Ishida, M., 127, 129, 219, 220, 223
Ittycheriah, P. I., 169, 171
Iverson, F., 71, 82
Ivie, G. W., 86, 90, 122, 205
Iwasa, J., 81, 95

Jacobson, M., 156, 157, 158, 173, 176, 177
Jager, K. W., 234, 235, 239, 241
James, J. D., 158

Janes, N. F., 11, 16, 17, 18, 19, 20
Jantz, O. K., 163, 170
Jenner, D. W., 200, 202
Jensen, S., 107, 113
Jerina, D. M., 207, 209
Johnson, A. L., 201, 202
Johnson, C. C., 132, 135
Johnson, H. E., 226, 228
Johnson, M. K., 228, 229
Jondorf, W. R., 128, 129
Jones, R. L., 72, 78, 81, 82, 95
Judah, J. D., 107, 113
Jukes, T. H., 134, 135

Kaae, R. S., 151, 153
Kallman, B. J., 107, 113
Kamimura, H., 23, 25
Kaplanis, J. N., 168, 171
Kapoor, I. P., 107, 110, 111, 113
Kapp, W. A., 144, 147
Karlson, P., 156, 157, 158, 168, 171
Kato, T., 54, 64
Katsuda, Y., 7, 19
Kauer, J. C., 201, 202
Kaul, R., 122, 123
Kay, K., 129, 227, 229, 231, 232, 234, 238, 240
Kaya, H., 18, 20
Kazantzis, G., 231, 234
Kearns, C. W., 8, 13, 15, 19, 20, 95, 99, 107, 112, 113, 127, 129, 187, 193, 201, 202, 218, 222
Keenan, G. L., 7, 19
Keith, J. O., 227, 228
Kemeny, T., 238, 241
Kenaga, E. E., 35
Keplinger, M., 239, 241
Kerb, U., 168, 171
Khalil, S. K. W., 169, 171
Khan, M. A. Q., 119, 123, 205, 209
Kido, H., 146, 147
Kilpatrick, J. W., 222, 223
Kilsheimer, J. R., 70, 94
Kimbrough, R., 144, 147
Kimmel, E. C., 15, 16, 17, 20
Kimura, T., 219, 223
Kindstedt, M. O., 107, 113
King, W. V., 99, 112
Kinoshita, Y., 54, 64
Kirby, P., 200, 202

Kitz, R. J., 82, 96
Klassen, W., 248, 250, 251, 257, 258
Klein, A. K., 107, 113, 118, 122, 123, 131, 134
Klempan, L., 128, 129
Klien, M., 238, 240
Klun, J. A., 153, 154
Knaak, J. B., 62, 85, 87, 90, 94, 206, 209
Knipling, E. F., 142, 147, 158
Koch, R. B., 127, 129
Kodaira, Y., 175, 176, 177
Kohli, 118, 122, 123, 205, 209
Kolbezen, M. M., 69, 71, 94
Kooy, H. J., Jr., 73, 95
Koranski, W., 128, 129
Koreeda, M., 169, 171
Korner, A., 175, 177
Korte, F., 118, 122, 123, 205, 207, 209
Kowalczyk, T., 62, 65
Kozuma, T. T., 149, 153
Kroger, M., 232, 234
Krueger, H. R., 62, 65
Krupka, R. M., 30, 31, 63, 80, 95
Ku, T. Y., 87, 89, 96
Kuhr, R. J., 85, 87, 90, 206, 209
Kulkarni, A. P., 201, 202
Kurihara, N., 190, 193
Kuyama, S., 176, 177
Kwalick, D. S., 232, 234

Laarman, J. J., 222, 223
Labler, 162, 170
LaBrecque, G. C., 142, 143, 146, 147
LaForge, F. B., 7, 8, 19
Laird, E. F., Jr., 248, 251
Lambrech, J. A., 69, 94
Laug, E. P., 107, 113
Lauger, P., 100, 112
Laws, E. R., Jr., 231, 234
Leeling, N. C., 87, 88
Lewis, C. T., 180, 193
Lewis, D. K., 77, 95
Liang, T. T., 118, 122
Lichtenstein, E. P., 118, 122
Lichty, R. W., 151, 153
Liebman, K. C., 201, 202
Lillies, J. N., 175, 177
Lindquist, D. A., 58, 64, 90, 96, 181, 193, 206, 208
Lipke, H., 107, 113, 218

Lloyd, C. J., 18, 20
Loh, A., 227, 228
Long, K. R., 234, 235
Longcore, J. R., 227, 228
Look, M., 77, 78, 95
Loos, J. L., 256, 258
Lord, K. A., 180, 183, 185, 192, 193, 222
Loschiavo, S. R., 152, 154
Louloudes, S. J., 168, 171
Lu, P.-Y., 107, 110
Luag, E. P., 122, 123
Lucier, G. W., 56, 57, 64
Luckman, W., 154, 258
Lupton, E. C., 81, 95
Lutz, H., 240, 241
Lutz-Ostertag, Y., 240, 241
Lykken, L., 62, 65

McBain, J. B., 39, 56, 63
McCarthy, J. F., 90, 94, 96
MacConnell, E., 174, 177
MacConnell, J. G., 156, 158
McDonald, W. E., 228, 229, 239, 241
McDonnell, C. C., 7, 19
Macek, K. J., 120, 122, 226, 228
McGovern, T. P., 151, 153
McHaffey, D. G., 146, 147
McIntosh, A. H., 180, 192
McLaughlin, A. I. G., 231, 234
Maclean, W., 157, 158
McMillan, W. W., 152, 153, 154
McNeil, J., 164, 170
Madsen, H. F., 256, 258
Magee, S., 54, 63
Mahfouz, A. M., 75, 95
Main, A. R., 32, 50, 59, 63, 64, 71, 82, 95,
    96
Maitra, N., 119, 123, 205, 209
Manning, D. T., 70, 94
March, R. B., 37, 49, 56, 58, 62, 63, 64, 71,
    77, 78, 95, 99, 112, 121, 123, 195, 221,
    223
Marfurt, T. A., 152, 154
Marilyn, J., 87, 96
Markin, T., 232, 234
Marrow, G. S., 107, 113
Martin, H., 100, 112, 181, 193
Matin, A. S. M. A., 180, 192
Matsuii, M., 7, 19
Matsumura, F., 59, 104, 113, 122, 123,

189, 193, 205, 209, 219, 223
Matthews, H. B., 122, 123
Mattson, A. M., 201, 202
Mattson, M., 169, 171
Mayer, M. S., 158
Mazur, A., 59, 64
Mehendale, H. M., 90, 96
Meifert, D. W., 146, 147
Meiniel, R., 240, 241
Meltzer, J., 73, 75, 80, 95
Mendel, J. L., 107, 113
Menn, J. J., 39, 51, 63, 160, 163, 170
Menzel, D. W., 225, 228
Menzer, R. E., 55, 56, 57, 64
Metcalf, R. L., 37, 49, 56, 58, 62, 63, 64,
    69, 70, 71, 73, 75, 76, 77, 78, 79, 81, 82,
    94, 95, 98, 99, 103, 104, 105, 107, 108,
    110, 111, 121, 123, 153, 154, 167, 206,
    209, 220, 223, 248, 252, 258
Meyer, A. S., 161, 170
Middleton, E. J., 168, 171
Milby, T. H., 232, 234
Miller, J. A., 166, 171
Miller, M. W., 227, 228
Miller, R. W., 167, 171
Miller, S., 99, 107
Minton, M. Y., 62, 65
Minyard, J. P., 157, 158
Miskus, R. P., 78, 95, 107, 113, 122, 123
Mitchell, I., 238, 240
Mitchell, W. C., 149, 153
Mitrovic, M., 145, 147
Miura, T., 165, 170
Miyamoto, S., 176, 177
Mofidi, Ch. M. H., 215, 216
Moore, J. B., 3, 19
Moore, L. A., 204, 208
Moorefield, H. H., 69, 80, 82, 94, 95, 96,
    201, 202
Mordue, J., 152, 153
Morello, A., 40, 41, 61, 63
Morgan, D. P., 107, 113
Morgan, E. D., 152, 153
Morgan, P. B., 146, 147
Mori, K., 162, 170
Mori, R., 170, 176
Mostafa, I. Y., 38, 63, 90, 96
Mounter, L. A., 62, 64
Mrak, E. M., 131, 132, 134
Mulder, R., 168, 178

Mulhern, B. M., 227, 228
Mukai, T., 9, 19
Mullins, L., 102, 112
Munakata, K., 152, 154
Munger, D. M., 90, 94, 96
Munson, S., 99, 112
Myers, R. O., 72, 78, 95

Nagasawa, M., 156, 158
Nagasawa, S., 146, 147
Nagatsu, J., 176, 177
Nair, J. H., 87, 96
Nakajima, E., 190, 193
Nakajima, M., 24, 25, 156, 158
Nakanishi, K., 169, 171
Nakanishi, M., 9, 19
Nakatsugawa, T., 56, 62
Narahashi, T., 12, 13, 14, 19, 20, 29, 99,
    100, 112, 113, 219, 223
National Academy of Sciences, U. S., 211,
    215, 216, 255, 258
Neal, J. W., 164
Neal, R. A., 39, 56
Neal, R. E., 101, 112
Needham, P. H., 8, 11, 19
Nelson, A. A., 238, 239, 240, 241
Nelson, M. J., 122, 123
Newallis, P. E., 103, 112
Niedermeyer, R. P., 62, 65
Nield, P., 127, 129
Nimmo, D. R., 226, 228
Nishimoto, S., 169, 171
Norris, J. R., 173
Novak, V. J. A., 160, 169
Nystrom, R. F., 107, 110, 113

O'Brien, R. D., 23, 25, 30, 36, 49, 56, 58,
    60, 62, 63, 64, 65, 70, 72, 95, 181, 189,
    193, 221, 223
Ogawa, S., 169, 171
Okey, A. B., 239, 241
Okhawa, H., 72, 95
Okhawa, R., 72, 95
Olson, W. H., 256, 258
Olson, W. P., 181, 183, 189, 193
O'Neil, J., 165, 170
Oonnithan, E. S., 85, 96, 122, 123, 206,
    209
Oosterbaan, R. A., 30, 63
Oppenoorth, F. J., 56, 64, 128, 129, 219,

    220, 221, 223
Ortiz, E., 201, 202
Oshima, Y., 54, 64
Osman, M. F., 79, 90, 95
Otto, H. D., 101, 110, 112

Pallota, A. J., 238, 240
Palm, P. E., 87, 96
Parke, D. V., 128, 129
Patil, K. C., 104, 113, 205, 209
Patterson, B. D., 169, 171
Paulson, G. D., 85, 86, 88, 96
Payne, L. K., 70, 94
Pence, R. J., 139, 140
Perry, A. S., 99, 107, 112, 121, 123, 201,
    202, 222
Peters, S., 238, 240
Peterson, J. E., 107, 113
Petrucelli, L., 238, 240
Petty, H. B., 248, 251
Philleo, W. W., 219, 223
Pickering, B. A., 42, 61, 63, 64
Pillmore, R. E., 18, 19
Pimentel, D., 218, 222, 225, 228, 248, 249,
    251
Plapp, F. W., 46, 60, 62, 63, 167, 171, 198,
    201, 202, 221, 223
Poonawalla, Z. T., 181, 182, 183, 184, 185,
    186, 187, 188, 189, 190, 191, 192, 193
Porter, R. D., 227, 228
Portig, J., 128, 129
Portnoy, C. E., 85, 86, 88, 96
Post, L. C., 168, 171
Potter, C., 8, 19
President's Scientific Advisory Committee,
    The, 248, 249, 251
Printy, G. E., 221, 223
Prior, P. F., 231, 234
Prouty, R. M., 227, 228
Proverbs, M. D., 142, 147
Pullman, D. A., 11, 18, 19

Quaife, M. L., 131, 134, 239, 241
Quinby, G. E., 107, 113
Quinstad, G. B., 165, 171
Quraishi, M. S., 84, 96, 126, 129, 171, 175,
    177, 180, 181, 182, 183, 184, 185, 186,
    187, 188, 189, 190, 191, 192, 193

Radomski, J. L., 121, 123, 204, 208, 232,

234, 239, 241
Randtke, A., 225, 228
Rasmussen, I. M., 101, 112
Ratcliffe, D. A., 227, 228
Ray, J. W., 219, 223
Reddy, G., 119, 123
Reed, W. T., 128, 129
Reimschnider, R., 101, 110, 112, 115
Reinert, M., 68, 94
Renwick, J. A., 157,158
Rey, A. A., 232, 234
Reynold, H. T., 79, 90, 95, 96, 206, 209, 248, 251
Rhead, M. M., 107, 113
Richards, A. G., 174, 177
Richardson, A., 118, 122, 205, 209
Riddiford, L. M., 163, 170
Rierson, D. A., 139, 140
Riley, R. C., 181, 193, 221, 223
Ringer, R. K., 226, 228
Riou, J.-Y., 175, 177
Roan, C. C., 107, 113
Roark, R. C., 7, 19
Robbins, W. E., 15, 20, 37, 63
Robeau, R., 165, 170
Roberts, D. V., 234, 235
Roberts, D. W., 175, 177
Roberts, M., 239, 241
Roberts, P. A., 72, 73, 95
Roberts, R. B., 78, 95
Robertson, A., 67, 94
Robinson, A., 118, 122, 205, 209
Robinson, J., 131, 134, 232, 234, 235, 239, 241
Robison, W. H., 107, 113
Robson, J. M., 142, 147
Rodin, J. O., 157, 158
Roelfs, W. L., 157, 158
Rogers, E. F., 101, 112
Rogoff, M. H., 175, 177
Röller, H., 161, 170
Romañuk, M., 160, 161, 163, 166, 169
Rosen, J. D., 119, 123, 205, 209
Rosner, L., 174, 177
Rosenthal, J., 13, 20
Rubens, J. F., 240, 241
Ruzicka, L., 7, 19
Ryan, A. J., 85, 96

Sacca, G., 99, 112, 146, 147

Sacher, R. M., 73, 95
Saito, T., 37, 38, 53, 63
Sakai, S., 23, 25
Samuels, A. J., 232, 234
Sanchez, F. F., 218, 222
Sandal, P. C., 169, 171
Sangha, G. K., 110, 111, 113
Sasaki, S., 169, 171
Satter, L. D., 90, 94, 96
Sawicki, R. M., 7, 8, 19, 222, 223, 245, 251
Schaefer, C. H., 78, 95, 165, 167, 170
Schermeister, L. J., 169, 171
Schechter, M. S., 9, 19
Schmidt, C. H., 181, 193
Schmeltz, I., 22, 25
Schmialek, P., 161, 170
Schneiderman, H. A., 161, 170
Schonbrod, R. D., 219, 223
Schoof, H. F., 222, 223
Schooley, D. A., 165, 171
Schwardt, H. H., 218, 222
Seagram, H. L., 226, 228
Sĕbesta, K., 175, 177
Sellers, L. G., 185, 193
Sham, F., 3, 19
Sharung, D. C., 201, 202
Shaw, W., 256, 257, 258
Shell Chemical Company, 244, 249, 251
Shepard, H. H., 7, 19
Sherman, M., 218, 222
Shih-Colman, C., 158
Shindo, H., 190, 193
Shinohara, H., 146, 147
Shishido, T., 60, 64
Shott, L. D., 144, 147
Shubik, P., 238, 241
Siddall, J. B., 160, 162, 163, 164, 165, 168, 170, 171
Silhacek, D. L., 158
Silverstein, R. M., 151, 153, 156, 157, 158
Simmons, J. H., 227, 228
Simmons, S. W., 98, 112
Sims, P., 128, 129, 207, 209
Sinclair, J. W., 227, 229, 231, 234
Sink, J., 107, 113
Skea, J., 226, 228
Slade, R. E., 127, 129
Slama, K., 160, 161, 162, 163, 166, 169
Sleeman, R. L., 7, 19
Smalley, H. E., 240, 241

Smith, C. N., 142, 143, 146, 153, 154
Smith, M. L., 216
Smith, R. F., 255, 258
Smith, S. H., 226, 228
Smyth, H. F., Jr., 87, 96
Snarey, M., 7, 19
Solomon, K. R., 167, 171
Sondgren, P., 134, 135
Sörm, F., 160, 161, 162, 163, 166, 169, 170
Spates, G. E., 164, 170
Spencer, E. Y., 41, 58, 61, 63, 64
Spurr, H. W., 90, 96
Staal, G. B., 162, 163, 164, 170
Stamm, M. D., 157, 158, 168, 171
Stafford, E. M., 146, 147
Standen, H., 127, 129
Stanovick, R. P., 94, 96
Stansbury, H. A., 70, 94
Stanton, R. H., 119, 123, 205, 208
Stark, K. J., 152, 153, 154
Staudinger, H., 7, 19
Stedman, E., 67, 94
Steiger, L. E., 165, 171
Steiner, L. F., 149, 153
Stempel, A., 68, 94
Stern, V. M., 255, 256, 258
Sternburg, J., 99, 107, 112, 113, 127, 129,
    187, 193, 218, 222
Stevenson, D. D., 239, 241
Stevenson, J. H., 7, 11, 18, 19
Stewart, D. K. R., 122, 123
Strother, A., 85, 96
Sullivan, L. J., 87, 90, 96, 206, 209
Sullivan, W. N., 13, 20
Sun, Y.-P., 181, 196, 198, 204
Sutherland, D. J., 119, 123, 205, 209
Suvak, L. N., 232, 234
Suzuki, S., 176, 177
Swain, G. G., 81, 95
Sweeney, T. R., 107, 113

Tabanova, S. A., 234, 235
Taft, R. W., 105, 115
Tahori, A. S., 195, 201, 202
Takahashi, M., 165, 170
Takahashi, N., 176, 177
Takemoto, T., 169, 171
Tallant, M. J., 87, 90, 96, 206, 208
Tamura, S., 176, 177
Tanada, Y., 173, 176

Tarjan, R., 238, 241
Tarrant, K. B., 134, 135
Tatton, J. O'G., 134, 135, 227, 228
Terracini, B., 238, 241
Terriere, L. C., 54, 64, 219, 223
Terry, P. H., 144, 146, 147
Thain, E. M., 7, 19
Thompson, A. C., 157, 158
Thompson, M. J., 161, 162, 168, 170, 171
Thorpe, E., 239, 241
Thorsteinson, A. J., 169, 171, 175, 177
Thrailkill, R. B., 151, 153
Todd, J. W., 167, 171
Tolman, N. M., 56, 62, 64, 65
Tomatis, L., 238, 241
Toscani, H. A., 152, 154
Tsuda, A., 9, 19
Tsukamoto, M., 58, 107, 113, 214, 216
Tucker, R. K., 227, 228
Tumlinson, J. H., 157, 158
Turner, C. R., 77, 95
Turnipseed, S. G., 167, 171
Turusov, V., 238, 241

Uchida, T., 62, 65
Udall, S., 226, 228
Udenfriend, S., 207, 209
Uebel, E. C., 161, 162, 170
Ueda, K., 18, 20
Ukeles, R., 225, 228
Ulland, B. M., 238, 240
Upholt, W. M., 107, 113
USDA, 244, 248, 249, 251

Vakenti, J. M., 256, 258
Valerio, M. G., 238, 240
Van Asperen, K., 127, 128, 129, 219, 221,
    223
Vandekar, M., 83, 122
Vardanis, A., 41, 61, 63
Varela-Alvarez, H., 107, 113
Velsicol Corporation, 116, 122
Versteeg, J. P., 231, 234
Viado, G. B., 56, 64
Vingiello, F. A., 103, 112
Vink, G. J., 62, 65
Vinopal, J. H., 62, 65
Vinson, S. B., 164, 165, 167, 170, 171, 218,
    222
Viray, M. S., 139, 140

Vité, J. P., 157, 158
Voerman, S., 62, 65

Wain, R. L., 100, 112
Wakerly, S. B., 77, 78, 95
Waldvogel, G., 168, 171
Walker, A. I. I., 239, 241
Walker, K. C., 107, 113
Walker, T. M., 226, 228
Walker, W. F., 163, 170
Walton, M. S., 107, 113
Ware, G. H., 228, 229
Waserman, D., 232, 234
Waserman, W., 232, 234
Waters, H. J., 243, 251
Watson, W. A., 58, 64
Watts, J. O., 122, 123
Weatherston, J., 157, 158
Weichert, R., 168, 171
Weiden, M. H. J., 70, 73, 77, 78, 79, 80,
    81, 83, 94, 95
Weil, C. S., 87, 96
Weisgerber, I., 118, 122, 123, 205, 209
Welch, R. M., 227, 229
Welle, H. B. A., 73, 75, 80, 95
Wellinger, K., 168, 171
Wenyon, E., 238, 241
Weseloh, R., 18, 20
White, R. W., 58, 64
White, W. C., 107, 113
Widmark, G., 227, 228
Wiemeyer, S. N., 227, 228
Wiggleworth, V. B., 160, 161, 170
Wilder, W. H., 78, 90, 165, 167, 170
Wilhelm, K., 83, 96
Wilkinson, C. F., 198, 201, 202
Wilkinson, J. D., 164, 170
Williams, C. H., 239, 240, 241
Williams, C. M., 160, 163, 170
Willy, W. E., 163, 170
Wilson, A., 234, 235
Wilson, A. R., Jr., 226, 228
Wilson, I. B., 70, 82, 94, 96

Wilson, M. C., 256, 258
Winbush, J. S., 131, 134
Wiseman, B. R., 152, 154
Winterigham, F. P. W., 15, 20
Winton, M. Y., 72, 73, 95
Witkop, B., 207, 209
Wohland, H. W., 128, 129
Wood, D. L., 156, 158
Woods, C. W., 145, 146, 147
Woodside, M. W., 87, 96
World Health Organization, 99, 112, 119,
    250, 251, 252
Wright, A. N., 62, 64
Wright, C. C., 139, 140
Wright, J. E., 164, 170
Wunderlich, J. A., 168, 171
Wurster, C. F., 225, 228

Yamamoto, I., 7, 19, 23, 24, 25, 39, 56, 63,
    72, 221, 223
Yamamoto, R., 7, 15, 16, 19, 23, 25
Yamamoto, R. T., 168, 171
Yamasaki, T., 219, 223
Yang, R. S. H., 62, 65
Yeager, J., 99, 112
Yip, G., 50, 63
Younger, R. L., 165, 170
Youatt, 226, 228
Yu, C. C., 70, 95
Yuan, C., 161, 170

Zachrison, C. H., 122, 123
Zaltman-Nirenberg, P., 207, 209
Zapp, J., Jr., 145, 147, 240, 241
Zarif, S., 118, 122, 205, 209
Zayed, S. M. A. D., 37, 38, 63, 90, 96
Zaylskie, R. G., 85, 86, 88, 96, 122, 123
Zehr, M. V., 85, 86, 88, 96
Zeid, M. M. I., 15, 20
Zeidler, O., 98, 112
Zellenmeyer, L., 232, 234
Zuberi, M. Y., 206

# SUBJECT INDEX

Abate, 29, 47
*Abies, balsamea,* 162
AC-24055, 152
10-Acetoxy-7-hexadecen-1-ol (gyptol), 157
12-Acetoxy-9-octadecen-1-ol (gyplure), 157
*p*-Acetoxyphenyl-methyl ketone, 150
Acetylcholine, 29, 30, 31
  acetylcholinesterase interactions, 29, 31
Acetylcholinesterase, 30, 31
*N*-Acetyl zectran, 78
Activation hormone, 160
*N*-Acyl, and other substitutions at carbamate N, 77
*Acyrthosiphon pisum,* 163
*Aedes aegypti,* 151, 169, 218, 222
*Aedes nigromaculis,* 78
*Aedes taeniorhynchus,* 218
Aerosols, use of pyrethrins in, 4
Africa, advantages of pesticides, 245, 246
  losses, due to agricultural pests, 245, 246
Alarm substance, 156
Aldicarb, 70, 84, 205, 206
  activation, 70, 205, 206
  sulfone, 206
  sulfoxide, 205, 206
Aldrin, 117, 198, 204, 205, 227
  advisory committee, 134
  ban on, 134
  epoxidation of, 204

Alkylating agents, 142, 143
  as sterilants, 142, 143
Allalotropic hormones, 160
Allethrins, 4, 8, 13
Alodan®, 121
Altocid, 164
  degradation of, 165
  metabolism of, 165
Altozar®, 166
American cockroach, 180, 181, 219
Aminocarb, 75
Aminopterin, 145
Amoy (China), manufacture, of pyrethrum joss sticks in, 3
*Amphamillon majalis,* 151
Anabasine, 22
*Anomis texana,* 255
*Anopheles, albimanus,* 222
  *atroparvus,* 222
  *culicifacies,* 222
  *gambiae,* 29, 214, 222
  *saccharovi,* 222
  *stephensi,* 214
  *stephensi mysorensis,* 215
*Anopheles* spp., DDT resistance in, 214
Antagonists, 198, 201
*Anthonomus grandis,* 138
*Anthonomus vestitus,* 255
Anthramycin, 145

271

Antifeedants, 152
Antimetabolites, as sterilants, 142
Antineoplastic agents, 146
*Aphidius nigriceps,* 164
*Aphis fabae,* 80
Apholate, 143
Arabica coffee, 246
Aramite®, 240
Arene epoxides, 207, 208
Argentina, protection of apples, with fentin
    acetate in, 152
*Argyotaenia sphaleropa,* 255
Arizona, cooperative Federal-State project
    in, 256
    integrated pest management services in,
    257
1-Arylimidazoles, as microsomal enzyme in-
    hibitors, 201
Aryloxylamines, 199
L-Asparaginase, 146
ATP-ase, and dieldrin resistance, 219
Autotoxin, 100
Azadirachtin, 152
Azaridines, as chemosterilants, 142, 143,
    146
Azinphosmethyl, 28, 52, 256
Azodrin®, 41, 56, 57

*Bacillus cereus,* 173
*Bacillus popilliae,* 173
*Bacillus thuringiensis,* 173, 174, 240
    α-exotoxin, 173, 174
    β-exotoxin, 173, 174
    γ-exotoxin, 173, 174
    δ-endotoxin, 173, 174
Balsam fir, *see Abies balsamea*
Bangladesh, agricultural production increase,
    due to malaria control in, 99
    *joie de vivre,* in the absence of malaria in,
    250
Bark beetles, 157
Barthrin, 9, 10
Base line, 213
Baygon®, *see* Propoxur
Baytex®, *see* Fenthion
Bean aphid, *see Aphis fabae*
*Beauveria bessiana,* 175, 176
Bedbug, 215
*Bemisia tabaci,* 245
Benzo-(*d*)-1,2,3-thiadiazoles, 200

1,4-Benzodioxan-2-carboxylate, 151
Benzyl benzoate, 153
BHC, 240
Bidrin®, 56, 57
Bioactivation, 204
Bioresmethrin, 9, 10
Biotrol®, 174
Birlane®, *see* Chlorfenvinphos
Black flies, methoxychlor, for the control of,
    109
*Blatella germanica, see* German cockroach
*Bluthyplectes curculionis,* 256
*Bombyx mori,* 168; *See also* Silkworm moth.
*Boophilus microplus,* 214
Brain hormone, 160
Brazil, coffee rust in, 246
Bromophos®, 46
*Bruchus pisorium,* 134
Bulan®, 101, 108
Butacarb, 73, 74
*m-tert*-Butylphenyl methylcarbamate, 69

Calcium arsenate, 138
California, profit increase, with private, inte-
    grated pest management firms, 257
California red scale, 212
Cal-O-Sil®, 139
Cameroun, cotton production, increase in,
    245
*Campoletis sonorensis,* 164
Canete Valley, Peru pesticides, and cotton
    production in, 255
Carbamates:
    alkylphenyl and alkoxyphenyl methylcar-
        bamates, 72
    metabolism of, 85
    mode of action of, 70
    multiring, 75
    resistance to, 215, 221
    selective toxicity of, 79, 80
    structure-activity correlation, 71
    symptoms of intoxication, 82, 83
    syn-anti isomerism, 79
    teratogenic effects of, 240
Carbaryl, 67, 69, 77, 83. *See also* Sevin®
Carbofuran, 76, 77, 84
    metabolism of, 90, 94
Carboxylesterase, 49, 220
*Cardiochiles nigriceps,* 164
Cecropia moth, 161

*Ceratitis capitata,* 149, 150
Ceylon, *see* Sri Lanka
Chemosterilants, 142
  alkylating agents as, 142, 143
  field use possibilities of, 146
  methods for testing of, 146
  non-alkylating agents as, 144
  undesirability of, 146
Chevron Research Corporation, 63
Chlorbenzilate, 238
Chlordane, 115
  ban on, 134
Chlordene, 115
Chlorfenvinphos, 42, 61
  *O*-dealkylation in, 61
Chlorhydrocarbons, *see* Chlorinated hydrocarbons
Chlorinated hydrocarbons, in coho salmon,
  226
  effect on:
    egg shell thickness, 227
    esterus in rats, 227
    man, 231
    MFO induction, 233
    reproduction, in beagle dogs, 227
      in mink, 226
    phytoplankton, 225
  in food chain, 225
  in grebes, 226
  pharmacodynamics of, 131
  resistance to, 214-215
  restrictions on, 131-134
  trophic concentration of, 226
  tumorigenic effects of, 238
Chloro-methylchlor, 111
1-(Chlorophenyl)-3-(2,6-difluorobenzoyl)
  urea, 167
Chlorpyriphos, 51
Chlorthion, 46
Cholinesterase, 29
  carbamates, reaction with, 70-71
  organophosphates, reaction with, 31-32
*Choristoneura fumiferana,* 78, 157
*Choristoneura occidentalis,* 77
(+)-*trans*-Chrysanthemic acid, 6
*Chrysanthemum cinerariaefolium,* 3
Cinerin I and II, 4-6
Cinerolone, 6
Ciodrin®, *see* Crotoxyphos
Clear Lake, California, DDD, trophic

concentration in, 226-227
Clones, in pyrethrum production, 4
*Cocculus trilobus,* 152
*Cochliomyia hominivorax,* 244. *See also*
  Screwworm
Cocoa, fungus disease affecting production,
  in tropical America, 245
  pests of, 245, 246
  production, in African countries, 245, 246
Coconut rhinoceros beetle, 150
Codling moth, 99, 174, 244. *See also Las-*
  *peyresia pomonella*
*Coelomyces* spp., 175
Coffee, pests of, 246
  production, in Africa, Brazil, and Sri Lan-
  ka, 246
Coho salmon, and chlorinated hydrocarbons,
  226
*Colletotrichum coffeanum,* 246
Colorado potato beetle, *see Leptinotarsa*
  *decemlineata*
Congo, pyrethrum production in, 3
*Conotrechulus nenuphar,* 22
Co-Ral®, *see* Coumaphos
Corn borer, resistance factor in, 153
Corn rootworm, fonofos, for the control of,
  39
Cotton boll weevil, 138
*Coturnix coturnix japonica,* 240
Coumaphos, 52
Crops, and threat of pest resistance, 215
Cross resistance, 214
Crotoxyphos, 42
Cruformate, 43
Cryolite, 139
Cue-lure, 150
*Culex fatigans,* 76, 215, 218, 219, 221. *See*
  *also Culex pipiens quinquefasciatus*
*Culex pipiens,* 167
*Culex pipiens quinquifasciatus,* 104, 167,
  215, 222. *See also Culex fatigans*
*Culex* spp., 201
*Culex tarsalis,* 168
Cuticle, back diffusion through, 181
  and insecticides, metabolism of, 187
    penetration of, 183-189
    penetration, dose dependence on rate of,
    182
  half time of, 181-182
  as resistance factor, 221

spread of chemicals on, 180
Cyanofenphos, 39
Cyclethrin, 9
Cyclodienes, activation of, 204
  advisory committee, 134
  epoxidation of, 204
  hearings, and ban on, 134
  metabolism of, 122-123
  photoproducts of, 204, 205
  resistance to, 214
  synthesis of, 115
Cyclopentadiene, 115
*Cynoglossum officinale,* 169
Cythion®, *see* Malathion
Cytochrome P450, 55, 195, 196

*Dacus cucurbitae,* 150
*Dacus dorsalis,* 149, 150, 151
Dahomey, cotton production increase, due
    to use of pesticides, 245
Dalmatia (Yugoslavia), pyrethrum produc-
    tion in, 3
Dasanit®, *see* Fensulfothion
DBP, 106, 107
DDA, 105, 106, 107, 218, 232
DDCN, 106, 107
DDD, 106, 107, 108, 214, 218, 219, 226,
    227
DDE, 106, 107, 218, 219, 227, 232
  and egg shell thickness, 227
DDMU, 106, 107
DDNU, 106, 107
DDOH, 106, 107
DDT, 180, 211, 214, 215, 218, 219, 220,
    221, 225, 226, 227, 228, 231, 232.
    *See also* Chlorinated hydrocarbons
iodo-DDT, 103
o-Cl-DDT, 102, 218
o-p-DDT, 218
DDT-md, 219
DDVP, *see* Dichlorvos
Deet, 151, 153
Deh-gene, 214
Dehydrojuvabione, 162
Demeton, 27, 47
  metabolism of, 48
  tautomerism, 48
Derris, 24
Desmethyl trichlorfon, 37
Destruxins, 175, 176

Desulfuration, of organophosphates, 35
DFP, 59
*Diabrotica* spp., 248
Diazinon, 51, 59, 219
Dibrom®, 40. *See also* Naled
Dibutyl phthalate, 153
Dicapthon, 44
Dichlorodihydrofarnesoate, 161
Dichlorvos, 37, 38, 40
Dicofol, 106, 107
Dieldrin, 117, 118, 119, 221, 244
  carcinogenicity, 239
  9-*syn*-hydroxy dieldrin, 118
  metabolism of, 122
  pentachloroketone of dieldrin, 118
  photodieldrin, 118
  resistance to, 214, 215
  *see also* Chlorinated hydrocarbons
Diethyl phosphate, 35
*N,N*-Diethyl-*m*-toluamide (deet), 151, 153
2,4-Dihydroxy-7-methoxy-1,4-benzoxazoli-
    none (6 MBOA), 153
Diiodooctadecane, 180
Diisopropyl-*O*-nitrophenyl thiophosphate,
    36
Diisopropyl phosphorofluoridate (DFP), 59
Dimecron®, 41
Dimetan, 68
*N*-Dimethylcarbamates, 68
Dimethylbenzanthracene (DMBA), 239
*O,S*-Dimethyl phosphoramidothioates, 53
Dimethyl phosphoric acid, 37
Dimethyl phthalate, 153
4-(3,3-Dimethyl-1-triazeno) acetanilide
    (AC 24055), 152
Dimetilan, 68
Dimilin®, *see* TH 6040
Dipel®, 174
*Diprion similis,* 156
Diptera, resistance in, 215
Dipterex®, *see* Trichlorfon
Disparlure, 158
*Distantiella theobroma,* 245
Disulfoton, 48, 51, 58
Di-Syston®, *see* Disulfoton
Dithiazolium salts, 145
Dithioburets, 145
Dithiophosphoric acid, derivatives, as pesti-
    cides, 43-51
DMC, 201, 205, 218

DNOC, 137
DNOCHP, 137
DNOSB, 137
*Dociostaurus maroccanus,* 168
Doom®, 175
Dri-Die®, 139
Dursban®, *see* Chlorpyrifos
Dyfonate®, *see* Fonofos

Ecdysone, 160
α-Ecdysone, 168
EDF, *see* Environmental Defense Fund
Egg shell thickness, and pesticides, 227
Egyptian cotton leafworm, *see Spodoptera
    littoralis*
*Empoasca solana,* 248
Endosulfan, 121, 122
Endrin, 119-120
    Δ-keto, 120
    hexachloraldehyde, 120
Entomogenous fungi, 175
Entomological Society of America, publica-
    cation, of approved common names
    of insecticides by, 35
Environmental Defense Fund (EDF), 133
Environmental Protection Agency, 133,
    158
    establishment of, 133
    and ban on DDT, 133
    and DDT hearings, 133-134
    registration of a pheromone for insect
        control, 158
*Epilachna varivestis,* 80, 108
EPN, 39
    oxidative metabolism of, 39, 60
Epoxide hydrases, role in inactivation of in-
    secticides, 207
*trans-trans*-10,11-Epoxy methyl farnesenate,
    161
Epoxy methyl octadecane (disparlure), 158
*Escherichia coli,* 175
Eserine, 67
Esterases, and organophosphates, 29
*Estigmene acrea,* 112
Ethoxychlor, 110
Ethyl dimethylphosphoramidocyanidate,
    59
Ethyl 3-isobutyl-2,2-dimethylcyclopropane
    carboxylate, 150
Eugenol, 151

European chafer, *see Amphamillon majalis*
European corn borer, *see Ostrinia nubilalis*
*Euxesta notata,* 222

FAO, *see* Food and Agriculture Organiza-
    tion, United Nations
*Farm Chemicals Handbook,* 35
Farnesol, 161
FDMC, 201
Federal Insecticide, Fungicide, and Rodenti-
    cide Act, 133
Fenchlorphos, *see* Ronnel
Fenitrothion, 28, 44, 60, 61
    desulfuration in, 56
Fensulfothion, 47
Fenthion, 28, 46
    thioether oxidation in, 58
Fentin acetate, 152
FIFRA, *see* Federal Insecticide, Fungicide
    and Rodenticide Act
Fire ant, imported, *see Solenopsis saevis-
    simia richteri*
Floodwater mosquitoes, resistance to insecti-
    cides, in California, 215
Fluenetil, 139
Fluorenyl methylcarbamates, 77
Fonofos, 39, 56
Food and Agriculture Organization, United
    Nations, 212, 245, 248
    working party, on pest resistance to pesti-
        cides, 212
    cost/benefit relationship, of pesticides and
        rice in India, 248
Fruit flies, sterilization of, 142
Furadan®, *see* Carbofuran

*Galleria mellonella,* 162
Gambusia fish, 112
Gardona, *see* Tetrachlorvinphos
Geigy Company, 68, 98
Geraniol, 151
German cockroach, 89, 215, 219
    resistance to insecticides, 215, 219
Gezira area, of Sudan, cotton reduction, due
    to white fly, 245
Ghana, cocoa production in, 245
Grapes, integrated control of pests of, 257
*Grapholitha molesta,* 152
Grasshoppers, and grazing lands, 246
    tolerance to DDT, 211

Green peach aphid, *see Myzus persicae*
Guthion®, *see* Azinphosmethyl
Gyplure, 157
Gypsy moth, check of spread, through
    pheromone baited traps, 157
  sex pheromone, mistaken identity of,
    157, 158
  true pheromone, 157, 158
Gyptol, 157

Half time, of penetration through inspect
    integument, 181
*Heliothis*, 215
*Heliothis armigera*, 245
Heliothis nuclear polyhedrosis virus, 173
*Heliothis viriscens*, 163, 167, 255
*Heliothis zea*, 152
*Hamileia vastatrix*, 246
Hempa, 144, 240
HEOM, 207
  selective toxicity to tsetse fly, 207
Heptachlor, 116-117
  ban on, 134
  epoxide, 117
Heterocyclic derivatives, of phosphorus
    compounds, 51
Hexachlorocyclohexane, 240. *See also*
    BHC; Lindane
Hexachlorocyclopentadiene, 115, 116
*cis*-7-Hexadecenyl acetate, 151
Hexalure, *see cis*-7-Hexadecenyl acetate
Hoffman-LaRoche 5503, 166
House fly:
  commercial registration, of pheromone,
    158
  Danish strain (Fc), 219
  genes, for resistance to DDT, 214
  Illinois strain L., 219
  Kdr gene in, 220
  R-Baygon strain, 221
  resistance to DDT, 218
  resistance (cross-tolerance) to JH analogs,
    167
  Rutger's A strain, 221
  sex attractant, 158
Human louse, 215
  conversion of DDT to DDA in, 106
  resistance to insecticides, 215
β-Hydroxyecdysone, 168
*Hypera postica*, 164, 167, 256

*Icerya purchasi*, 243
Imidan®, 52
Imported fire ant, *see Solenopsis saevissima
    richteri*
India, 244, 245, 248, 250, 254
Inokosterone, 168
Inorganic insecticides, 138
Insect development inhibitors, 163
Insect growth regulators, 163
Insecticides, economics of use, 245
  patterns of use, 244
Iran, pyrethrum use in 2000 B.C. in, 3
Isodrin, 119
Isolan, 68, 80
Isopropyl-11-methoxy,3,7,11-trimethyl-
    *trans*-2-*trans*-4-dodecadienoate, 162,
    164
Ivory Coast, 68, 80

Japanese quail, *see Coturnix coturnix japon-
    ica*
Jasmolin, 4, 6
*Jasus lalandei*, 168
Joss sticks, 3
Juvabione, 162
Juvenile hormone, 160
  analogs, 160
  effect, on insects, 163
    on parasitoids, 164
  inactivation, in beetles, 207
    in house fly, 167
  use, in insect control, 163, 164
Juvenoids, 163
  use of the term, 163

Kenya, pesticide use in Arabica coffee, 246
  pyrethrum production in, 3
Kikuthrin, 9, 10
Knockdown r-gene (Kdr), in house fly, 220
Korlan®, *see* Ronnel

Lake Michigan, fish and DDT in, 226
Lake Superior, fish and DDT in, 246
Landrin®, 74, 84
Lannate®, *see* Methomyl
*Laspeyresia pomonella*, 99, 174, 244
Latin America, crop losses, 246
  grasshoppers and grasslands, 246
Law-Williams mixture, 161
Lead arsenate, 138

*Leptinotarsa decemlineata,* 73, 138
*Leucophaea maderae,* 13
*Leucoptera coffeina,* 246
Lilly 18947, 199
Lindane, 125, 180
  hematologic reaction to, 129
  metabolism, 127-128
  mode of action, 127
  resistance to, 214
Locusts, 152
  and insecticides, 247
*Lonchocarpa,* 24

*Macrosiphum euphorbiae,* 164
Madeira cockroach, *see Leucophaea mader-*
  *ae*
Malaria, 250, 251
  in Bangladesh, 250
  in India, 250
  *joie de vivre,* after control of, 250
  in Sri Lanka, 250
Malathion, 29, 49, 56, 78
  metabolism, in insects and man, 49, 62
  resistance to, 220
Malathionase, 50
*Manduca sexta,* 168
Mariana Islands, eradication of *Dacus dor-*
  *salis* from, 149
6-MBOA, 153
Mediterranean fruit fly, *see Ceratitis capi-*
  *tata*
Melamines, 145
*Melia azadirachta,* 152
*Melia azedarch,* 152
*Melia indica,* 152
Meobal, 73, 74, 84
Meobam, 76
Mercaptophos, *see* Demeton
Metacil®, 206
*Metarrhizium anisopliae,* 175
Meta-Systox-R®, *see* Oxydematonmethyl
Methamidophos, 53
Methane sulfones, as sterilants, 142
Methiochlor, 111
Methomyl, 80, 83, 85
Methoxychlor, 108, 214
Methoxy citronellal, 166
Methoxy-methiochlor, 111
2-Methyl-1,2-bis(3-pyridyl) propane, 200
N-Methylcarbamoyl oximes, 78

Methyl-ethoxychlor, 111
Methyl eugenol, 150, 151
Methyl parathion, 44
Methyl phosphoric acid, 37
Metyrapone, 200
Mevinphos, 28, 41
  isomers of, 41, 61
  metabolism of *cis* isomer, 61
Mexican bean beetle, *see Epilachna varivestis*
Micro-Cel-C®, 139
Mirex, 121, 122
MGK 264, 199, 201
Microsomal oxidative system, 55, 195, 196
Microtrol, 174
Milky disease, of Japanese beetle, 173
Mitin FF, 100
Mixed-function oxidases, 55, 195, 196
  reactions of, 197, 198
Mobam®, 76
Monitor®, *see* Methamidophos
Monocrotophos, 41
Multiresistance, 214
*Musca domestica, see* House fly
Muscalure, 158
Mustard gas, as sterilant, 142, 145
Mutagens, as sterilants, 142
*Myzus persicae,* 248

Naled, 40
Naphthalene, 207
1-Naphthyl-3-butynyl ether, 73
Naphthyl propnyl ether, 198, 199
Negative temperature coefficient, in DDT,
  100
  in pyrethrum, 13
Neo-Pynamin®, 9, 10
Neostigmine, 68
NIA 16824, 201
Niagara 23509, 166
*Nicotiana rustica,* 22
*Nicotiana tabacum,* 22
Nicotine, 22, 256
Nicotinoids, 22
  mode of action, 23
Nomenclature, of organophosphorus com-
  pounds, 33
Nornicotine, 22

OMPA, *see* Schradan
Organophosphorus compounds:

activation of, 55
cholinesterase, reaction with, 31-32
cross resistance in, 220
degradation of, 59-62
delayed neurotoxic action of, 227
effect on man of, 233
and electromyographic studies, 234
mode of action of, 29-32
nomenclature of, 33
reactions of, 35
resistance to, 220
    role of hydrolysis in, 220
tautomeric conversions in, 43, 48
Organothiocyanates, 138
Oriental fruit fly, see Dacus dorsalis
Oriental fruit moth, see Grapholitha molesta
Oryctes rhinoceros, 150
Oryzaephilus surinamensis, 152
Osborne-Mendel rats, 238, 239
Ostrinia nubilalis, 174
9-Oxodec-trans-2-enoic acid, 156, 169
Oxodemetonmethyl, 48, 58

Pakistan, 244, 249, 254
    cost/benefit ratio, and use of insecticides
        on sugar crop, 249
Paper factor, 161, 163
Paraoxon, 31, 35, 59, 60
Parathion, 44, 56, 60, 240, 257
Paris green, 138
PCB, see Polychlorinated biphenyls
PCCH, see Pentachlorocyclohexene
Pediculus humanus, see Human louse
γ-1,3,4,5,6-Pentachlorocyclohexene, 128
Permethrin, 10, 11
Persia, see Iran
Perthane, resistance to, 214
Peru, insecticide use, and integrated control
        in Canete Valley, 255
Pestox, see Schradan
Pests, control, by conventional insecticides,
        244
    losses due to, differing views, 243-250
        need for the assessment of long term,
            243
Phenethyl butyrate, 151
Phenothrin, 10, 11
Pheromones, 156
Philippines, 234
Philips-Duphar, 167

Phorate, 27
Phosalone, 257
Phosdrin®, see Mevinphos
Phosmet, 52
Phosphamidon, 41
Phosphonic acid, derivatives, as insecticides,
        37
Phosphoramides, 144, 146
Phosphoric acid, derivatives, as insecticides,
        40
Photoaldrin, 117
Photodieldrin, 117
Photoheptachlor, 117
Phthalthrin, 9, 10
Physa snail, 112
Physostigma venenosum, 67
Physostigmine, 67, 69
Polychlorinated biphenyls, 226
Polyresistance, 214
Ponasterone, 169
Popillia japonica, 151, 175
Porthetria dispar, 18, 99, 174
Prolan®, 101, 108
Promecarb, 73, 74, 83
Propoxur, 85
10-n-Propyl-trans-5,9-tridecadienyl acetate,
        151
Prothoracic gland hormone, 160
Prothoracotropic hormone, 160
Pseudococcus spp., 246
Pyramat, 68
(+)-trans-Pyrethric acid, 6
Pyrethrins, relative toxicities, 8
    resistance to, 221
    structures, 4
    synergism, 8
Pyrethroids:
    metabolism, 14-17
    mode of action, 12
    nerve interactions, 13, 14
    resistance to, 221
    structure, 9, 10
    structure-activity relationship, 11
    synthetic, 9, 10
Pyrethrolone, 6
Pyrolan, 68
Pyrophosphoric acid, derivatives, as insecti-
        cides, 52
Pyrrhocoris apterus,
        162

Queen substance, 156, 169

Rabon®, *see* Tetrachlorvinphos
Radiomimetic chemicals, as sterilants, 142
RE 11775, 78
Regression line, 213
Repellents, 152
Resistance, 211, 214
  behavioristic, 222
  as crop threat, 215
  DDT, 214, 215
  dieldrin, 214
  effect, on malaria eradication programs, 215
  genetics of, 213, 214
  in insects, agricultural importance of, 215
    medical importance of, 215
  mechanisms, types of, 218
  monofactorial, 213
  polyfactorial, 214
  *see also individual compounds*
Resmethrin, 9
Rethrins, 7
Rethrolones, 7
*Rhodnius,* 161
Ronnel, 28, 46, 60, 62
Rotenoids, 24
Ruelene®, *see* Cruformate
*Ryania speciosa,* 24

Sabadilla, 24
Safrole, 199
*Sahlbergella singularis,* 245
Saligenin phosphorus esters, 54
Salithion, 54
Saltmarsh caterpillar, *see Estigmene acrea*
San Jose scale, resistance to lime sulfur, 212
*Schistocerca gregaria,* 152
*Schoenocaulon officinale,* 24
Schradan, 27, 58
Screwworm, 142. *See also Cochliomyia hominivorax*
Scrub typhus, 153
Sesamin, 199
Sevin®, *see* Carbaryl
Silkworm moth, 157
*Sitophilus granarius,* 249
SKF 525 A, 199, 204
*Solenopsis saevissimia richteri,* 121
Southern armyworm, *see Spodoptera*

  *eridania*
*Spodoptera eridania,* 80
*Spodoptera frugiperda,* 152
*Spodoptera littoralis,* 77, 152, 248
Spotted root fly, *see Euxesta notata*
Spruce budworm, *see Choristoneura fumiferana*
Sri Lanka, malaria in, 250
  replacement of coffee by tea in, 247
Stauffer 20458, 166
*Streptomyces mobarensis,* 176
Strobane®, 238
Substituted phenylmethyl carbamates, 72-75
Sudan, cotton losses, due to insect pests, 245
Sumithion®, *see* Fenitrothion
Supracide®, 52
Surecide®, *see* Cyanofenphos
Synergists, 198, 199-201
Systam, *see* Schradan
Systox®, *see* Demeton

Tabun, 59
Tamaron, *see* Methamidophos
Tanzania, pyrethrum production in, 3
Tchad, importance of cotton in, 245
TDE, *see* DDD
Tea, as a replacement of coffee in Sri Lanka, 246
Temephos, *see* Abate®
Temik®, *see* Aldicarb
Tepa, 143
  mutagenic properties of, 240
Tepp, 27, 28, 53, 59
Termites, 116, 134, 139
Tetrachlorvinphos, 42
Tetramethrin, 9, 10
*Tetranychus* spp., 80
*Tetranychus urticae,* 248
  resistance in, 215
Tetron®, *see* Tepp
TH 6040, 167
Thanite®, 200
*Therioaphis maculata,* 243, 256
Thimet®, *see* Phorate
Thiodan®, *see* Endosulfan
Thiometon, 51
Thiophosphoric acid, derivatives as insecticides, 43-51

Thiotepa, 144
Thompson-Hayward Chemical Company, 167
Thuricide®, 174
Thuron Industries, 158
TOCP, *see* Tri-*o*-tolyl phosphate
Tolerance, natural to insecticides, 211
*O*-Tolyl seligenin phosphate, 59
Toxaphene, 120, 244
    resistance to, 215
Toxic factor, in the blood of treated insects, *see* Autotoxin
*s*-Triazine, 146
*Tribolium castaneum,* 167
*Tribolium confusum,* 152
Trichlorethanol, 37
Trichlorfon, 27, 28, 37, 38, 256
    metabolism of, 37-38
*Trichoplusia ni,* 174
*m*-Trimethylammoniumphenyl methylcarbamate, 67, 69
Tri-*o*- tolyl phosphate, 58, 59
*Trombicula* spp., 153
Tropical America, decline of cocoa production in, 245
Tsetse fly, and insecticides, 247
    selective toxicity of HEOM to, 207
Turkey, resistance of cotton pests in, 247
Typhus, 250

UC 10854, 72
Uganda, pyrethrum production in, 3

Union Carbide Corporation, 69, 70
Upper Volta, cotton production increase in, 245
USA, cost/benefit equations, 248-249
    important crops, and their protection, 244

Vapona, *see* Dichlorvos
*Vespula* spp., 150

WARF antiresistant, 199
West Africa, cotton yield, increase in, 245
Western spruce budworm, *see Choristoneura occidentalis*
Wireworm, DDT, for control of, 99
    fonofos, for control of, 39
World Health Organization, U.N., 29

*Xanthomonas malvacearum,* 248
Xenobiotic, 55

Yellow jacket, 150
Yellow mealworm, *see Tenebrio molitor*
Yugoslavia, pyrethrum production in, 3

Zaire, importance of agricultural exports to, 245
Zectran®, 75, 78, 84
Zoecon 512, *see* Altozar®
Zolone®, *see* Phosalone
ZR-619, 166
ZR-777, 166